Carbon Nanotube Science

Synthesis, Properties and Applications

Carbon nanotubes represent one of the most exciting research areas in modern science. These molecular-scale carbon tubes are the stiffest and strongest fibres known, with remarkable electronic properties, and potential applications in medicine, sensing devices and a wide range of other fields. Cutting through the plethora of information available, *Carbon Nanotube Science* is the most concise, accessible book for the field, presenting the basic knowledge graduates and researchers need to know.

Based on the successful *Carbon Nanotubes and Related Structures*, this new book focuses solely on carbon nanotubes, covering the major advances made in recent years in this rapidly developing field. Chapters focus on electronic properties, chemical and biomolecular functionalization, nanotube composites and nanotube-based probes and sensors. The book begins with a comprehensive and up-to-date discussion of synthesis, purification and processing methods. With its full coverage of the state of the art in this active research field, this book will appeal to researchers in a broad range of disciplines, including nanotechnology, engineering, materials science, chemistry and physics.

Peter J F Harris is Manager of the Centre for Advanced Microscopy at the University of Reading, where he is involved in a wide range of projects in both the physical and biological sciences. His personal research interests mainly involve the application of high-resolution TEM to carbon materials. He is a member of the Editorial Advisory Boards both for the *Journal of Physics: Condensed Matter* and *Carbon*.

Carbon Nanotube Science

Synthesis, Properties and Applications

PETER J. F. HARRIS
University of Reading, UK

CAMBRIDGE
UNIVERSITY PRESS

CAMBRIDGE UNIVERSITY PRESS
Cambridge, New York, Melbourne, Madrid, Cape Town,
Singapore, São Paulo, Delhi, Tokyo, Mexico City

Cambridge University Press
The Edinburgh Building, Cambridge CB2 8RU, UK

Published in the United States of America by Cambridge University Press, New York

www.cambridge.org
Information on this title: www.cambridge.org/9780521535854

First published 2009
First paperback edition 2011

A catalogue record for this publication is available from the British Library

ISBN 978-0-521-82895-6 Hardback
ISBN 978-0-521-53585-4 Paperback

Contents

Preface

This book was originally conceived as a second edition of my earlier work *Carbon nanotubes and related structures: new materials for the twenty-first century* (Cambridge University Press, 1999). However, the field has expanded rapidly since 1999, and the tale grew in the telling, to the point where I realized I had essentially written a new book. The new title reflects this, as well as the fact that most of the material concerned with 'related structures' has been omitted: the book now focuses almost entirely on carbon nanotubes themselves. As with the first book, I have benefited enormously from the freely given assistance of colleagues from around the world, many of whom have also provided copies of images and preprints. The following list almost certainly fails to include all who have helped me, so I apologize for any omissions. I also stress that any errors which remain in the book are my responsibility alone.

I wish to thank: Pulickel Ajayan, Lizzie Brown, Marko Burghard, Hui-Ming Cheng, Hongjie Dai, Walt De Heer, Cees Dekker, Chris Ewels, John Gallop, Jason Hafner, Michael Holzinger, Martin Hulman, Kaili Jiang, Hiromichi Kataura, Ian Kinloch, Ralph Krupke, Alan Lau, Cheol Jin Lee, Jannik Meyer, Geoff Mitchell, Pasha Nikolaev, Henk Postma, Zhifeng Ren, Daniel Resasco, Andrew Rinzler, Milo Shaffer, Wenhui Song, Kazu Suenaga, Sander Tans, Kenneth Teo, Edman Tsang, Daniel Ugarte, Bruce Weisman and Karen Winey.

I would also like to thank Cambridge University Press for their encouragement and patience.

Most importantly, I want to thank my wife, Elaine, and daughters Katy and Laura for their continuing love and support.

Peter Harris,
Twyford, November 2008

1 Introduction

Carbon in its miscellaneous forms has been used in art and technology since prehistoric times (1.1–1.4). Some of the earliest cave paintings, at Lascaux, Altamira and elsewhere, were produced using a mixture of charcoal and soot. Charcoal, graphite and carbon black (a pure form of soot) have been used as drawing, writing and printing materials ever since: photocopier toner is largely composed of carbon black. Coal and charcoal, of course, have been used as fuels for millennia, and charcoal played an important role in what might be considered humankind's first technology, the smelting and working of metals. Charcoal was used in this way right up to the eighteenth century, when it began to be replaced by coke, a development which helped to stimulate the Industrial Revolution. With the development of the electrical industry in the late nineteenth century, a demand developed for graphite. The American Edward Acheson is credited with producing the first synthetic graphite in 1896. In the twentieth century, the importance of activated carbon in purifying air and water supplies grew steadily, and the invention of carbon fibres in the 1950s provided engineers with a new lightweight, ultra-strong material. Diamonds, like graphite, have been known since antiquity, but until quite recently were only used decoratively. The development of a commercial synthetic method at General Electric in the 1950s opened the way for the industrial use of diamonds.

The history of carbon science is littered with illustrious names. Antoine Lavoisier, in a famous experiment in 1772, proved that diamonds are a form of carbon by demonstrating that they produce nothing but carbon dioxide on combustion. Carl Wilhelm Scheele carried out a similar experiment with graphite in 1779; before that time graphite had been thought to be a form of lead. Humphrey Davy and Michael Faraday carried out extensive studies of combustion at the Royal Institution. Davy correctly ascribed the yellow incandescence of a flame to glowing carbon particles, while Faraday, in his famous series of lectures on 'The chemical history of a candle', used a burning candle as the starting point for a wide-ranging dissertation on natural philosophy (1.5). The structure of diamond was one of the first to be solved using X-ray diffraction, by William and Lawrence Bragg in 1913 (1.6), while nine years later John D. Bernal solved the structure of graphite (1.7). In 1951 Rosalind Franklin demonstrated the distinction between graphitizing and non-graphitizing carbons (1.8, 1.9), and at about the same time Kathleen Lonsdale made important contributions to the study of diamonds, both natural and synthetic (e.g. 1.10).

By the early 1980s, however, carbon science was widely considered to be a mature discipline, unlikely to yield any major surprises, let alone Nobel Prizes. That the situation is so different today is due, in large measure, to the synthesis in 1985 by Harry Kroto of the University of Sussex and Richard Smalley of Rice University, and their colleagues, of the first all-carbon molecule, buckminsterfullerene (1.11). It was this discovery which led to the synthesis of fullerene-related carbon nanotubes and which made carbon science suddenly so fashionable.

1.1   Buckminsterfullerene

Neither Kroto nor Smalley were traditional carbon scientists, but both had a strong interest in synthesizing carbon clusters. Kroto's interest arose from a long standing fascination with the chemical species that are found in the interstellar medium. He believed that small carbon clusters or molecules might be responsible for some of the unexplained features in the spectra recorded by astronomers. Smalley's motivation was more down to earth. For many years he had been working on the synthesis of inorganic clusters using laser-vaporization, with the aim of producing new semiconductors or catalysts. Carbon clusters produced in a similar way might also have valuable properties.

The now-famous series of experiments involved vaporizing graphite using a Nd:YAG laser. The distribution of carbon clusters in the gas-phase was then determined using mass spectrometry. This produced an extremely striking result. In the distribution of gas-phase carbon clusters, detected by mass spectrometry, clusters containing 60 carbon atoms were by far the dominant species. This dominance became even more marked under conditions that maximized the amount of time the clusters were 'annealed' in the helium. There was no immediately obvious explanation for this since there appeared to be nothing special about open structures containing 60 atoms. The eureka moment came when they realized that a closed cluster containing precisely 60 carbon atoms would have a structure of unique stability and symmetry, as shown in Fig. 1.1. The unwieldy name which they gave to this structure, buckminsterfullerene, honoured the visionary designer of geodesic domes, Richard Buckminster Fuller. The discovery of C_{60}, announced in *Nature* in November 1985 (1.11), had an enormous impact (1.12–1.15).

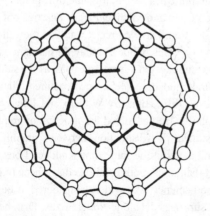

Fig. 1.1 C_{60}: buckminsterfullerene.

At first, however, further progress was slow. The main reason was that the amount of C_{60} produced in the Kroto–Smalley experiments was minuscule. If C_{60} were to become more than a laboratory curiosity, some way must be found to produce it in bulk. Eventually, this was achieved using a technique far simpler than that of Kroto and Smalley. Instead of a high-powered laser, Wolfgang Krätschmer of the Max Planck Institute at Heidelberg, Donald Huffman of the University of Arizona and their co-workers used a simple carbon arc to vaporize graphite, again in an atmosphere of helium, and collected the soot which settled on the walls of the vessel (1.16). Dispersing the soot in benzene produced a red solution which could be dried down to produce beautiful plate-like crystals of 'fullerite': 90% C_{60} and 10% C_{70}. Krätschmer and Huffman's work, published in *Nature* in 1990, showed that macroscopic amounts of solid C_{60} could be made using methods accessible to any laboratory, and it stimulated a deluge of research. Carbon nanotubes are perhaps the most important fruits of this research.

1.2 Fullerene-related carbon nanotubes

Sumio Iijima, an electron microscopist then working at the NEC laboratories in Japan, was fascinated by the Krätschmer–Huffman *Nature* paper. Ten years earlier he had used transmission electron microscopy to study soot formed in a very similar arc-evaporation apparatus to that used by Krätschmer and Huffman (1.17, 1.18). He found that the soot contained a variety of novel carbon architectures including tightly curved, closed nanoparticles and extended hollow needles. Might such particles also be present in the K–H soot? Initial high-resolution TEM studies were disappointing: the soot collected from the walls of the arc-evaporation vessel appeared almost completely amorphous, with little obvious long-range structure. Eventually, Iijima gave up sifting through the wall soot, and turned his attention to the hard, cylindrical deposit which formed on the graphite cathode after arc-evaporation. Here his efforts were finally rewarded. Instead of an amorphous mass, the cathodic soot contained a whole range of novel graphitic structures, the most striking of which were long hollow fibres, finer and more perfect than any previously seen. Iijima's beautiful images of carbon nanotubes, shown first at a meeting at Richmond, Virginia in October 1991, and published in *Nature* a month later (1.19), prompted fullerene scientists the world over to look again at the used graphite cathodes, previously discarded as junk.

A typical sample of the nanotube-containing cathodic soot is shown at moderate magnification in Fig. 1.2(a). As can be seen, the nanotubes are accompanied by other material, including nanoparticles (hollow, fullerene-related structures) and some disordered carbon. The nanotubes range in length from a few tens of nm to several μm, and in outer diameter from about 2.5 to 30 nm. At high resolution the individual layers making up the concentric tubes can be imaged directly, as in Fig. 1.2(b).

Virtually all of the tubes produced using the arc-evaporation method are closed at both ends with caps which contain pentagonal carbon rings. The structural relationship between nanotubes and fullerenes can be illustrated by considering the two 'archetypal' carbon nanotubes that can be formed by cutting a C_{60} molecule in half and placing a graphene cylinder between the two halves. Dividing C_{60} parallel to one of the three-fold axes results in

(a)

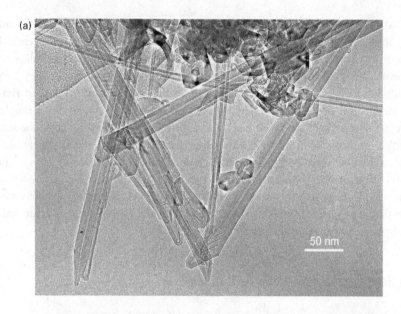

50 nm

(b)

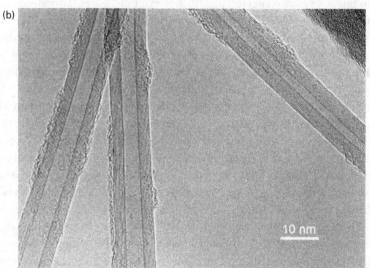

10 nm

Fig. 1.2   (a) A TEM image of multiwalled carbon nanotubes produced by arc-evaporation. (b) A higher magnification image of individual tubes.

the zig zag nanotube shown in Fig. 1.3(a), while bisecting C_{60} along one of the five-fold axes produces the armchair nanotube shown in Fig. 1.3(b). The terms 'zig zag' and 'armchair' refer to the arrangement of hexagons around the circumference. There is a third class of structure in which the hexagons are arranged helically around the tube axis (see Chapter 3). In practice, the caps are rarely hemispherical in shape, but can have a variety of morphologies; a typical example is shown in Fig. 1.4. More complex cap structures are often observed, owing to the presence of heptagonal as well as pentagonal carbon rings (1.21).

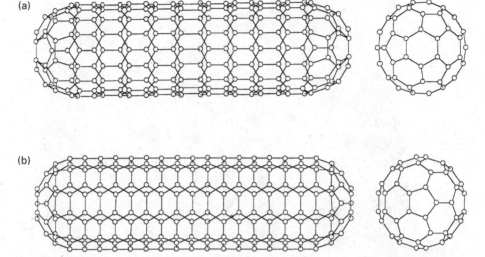

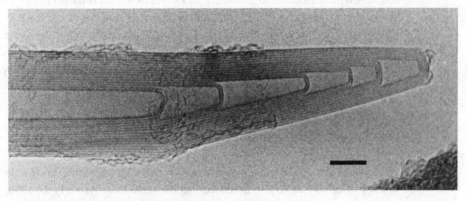

Fig. 1.3 Drawings of the two nanotubes that can be capped by one-half of a C_{60} molecule (1.20). (a) Zig zag $(9, 0)$ structure, (b) armchair $(5, 5)$ structure (see Chapter 5 for an explanation of the indices).

Fig. 1.4 An image of a typical multiwalled nanotube cap. Scale bar 5 nm.

1.3   Single- and double-walled nanotubes

Nanotubes of the kind described by Iijima in 1991 invariably contain more than one graphitic layer, and generally have inner diameters of around 4 nm. In 1993, Iijima and Toshinari Ichihashi of NEC, and Donald Bethune and colleagues of the IBM Almaden Research Center in California independently reported the synthesis of single-walled nanotubes (1.22, 1.23). This proved to be an extremely important development, since the single-walled tubes appeared to have structures that approximate to those of the 'ideal' nanotubes shown in Fig. 1.3. They proved to have extraordinary properties, and today there are more papers published on single-walled tubes than on their multiwalled counterparts. An important advance came in 1996 when Smalley's group described the synthesis of single-walled

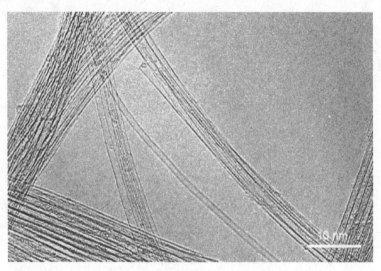

Fig. 1.5   A typical image of single-walled nanotubes. Courtesy Kazu Suenaga.

tubes using laser-vaporization (1.24). They can also be made catalytically, as discussed below. A typical image of single-walled nanotubes (SWNTs) is shown in Fig. 1.5. It can be seen that the appearance is quite different from that of samples of multiwalled nanotubes (MWNTs). The individual tubes have very small diameters (typically ~1 nm), and are often curved rather than straight. They also have a tendency to form bundles or 'ropes'. Methods for producing double-walled carbon nanotubes (DWNTs) in high yield using arc-evaporation have also now been developed (1.25).

1.4   Catalytically produced carbon nanotubes

The production of filamentous carbon by catalysis had been known long before Iijima's discovery of fullerene-related carbon nanotubes. As early as 1890, P. and L. Schultzenberger observed the formation of filamentous carbon during experiments involving the passage of cyanogen over red-hot porcelain (1.26). Work in the 1950s established that filaments could be produced by the interaction of a wide range of hydrocarbons and other gases with metals, the most effective of which were iron, cobalt and nickel. Probably the first electron micrographs showing tubular carbon filaments appeared in a 1952 paper by Radushkevich and Lukyanovich in the *Russian Journal of Physical Chemistry* (1.27). Serious research into the catalytic formation of carbon filaments began in the 1970s when it was appreciated that filament growth could constitute a serious problem in the operation of nuclear reactors, and in certain chemical processes. The most extensive programme of research was carried out in the 1970s by Terry Baker and his colleagues at the United Kingdom Atomic Energy Authority's laboratories at Harwell, and later in the USA (e.g. 1.28). This group were concerned with filamentous carbon growth in the cooling circuits of gas-cooled nuclear reactors. Thus, Baker's work was primarily motivated by the need to avoid filamentous carbon growth in these cooling pipes. A few workers, however, recognized that it might be possible to produce useful carbon nanofibres by catalysis. Notable among these was Morinobu Endo of Shinshu University in Japan, who proposed in 1988 (1.29) that catalytic carbon nanotubes could represent an alternative to conventional carbon fibres, which are produced by pyrolysing strands of polymer or pitch.

Following Iijima's paper in 1991, interest in catalytically produced carbon nanotubes exploded. As well as multiwalled nanotubes, single-walled and double-walled tubes can now also be produced by catalysis (e.g. 1.30, 1.31). The production of nanotubes catalytically is attractive for many reasons. Perhaps most importantly, it is much more amenable to scale-up than arc-evaporation. A number of processes have now been developed for the bulk synthesis of single-walled tubes using catalysis (e.g. 1.32, 1.33). Catalytic methods also allow the controlled growth of tubes on substrates, for applications in display technology and other areas. The main drawback is that, at least for multiwalled nanotubes, the quality of tubes produced in this way is poorer than for those produced by arc-evaporation.

1.5   Who discovered carbon nanotubes?

If carbon nanotubes have been known since the 1950s, if not earlier, why is Iijima's work considered so important? And who, in fact, should be credited with discovering carbon nanotubes? This has often been debated at nanotube conferences, and in the literature (1.34, 1.35), and the uncertainty surrounding the question of who actually discovered nanotubes probably explains why no Nobel Prizes have yet been awarded in this area. In discussing this topic, some researchers have attempted to downplay the importance of Iijima's work. It has even been stated that Iijima simply took better micrographs than anyone else! In the view of the present author this is wrong. Iijima's 1991 work is undoubtedly responsible for the current explosion of interest in carbon nanotubes, and for good reason. The nanotubes he prepared were far more perfect than those that had been previously produced catalytically, and differed from them in being all-carbon structures, closed at both ends, and not 'contaminated' with catalyst particles. Moreover, all the evidence suggests that arc-produced tubes have superior properties to catalytically-grown ones. This apparently remains the case even after the catalytic tubes have been annealed in an attempt to remove the defects (see Section 7.1.2). Thus it can be argued that fullerene-related carbon nanotubes are different in kind from catalytically produced tubes. It is important to note, however, that this distinction only seems to apply to multiwalled tubes. Tubes with one or two layers seem to have similar properties no matter how they are produced.

If we accept, then, that fullerene-related carbon nanotubes are a different species from catalytically produced tubes, is there evidence that the fullerene-related variety were known before 1991? The answer is yes, although the potential importance of these structures was not recognized. To begin, there are Iijima's studies of carbon films carried out in the late 1970s and early 1980s, which were mentioned above. For this work he prepared specimens of arc-evaporated carbon using an apparatus of the type commonly employed to make carbon support films for electron microscopy. The method he used would have differed slightly (but significantly) from the Krätschmer–Huffman technique in that the chamber would have been evacuated rather than filled with a small pressure of helium. The resulting films were largely amorphous, but contained small, partially graphitized regions which contained some unusual structures. These structures included discrete graphitic particles apparently made up of concentric closed shells, tightly curved around a central cavity. One of these structures, reproduced as Fig. 5(a) in his 1980 *Journal of Microscopy* paper (1.17), is clearly a nanotube, and Iijima confirmed its tubular nature using tilting experiments. But he did not explore these structures in detail, and suggested that the curved layers were probably due to sp^3 bonding, rather than, as we now believe, the presence of pentagonal rings.

There are other examples of structures resembling 'Iijima-type' multiwalled nanotubes scattered throughout the pre-1991 carbon literature. In some cases these structures might be contaminants on the carbon films used to support the samples (1.36). It has also been claimed that an image in a 1976 paper by Agnes Oberlin and colleagues contains an image of a single-walled tube (Fig. 11 of ref. 1.37). While this may be true, the authors did not recognize its significance at the time.

Work by theoreticians also anticipated the discovery of carbon nanotubes. For example, Patrick Fowler of Exeter University described theoretical studies of small cylindrical fullerene molecules in early 1990 (1.38). Two groups of American theorists, one at the Naval Research Laboratory, Washington DC (1.39), and one at the Massachusetts Institute of Technology (1.40) submitted papers on the electronic properties of fullerene tubes just a few weeks before Iijima's paper appeared in *Nature*. Last, but not least, the highly imaginative British chemist David Jones, under his pen-name Daedalus, ruminated about rolled-up tubes of graphite in The *New Scientist* in 1986 (1.41).

1.6   Carbon nanotube research

Interest in carbon nanotubes, which took off following the publication of Iijima's 1991 paper, continues to grow. This can be seen in the rise in the annual total of papers on nanotubes, up from 886 in 2000 to 5406 in 2007 (Fig. 1.6). During the same period the number of papers on fullerenes has remained fairly static at about 700 per year. The contrast between the number of papers on carbon nanotubes and the number on fullerenes of course reflects the far greater potential for practical applications of nanotubes. Despite the continuing growth of interest in carbon nanotubes, however, there are signs that the number of major breakthroughs in the field is falling. Thus, in 2007 there was just one paper on carbon nanotubes in *Nature* and two in *Science*, compared with nine and ten respectively in 1999. As far as the origin of papers is concerned, the United States leads the way, as might be expected, with 27% of the total of carbon nanotube papers published in 2007. Perhaps surprisingly, this is closely followed by China, with 26%, demonstrating a major investment in carbon science and nanotechnology in that country. About 10% of nanotube papers originated in Japan and 7% from South Korea. Of the European countries, Germany contributed about 6% of the total, the United Kingdom about 5% and France 4%.

Perhaps the largest volume of research into nanotubes has been devoted to their electronic properties. The theoretical work that pre-dated Iijima's discovery has already

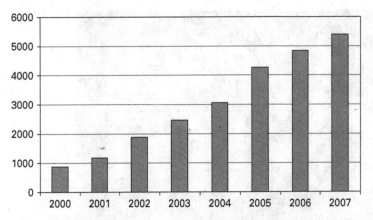

Fig. 1.6   The number of papers on carbon nanotubes published annually, from 2000 to 2007. Data from ISI Web of Knowledge.

been mentioned. A short time after the publication of Iijima's 1991 letter in *Nature*, theoretical studies appeared which demonstrated that the electronic properties of nano-tubes were a function of both tube structure and diameter. These remarkable predictions stimulated huge interest, but attempting to determine the electronic properties of nano-tubes experimentally presented great difficulties. By the late 1990s, however, studies were appearing which confirmed these predictions, and the first nanotube-based electro-nic devices began to be produced. The *Science* cover picture shown in Fig. 1.7(a) accompanied a 2001 article on nanotube logic circuits by Cees Dekker of Delft University of Technology and colleagues (see p. 167). The second major area of carbon nanotube research has revolved around their mechanical properties. Again, theory came slightly ahead of experiment, but ingenious measurements using electron microscopy and scanning probe microscopy soon confirmed the theoretical predictions: carbon nanotubes are the stiffest and strongest fibres known. These properties, coupled with their low density, mean that carbon nanotubes are the only fibres suitable for producing a 'space elevator', an Earth-to-space cable first proposed by Arthur C. Clarke (1.42), as illustrated in Fig. 1.7(b). On a more down-to-earth level, the outstanding mechanical properties of carbon nanotubes are beginning to find applications in a whole range of areas, from sports equipment to automobiles. In such applications it is almost always necessary to encapsulate the tubes in a matrix, and the production of carbon nanotube composites is becoming a major field in itself. As well as their electronic and mechanical properties, many other aspects of carbon nanotubes have captivated researchers, and, as will become clear in this book, nanotube research has developed in an amazingly wide range of directions. Some aspects of carbon nanotube research have even begun to impinge on popular culture. While Arthur C. Clarke was unaware of nanotubes when he speculated about space elevators, contemporary science fiction writers are alive to the possibilities offered by these new materials. Thus, in *Rollback* by Robert J. Sawyer (1.43) there is a glistening carbon nanotube tower called the Spire of Hope, while *River of Gods* by Ian McDonald (1.44) features a 'no-maintenance domestic scale carbon nanotube solar power generator', although how this actually works is not specified.

(a)

Fig. 1.7   Carbon nanotube cover art. (a) *Science*, 9 November 2001 issue, showing a nanotube-based electronic circuit. (b) *American Scientist*, July–August 1997 issue, with an illustration of a nanotube space elevator.

1.7   Scope of the book

The next two chapters cover the synthesis of carbon nanotubes. Chapter 2 is concerned with non-catalytic methods, primarily arc-evaporation and laser-vaporization, while catalytic synthesis is discussed in Chapter 3. Methods of purifying nanotubes and of processing them into defined forms and arrangements are covered in the following chapter. In Chapter 5, theoretical approaches to the analysis of nanotube structure are outlined, and experimental observations described. One of the major growth points of nanotube science, namely research into their electronic properties, is considered in Chapter 6, while other physical properties, including mechanical, are discussed in the following chapter. Chapter 8, entitled 'Chemistry and biology of nanotubes' covers the chemical and biomolecular functionalization of carbon nanotubes and their interaction with biological systems. Chapters 9–11 discuss the incorporation of nanotubes into polymer and other matrices; the filling of nanotubes with foreign materials and the production and properties of heterogeneous nanotubes; and the use of nanotubes in

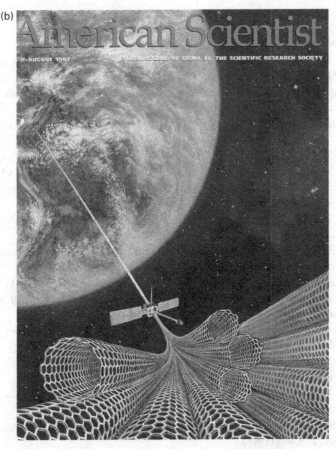

Fig. 1.7 (cont.)

sensing and imaging devices respectively. The final chapter summarizes the achievements and failings of carbon nanotube research.

References

(1.1) J. Emsley, *Nature's Building Blocks: an A–Z Guide to the Elements*, Oxford University Press, Oxford, 2003, p. 93.

(1.2) F. Derbyshire, M. Jagtoyen and M. Thwaites, *Porosity in Carbons*, ed. J. W. Patrick, Edward Arnold, London, 1995, p. 227.

(1.3) G. Collin, 'On the history of technical carbon', *CFI-Ceramic Forum Int.*, **77**, 28 (2000).

(1.4) P. J. F. Harris, 'On charcoal', *Interdisc. Sci. Rev.*, **24**, 301 (1999).

(1.5) M. Faraday, *The Chemical History of a Candle*, Collier, New York, 1962.

(1.6) W. H. Bragg and W. L. Bragg, 'The structure of diamond', *Proc. Roy. Soc. A*, **89**, 277 (1913).

(1.7) J. D. Bernal, 'The structure of graphite', *Proc. Roy. Soc. A*, **106**, 749 (1924).

(1.8) R. E. Franklin, 'Crystallite growth in graphitizing and non-graphitizing carbons', *Proc. Roy. Soc. A*, **209**, 196 (1951).

(1.9) P. J. F. Harris, 'Rosalind Franklin's work on coal, carbon, and graphite', *Interdisc. Sci. Rev.*, **26**, 204 (2001).

(1.10) K. Lonsdale, 'Diamonds, natural and artificial', *Nature*, **153**, 669 (1944).

(1.11) H. W. Kroto, J. R. Heath, S. C. O'Brien *et al.*, 'C_{60}: buckminsterfullerene', *Nature*, **318**, 162 (1985).

(1.12) J. Baggott, *Perfect Symmetry: the Accidental Discovery of Buckminsterfullerene*, Oxford University Press, Oxford, 1994.

(1.13) H. Aldersley-Williams, *The Most Beautiful Molecule*, Aurum Press, London, 1995.

(1.14) H. W. Kroto, 'Symmetry, space, stars and C_{60}' (Nobel lecture), *Rev. Mod. Phys.* **69**, 703 (1997).

(1.15) R. E. Smalley, 'Discovering the fullerenes' (Nobel lecture), *Rev. Mod. Phys.* **69**, 723 (1997).

(1.16) W. Krätschmer, L. D. Lamb, K. Fostiropoulos *et al.*, 'Solid C_{60}: a new form of carbon', *Nature*, **347**, 354 (1990).

(1.17) S. Iijima, 'High resolution electron microscopy of some carbonaceous materials', *J. Microscopy*, **119**, 99 (1980).

(1.18) S. Iijima, 'Direct observation of the tetrahedral bonding in graphitized carbon black by high-resolution electron microscopy', *J. Cryst. Growth*, **50**, 675 (1980).

(1.19) S. Iijima, 'Helical microtubules of graphitic carbon', *Nature*, **354**, 56 (1991).

(1.20) M. Ge and K. Sattler, 'Scanning tunnelling microscopy of single-shell nanotubes of carbon', *Appl. Phys. Lett.*, **65**, 2284 (1994).

(1.21) S. Iijima, T. Ichihashi and Y. Ando, 'Pentagons, heptagons and negative curvature in graphitic microtubule growth', *Nature*, **356**, 776 (1992).

(1.22) S. Iijima and T. Ichihashi, 'Single-shell carbon nanotubes of 1-nm diameter', *Nature*, **363**, 603 (1993).

(1.23) D. S. Bethune, C. H. Kiang, M. S. de Vries *et al.*, 'Cobalt-catalysed growth of carbon nanotubes with single-atomic-layer walls', *Nature*, **363**, 605 (1993).

(1.24) A. Thess, R. Lee, P. Nikolaev *et al.*, 'Crystalline ropes of metallic carbon nanotubes', *Science*, **273**, 483 (1996).

(1.25) J. L. Hutchison, N. A. Kiselev, E. P. Krinichnaya *et al.*, 'Double-walled carbon nanotubes fabricated by a hydrogen arc discharge method', *Carbon*, **39**, 761 (2001).

(1.26) P. Schultzenberger and L. Schultzenberger, 'Sur quelques faits relatifs à l'histoire du carbone', *C. R. Acad. Sci., Paris*, **111**, 774 (1890).

(1.27) L. V. Radushkevich and V. M. Lukyanovich, 'On the carbon structure formed during thermal decomposition of carbon monoxide in the presence of iron' (in Russian), *Zh. Fizich. Khim.*, **26**, 88 (1952).

(1.28) R. T. K. Baker and P. S. Harris, 'The formation of filamentous carbon', *Chem. Phys. Carbon*, **14**, 83 (1978).

(1.29) M. Endo, 'Grow carbon fibres in the vapour phase', *Chemtech*, **18**, 568 (1988) (September issue).

(1.30) H. J. Dai, A. G. Rinzler, P. Nikolaev *et al.*, 'Single-wall nanotubes produced by metal-catalyzed disproportionation of carbon monoxide', *Chem. Phys. Lett.*, **260**, 471 (1996).

(1.31) E. Flahaut, A. Peigney, C. Laurent *et al.*, 'Synthesis of single-walled carbon nanotube–Co–MgO composite powders and extraction of the nanotubes', *J. Mater. Chem.*, **10**, 249 (2000).

(1.32) P. Nikolaev, M. J. Bronikowski, R. K. Bradley *et al.*, 'Gas-phase catalytic growth of single-walled carbon nanotubes from carbon monoxide', *Chem. Phys. Lett.*, **313**, 91 (1999).

(1.33) D. E. Resasco, W. E. Alvarez, F. Pompeo *et al.*, 'A scalable process for production of single-walled carbon nanotubes (SWNTs) by catalytic disproportionation of CO on a solid catalyst', *J. Nanoparticle Res.*, **4**, 131 (2002).

(1.34) H. P. Boehm, 'The first observation of carbon nanotubes', *Carbon*, **35**, 581 (1997).

(1.35) M. Monthioux and V. L. Kuznetsov, 'Who should be given the credit for the discovery of carbon nanotubes?', *Carbon*, **44**, 1621 (2006).

(1.36) P. J. F. Harris, 'Carbonaceous contaminants on support films for transmission electron microscopy', *Carbon*, **39**, 909 (2001).

(1.37) A. Oberlin, M. Endo and T. Koyama, 'Filamentous growth of carbon through benzene decomposition', *J. Cryst. Growth*, **32**, 335 (1976).

(1.38) P. W. Fowler, 'Carbon cylinders: a new class of closed-shell clusters', *J. Chem. Soc., Faraday Trans.*, **86**, 2073 (1990).

(1.39) J. W. Mintmire, B. I. Dunlap and C. T. White, 'Are fullerene tubules metallic?', *Phys. Rev. Lett.*, **68**, 631 (1992).

(1.40) M. S. Dresselhaus, G. Dresselhaus and R. Saito, 'Carbon fibers based on C_{60} and their symmetry', *Phys. Rev. B*, **45**, 6234 (1992).

(1.41) D. E. H. Jones (Daedalus), *New Scientist*, **110**, (1505), 88 and (1506), 80 (1986).

(1.42) A. C. Clarke, *The Fountains of Paradise*, Gollancz, London, 1979.

(1.43) R. J. Sawyer, *Rollback*, Tor, New York, 2007.

(1.44) I. McDonald, *River of Gods*, Pyr, New York, 2006.

2 Synthesis I: arc- and laser-vaporization, and heat treatment methods

As we saw in the last chapter, the excitement surrounding carbon nanotubes was originally sparked by Iijima's production of highly perfect multiwalled tubes using arc-evaporation in 1991. Although the catalytic production of carbon tubules had been known for decades, the structures discovered by Iijima displayed a degree of perfection much greater than those seen in catalytic tubes. The first synthesis of single-walled nanotubes in 1993 also involved arc-evaporation, this time with metal-impregnated electrodes. Arc-evaporation remains an important method of nanotube synthesis, and will be discussed in detail in this chapter. The chapter begins with a description of the practical aspects of the arc synthesis of multiwalled nanotubes. A summary of the various models that have been put forward for the growth of multiwalled tubes in the arc is then given. This is followed by a discussion of the production of multiwalled nanotubes by high-temperature heat treatment of disordered carbon. The synthesis of single-walled carbon nanotubes by arc-evaporation and by laser-vaporization is then covered, and the possible mechanisms of nanotube formation in these processes summarized. Finally, the arc synthesis of double-walled nanotubes is described.

2.1 Production of multiwalled nanotubes by arc-evaporation

2.1.1 Early work

The original method used by Iijima to prepare nanotubes (2.1) differed slightly from the Krätschmer–Huffman technique for C_{60} production in that the graphite electrodes were held a short distance apart during arcing, rather than being kept in contact. Under these conditions, some of the carbon which evaporated from the anode recondensed as a hard cylindrical deposit on the cathodic rod. It was the central part of this deposit that Iijima found to contain both nanotubes and nanoparticles. But the yield was rather poor in these initial experiments. A significant advance came in July 1992 when Thomas Ebbesen and Pulickel Ajayan, working at the same Japanese laboratory as Iijima, discovered that increasing the pressure of He in the arc-evaporation chamber dramatically improved the yield of nanotubes formed in the cathodic soot (2.2). The discussion that follows draws considerably on the excellent reviews which have been given by Ebbesen (2.3, 2.4).

A variety of different arc-evaporation reactors have been employed for nanotube synthesis, but a stainless steel vacuum chamber with a viewing port is probably the

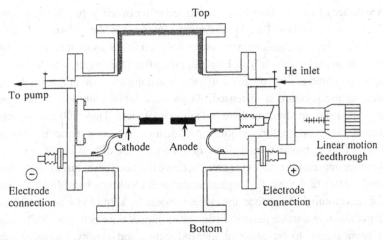

Fig. 2.1 A schematic illustration of arc-evaporation apparatus for the production of fullerenes and nanotubes (adapted from ref. 2.5). Although not shown here, it is usual for the electrodes to be water-cooled.

most commonly-used type. A typical example is illustrated in Fig. 2.1. A glass-dome chamber of the kind used in the original Krätschmer–Huffman experiments is not ideal, since this does not easily allow for the separation of the rods to be adjusted during discharge. The chamber must be connected both to a vacuum line with a diffusion pump, and to a He supply. A continuous flow of He at a given pressure is usually preferred over a static atmosphere of the gas. The electrodes are two graphite rods, usually of high-purity, although there is no evidence that exceptionally pure graphite is necessary. Indeed, nanotubes have been successfully produced using very impure forms of carbon, such as coal, as electrodes (see later). Typically, the anode is a long rod approximately 6 mm in diameter and the cathode a much shorter rod 9 mm in diameter. Efficient water-cooling of the cathode has been shown to be essential in producing good quality nanotubes, and the anode is also frequently cooled. The position of the anode should be adjustable from outside the chamber, so that a constant gap can be maintained during arc-discharge. A voltage-stabilized DC power supply is normally used, and discharge is typically carried out at a voltage of 20 V. The current depends on the diameter of the rods, their separation, the gas pressure and so on, but is usually in the range 50–100A.

When the pressure is stabilized, the voltage should be turned on. At the start of the experiment the electrodes should not be touching, so no current will initially flow. The movable anode is now gradually moved closer to the cathode until arcing occurs. When a stable arc is achieved, the gap between the rods should be maintained at approximately 1 mm or less; the rod is normally consumed at a rate of a few mm per minute. When the rod is consumed, the power should be turned off and the chamber left to cool before opening. The rate of nanotube synthesis in the arc-evaporation process is quite high: deposits are typically generated at rates of 20–100 mg min^{-1} (2.6). However, the discharge can only be sustained for a few minutes.

A number of factors have been shown to be important in producing a good yield of high-quality nanotubes. Perhaps the most important is the pressure of the He in the evaporation chamber, as demonstrated by Ebbesen and Ajayan in their 1992 paper (2.2). This is illustrated graphically in Fig. 2.2, taken from this paper, which shows nanotube samples prepared at 20, 100 and 500 torr. A striking increase in the number of tubes is evident as the pressure is increased. At pressures above 500 torr there is no obvious change in sample quality, but there is a fall in total yield. Thus, 500 torr appears to be the optimum He pressure for nanotube production. Note that these conditions are not optimum for C_{60} production, which requires a pressure of below 100 torr.

Another important factor in the arc-discharge method is the current, as demonstrated in several studies (2.7, 2.8). Too high a current will result in a hard, sintered material with few free nanotubes. Therefore, the current should be kept as low as possible, consistent with maintaining a stable plasma. Efficient cooling of the electrodes and the chamber has also been shown to be essential in producing good quality nanotube samples and avoiding excessive sintering. If arc-discharge has been carried out correctly, a cylindrical and homogenous deposit should form on the cathode. This consists of a hard outer shell, consisting of fused material and a softer fibrous core that contains discrete nanotubes and nanoparticles. These can be extracted by cutting open the outer shell. Some indication of the quality of the nanotube samples be gained by a simple physical examination of the carbon. A poor sample containing few nanotubes will generally have a powdery texture, while good quality material can be smeared to produce sheet-like flakes with a grey metallic lustre (note, however, that gloves should be worn when handling the carbon: see Section 2.1.4 below).

2.1.2   The arc-evaporation technique: further developments

There have been a number of variations on the 'classic' arc-evaporation method since the original work in the early 1990s. Several groups have experimented with using alternatives to He for arc-evaporation. These alternative gases include H_2 (2.9–2.11), N_2 (2.12), CF_4 (2.13) and organic vapours (2.14). Some of these experiments have produced interesting results. For example, in a paper published in *Nature* in 2000 (2.11), Iijima and colleagues claimed to have produced the 'smallest possible carbon nanotube'. The tube, with a diameter of 0.4 nm, was the innermost shell of a multiwalled nanotube in a sample produced by arc-evaporation under H_2. It was suggested that the tube had a (3, 3) armchair structure and was capped with half of a C_{20} molecule. Although smaller nanotubes are theoretically possible, the (3, 3) structure is believed to be the narrowest that could have any realistic stability. It was suggested that the H_2 atmosphere facilitated the formation of the semi-C_{20} dodecahedra by terminating dangling bonds with hydrogen. Addition of carbon species to these stabilized 'seeds' would then result in the growth of (3, 3) tubes. In the same issue of *Nature*, a group from the Hong Kong University of Science and Technology reported the pyrolytic synthesis of SWNTs with a similar diameter (2.15). In a later paper it was claimed that these tubes exhibited superconductivity, as discussed in Chapter 6 (p. 164). In general, there is little evidence that using alternative gases produces major benefits in terms of nanotube yield, although workers

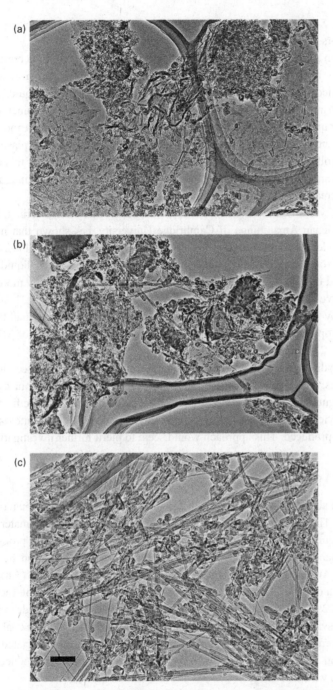

Fig. 2.2   Micrographs showing the effect of helium on the yield of nanotubes in arc-evaporation
experiments, from the work of Ebbesen and Ajayan (2.2). Samples prepared at (a) 20 torr,
(b) 100 torr and (c) 500 torr.

from the Fuji Xerox company in Japan (2.14) claim that the use of hexane or other organic vapours can dramatically increase the yield of MWNTs. Another variant on the arc-evaporation process involves carrying out the arcing in a liquid, thus removing the requirement for a vacuum chamber altogether. It seems that the first experiments of this kind were carried out in 2000 by Alex Zettl's team from Berkeley using liquid N_2 (2.16). Their method involved inserting a graphite anode into an open container of liquid nitrogen containing a short copper or graphite cathode. The electrodes were momentarily brought into contact and an arc was struck. Nanotubes formed in the arc plasma region dropped and collected on the bottom of the vessel. A major advantage of this approach over arc-discharge under a gas is that the process can be operated continuously, thus potentially increasing the yield of nanotubes many fold.

Subsequent work, notably by workers from Tsing Hua University, Beijing, and by a team led by Gehan Amaratunga of Cambridge University, has shown that nanotubes and nanoparticles can also be made by arc-discharge under water (2.17–2.21). The Amaratunga group compared the quality of nanotubes produced under liquid N_2 with those produced in water (2.20) and found that the latter were significantly more perfect. This group have also discussed the production of carbon onions (2.20, 2.21) and nano-horns (2.22) by arc-discharge under water. Other researchers have described the development of an optoelectronically automated system for the arc-synthesis of MWNTs in solution (2.23, 2.24).

Another modification of the original arc-evaporation method involved the use of magnetic fields. The Fuji Xerox group showed in 2002 that situating four cylindrical Nd–Fe–B magnets around the electrodes, to form a symmetrical magnetic field, greatly increased the yield of nanotubes (2.25). Soots containing up to 97% of nanotubes could apparently be produced. This approach would seem to merit further investigation.

2.1.3   Alternatives to graphite

As mentioned above, arc-evaporation can be carried out with electrode materials other than graphite. There has been some interest in using coal as an electrode material, since this would reduce the cost of raw material approximately ten-fold. The production of multiwalled nanotubes from coal using arc-discharge was first explored by Michael Wilson of the Australian National University and his co-workers in 1993 (2.8), and has been studied in detail by Jason Qiu and colleagues of Dalian University of Technology, China (2.26–2.29). It is not clear at present whether using coal as an electrode material for arc-discharge synthesis of nanotubes offers any real advantages. Although coal is cheap, the saving is probably insignificant compared with the labour costs associated with the arc-evaporation method. There is also the problem of contamination produced by the non-carbon constituents of coal.

2.1.4   Safety considerations for the arc-evaporation method

This section considers briefly the safety issues associated with the arc-evaporation synthesis of nanotubes. It is clearly important to check the machine for short circuits before carrying out arc-discharge and, if a vacuum chamber is being used, the vacuum should be tested for leaks before introducing the inert gas. Since most chambers will have a viewing port, care must also be taken to protect the operator's eyes from the intense light of the arc using a high-density optical glass filter. The soot produced by arc-evaporation, particularly that which condenses on the walls of the chamber, is extremely light and can easily become airborne. Precautions should therefore be taken to avoid inhalation. For this reason, it is recommended that the entire arc-discharge apparatus is enclosed in a fume hood. A mask should also be worn when opening the chamber, and it is advisable to wear gloves when handling the fullerene-related materials.

Carrying out arc-evaporation under liquids appears to be a less alarming procedure than one might expect. If carried out correctly, the discharge should not cause the liquid to dramatically vaporize or bubble violently. In the case of arc-discharge in water, some CO vapour may be released, which must be extracted. Otherwise the process is relatively mild and, with proper circulation of the liquid, can be run for several hours at a time.

The toxicity of carbon nanotubes is discussed in Chapter 8.

2.2   Growth mechanisms of multiwalled nanotubes in the arc

2.2.1   General comments

Before discussing theories of the growth mechanisms of MWNTs in the arc, it is worth considering the influence of tube structure on growth. Many of the comments here may also apply to other types of nanotube growth.

Iijima pointed out in his 1991 *Nature* paper (2.1) that the growth of tubes with a helical structure would seem to be favoured, since such tubes have a repetitive step at the growing edge. This situation, illustrated in Fig. 2.3, is rather similar to the emergence of a screw dislocation from a crystal surface. Armchair and zigzag nanotubes do not possess such a favourable growth structure and would require the repeated nucleation of a new ring of hexagons. This suggests that helical nanotubes should be much more commonly observed than armchair and zigzag tubes, and the experimental evidence tends to confirm this.

A further, very basic, question concerning the growth mechanism is why tubes remain open during growth. An early view, put forward by Smalley and colleagues, was that the electric field in the arc may be important in keeping tubes open during growth (e.g. 2.30, 2.31). If correct, this would help to explain why nanotubes are never found in the soot which condenses on the walls of the arc-evaporation vessel, but only on the cathode. However, calculations indicated that field-induced lowering of the open tip energy was not sufficient to stabilize the open configuration except for unrealistically high fields (2.32, 2.33). Therefore, a refined model was developed in which adatom 'spot-welds' between layers help to stabilize the open tip conformation against closure (2.34).

An alternative explanation for the phenomenon of open-ended growth is that the interactions between adjacent concentric tubes can stabilize open tubes (2.35, 2.36). A detailed analysis of the interaction of two adjacent tubes was carried out by

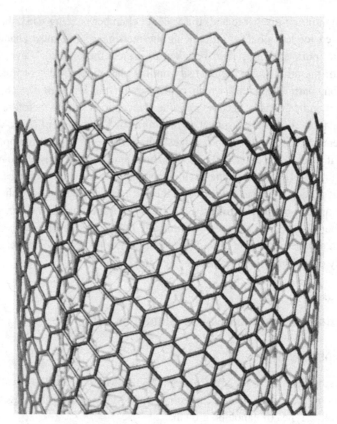

Fig. 2.3   A drawing of two concentric helical tubes showing the presence of steps at the growing edges (2.3).

Jean-Christophe Charlier and colleagues using molecular dynamics simulations (2.35). They considered a $(10,0)$ tube inside an $(18,0)$ tube, and found that bridging bonds formed between the edges of the two tubes. At high temperatures (3000 K), the configuration of the lip–lip bonding structures was found to fluctuate continuously. It was suggested that this fluctuating structure would provide active sites for the adsorption and incorporation of new carbon atoms, thus enabling the tube to grow. A final possibility is simply that nanotube growth is a kinetic process, and that the growth of an open tube is more kinetically favourable than closure.

As far as the detailed mechanism of MWNT growth in the arc is concerned, three types of model have been put forward, which could be labelled 'gas', 'solid' and 'liquid'. These are now discussed in turn.

2.2.2   Vapour phase growth

Most early theories of nanotube formation in the arc assumed that nucleation and growth occurred as a result of direct condensation from the vapour, or plasma, phase. It was also thought that the electric field of the arc played an essential role in inducing the 'one-dimensional' growth that leads to the formation of tubes (2.37). Probably the most detailed analysis of the gas phase nucleation and growth of MWNTs in the arc was given by Eugene Gamaly, an expert on plasma physics, and Thomas Ebbesen in 1995 (2.38). These authors began by assuming that the nanotubes and nanoparticles form in the region of the arc next to the cathode surface. They then analysed the density and velocity distribution of carbon vapours in this region, taking into account the temperature and the properties of the arc, in order to develop their model. They suggested that in this layer of carbon vapour there will be two groups of carbon particles with different velocity distributions. This idea is central to their growth model. One group of carbon particles will have a Maxwellian, i.e. isotropic, velocity distribution corresponding to the temperature of the arc (~3700 °C). The other group is composed of ions accelerated in the gap between the positive space charge and the cathode. The velocity of these carbon particles will be much greater than those of the thermal particles, and in this case the flux will be directed rather than isotropic. The process of nanotube (and nanoparticle) formation is considered to occur in three stages. In the first stage the isotropic velocity distribution results in the formation of approximately equiaxed structures such as nanoparticles. As the current becomes more directed, open structures begin to form which Gamaly and Ebbesen consider to be the seeds for nanotube growth. In the second stage, a stream of directed carbon ions flows in a direction perpendicular to the cathode surface, resulting in rapid tube growth. Finally, instabilities in the arc discharge lead to abrupt termination of nanotube growth by the formation of caps.

A variation of the vapour phase growth model has been given by Oleg Louchev and colleagues (2.39, 2.40). Here, the key process is not the direct condensation of carbon atoms onto a growing edge, but the adsorption of atoms onto a nanotube surface followed by surface diffusion to the growth edge. The kinetics of nanotube growth in this model have been analysed in detail (2.40). Han Zhang and colleagues of Peking University have extended the Louchev model by considering heptagon formation at the growing edge (2.41).

2.2.3 Liquid phase growth

The liquid phase model of multiwalled nanotube growth was put forward by Walt De Heer of Georgia Tech and colleagues in 2005 (2.42). These workers studied MWNTs formed on the surfaces of columns within the cathodic deposit. They found that these tubes were often decorated with beads of amorphous carbon, as shown in Fig. 2.4. The appearance of these beads was suggestive of solidified liquid droplets, and this led them to conclude that liquid carbon played a central role in nanotube nucleation and growth. Based on their observations, and on the known properties of liquid carbon, they proposed the following nanotube formation scenario. When arc-discharge is initiated, the carbon anode is locally heated by electron bombardment from the cathode, causing the surface to locally liquefy and liquid carbon globules to be ejected from the anode. Initially, because of the high vapour pressure of liquid carbon, the surface of a globule will evaporatively cool very rapidly. However, the cooling of the interior of the globule occurs much more slowly, and this causes the liquid carbon to supercool. It is within this supercooled liquid carbon that carbon nanotubes and nanoparticles are envisaged to homogeneously nucleate and grow.

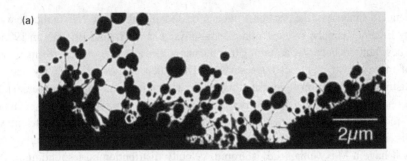

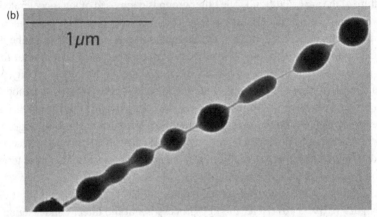

Fig. 2.4   Micrographs from the work of De Heer *et al.* showing arc-produced MWNTs covered with beads of amorphous carbon (2.42).

2.2.4   Solid phase growth

The solid phase theory of MWNT growth by arc-evaporation (2.43) was first put forward by the present author and his colleagues in a paper published in 1994 (2.44). In this paper, fullerene soot was heated to approximately 3000 °C in a positive-hearth electron gun. This resulted in the formation of single-walled cones and tubes, as discussed in Section 2.3 below. The observation that nanotube-like structures can be produced by high-temperature heat treatment of fullerene soot prompted us to put forward a solid-state model of nanotube growth, in which fullerene soot is an intermediate product. The model can be summarized as follows. In the initial stages of arc-evaporation, carbon in the vapour phase (consisting largely of C_2 species) condenses onto the cathode as a fullerene soot-like material. This condensed carbon then experiences extremely high temperatures as the arcing process continues, resulting in the formation firstly of nanotube 'seeds' and then of multiwalled nanotubes. Growth terminates when the supply of carbon is exhausted or when arcing finishes. The model is illustrated in Fig. 2.5. It is recognized that this model is incomplete. In particular, it has not been established why the fullerene soot evolves into carbon nanotubes rather than nanoparticles when heated in the arc. Experiments on the high-temperature heat treatment of fullerene soot and other carbons, described below, show that nanoparticles are the usual product of such a treatment. It may

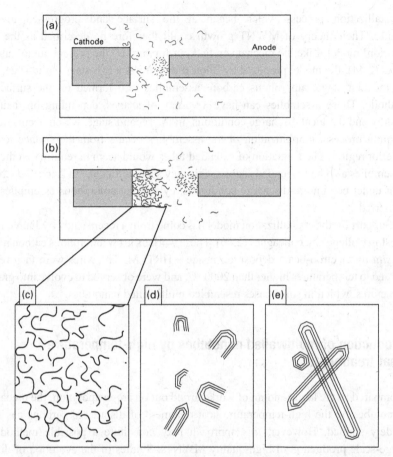

Fig. 2.5   A schematic illustration of the solid phase growth model for multiwalled carbon nanotubes. (a) Electron bombardment from the cathode causes heating of the anode surface, and evaporation of C_2 and other species. These rapidly coalesce into fullerene soot fragments. (b) Some of the fullerene soot condenses onto the cathode, with the remainder being deposited on the walls of the vessel. (c)–(e) Enlarged views of the interior of the cathodic deposit, showing the transformation of fullerene soot into firstly open-ended 'seed' structures and then multiwalled nanotubes and nanoparticles (2.43).

be that the electric field of the arc helps to promote tube growth, or that kinetic factors are involved.

2.2.5   The crystallization model

In 2003, Dan Zhou and Lee Chow of the University of Central Florida described detailed high-resolution TEM observations of defective nanotube-like structures produced by arc-evaporation (2.45). Among other structures, they found complex, branching forms (see also Section 5.3.5). These observations suggested to the authors that the formation and growth of multiwalled nanotubes does not proceed from one end to the other, as assumed in most previous theories, but that the formation of the tubes was a crystallization process which began at the surface and progressed toward the centre. Their theory of MWNT growth could therefore be described as the 'crystallization' model. Like the mechanism first put forward by the present author and colleagues (2.44), the model proposed by Zhou and Chow is a two-stage, solid-state, process. In the first stage, amorphous carbon 'assemblies' are formed on the surface of the cathode. These assemblies can have a variety of shapes, depending on their surface energy and the local discharge conditions. In the second stage, which occurs during the cooling process, graphitization of the assemblies occurs from the surface toward the interior region. The formation of extended tubes would seem to require that the original assemblies also had extended shapes. This may seem unlikely, but Zhou and Chow state that under certain arc-discharge conditions, cylindrical amorphous assemblies may be preferred.

Support for the crystallization model has come from a recent study by Jianyu Huang of Boston College and colleagues (2.46). These workers grew amorphous carbon nanowires *in situ* by electron-beam deposition inside a HRTEM. The wires were then resistively heated to temperatures higher than 2000 °C, and were observed to evolve into graphitized structures, which in some cases resembled multiwalled nanotubes.

2.3   Production of multiwalled nanotubes by high-temperature heat treatments

Compared to the large amount of work carried out on arc-evaporation, the preparation of nanotubes by the high-temperature heat treatment of disordered carbons has not been widely studied. However, this approach may constitute a possible method for the large-scale production of high-quality MWNTs. Studies of the evolution of disordered carbon into multiwalled nanotubes may also provide insights into the nucleation and growth mechanism. The first experiments in this area involved heat treatments of fullerene soot. As already mentioned, fullerene soot is the light, fluffy carbon that forms on the walls of the evaporation vessel during fullerene synthesis. It is this soot which contains the C_{60}, C_{70} and higher fullerenes, which can be extracted using organic solvents. A typical micrograph of fullerene soot is shown in Fig. 2.6. The structure of the soot is highly disordered, consisting of curved carbon fragments in which both pentagons and heptagons are distributed randomly throughout a hexagonal network, producing continuous curvature (2.44, 2.47, 2.48).

High-temperature heat treatments of fullerene soot were first carried out by the Oxford group (2.44) and by Walt De Heer and Daniel Ugarte, then at the Ecole Polytechnique Fédérale de Lausanne in Switzerland (2.49, 2.50). The results were somewhat different. The Oxford group found that heat treatment produced a structure apparently made up of large pores which were often extended in shape, resembling large-diameter single layer nanotubes, as can be seen in the micrograph shown in Fig. 2.7(a). Like nanotubes, the extended pores were almost invariably closed, and exhibited a variety of capping morphologies. In some cases features were observed which are thought to be indicative of the presence of seven-membered carbon rings. The extended pores were usually

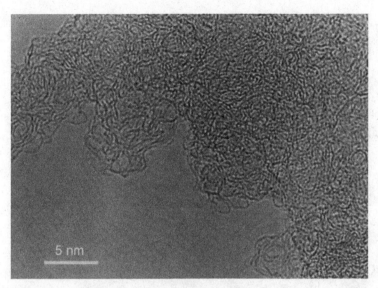

Fig. 2.6   A high-resolution electron micrograph of fullerene soot.

bounded by single carbon layers, although multilayer structures were also present. The precise mechanism of the transformation of fullerene soot into nanotube-like structures is not known, but may involve rearrangements such as the Stone–Wales mechanism (see p. 115).

De Heer and Ugarte found that high-temperature heat treatments generally tended to transform the fullerene soot into small, graphitic nanoparticles rather than nanotubes. However, rather short multiwalled nanotubes were occasionally observed in the heat-treated soot. An example is shown in Fig. 2.7(b) (2.49). These observations led the present author and his colleagues to propose the solid phase model of MWNT growth discussed above.

Robert Chang and colleagues from Northwestern University, in a paper published in 2000, described further studies of the synthesis of MWNTs by high-temperature heat treatment (2.51). They began by studying the annealing of fullerene soot made by the arc evaporation of graphite at 450 torr of He. The soot was activated in a CO_2 atmosphere at 850 °C to increase its surface area and then heated at 2200–2400 °C in a graphite resistance furnace. In the case of pure soot, this heat treatment mainly resulted in the formation of nanoparticles, as observed in the work of de Heer *et al.* When the soot was mixed with amorphous boron, however, heat treatment resulted in the growth of nano-tubes several microns in length. Similar experiments were then carried out on other disordered carbon materials: ball milled graphite, carbon black, and sucrose carbon. Heat treatment of ball milled graphite or carbon black did not produce any nanotubes, even when doped with boron. However, annealing sucrose carbon with boron did produce multiwalled nanotubes, albeit in relatively low yield. These experiments provide further evidence that nanotube growth is a solid-state process. In their initial study the Chang group succeeded in preparing nanotubes from fullerene soot and sucrose carbon, but not

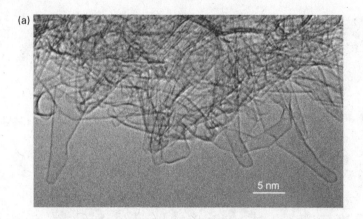

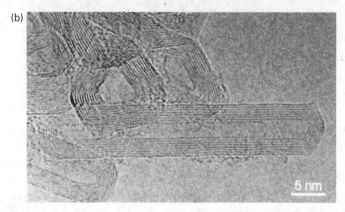

Fig. 2.7   Nanotube structures observed in fullerene soot following high-temperature heat treatment. (a) Tube-like structures in soot heated to approximately 3000 °C in a positive-hearth electron gun (2.44), (b) short MWNT in soot heated to 2400 °C, from work by De Heer and Ugarte (2.49).

from carbon black. Subsequent work showed that carbon black could also be used as a precursor for nanotube synthesis (2.52–2.54). In this study an arc-evaporation unit was modified so that it could be employed as a high-temperature furnace.

There are other examples of the solid-state synthesis of MWNTs. Lewis Chadderton and Ying Chen of the Australian National University reported the production of nanotubes by thermal annealing of mechanically milled graphite powder at temperatures around 1400 °C (2.55). Again, the formation of nanotubes at such low temperatures must have been a solid-state process, and the authors emphasize the importance of surface diffusion in the transformation. They also suggest that the nanotube growth may have been partly catalysed by impurity particles from the milling process. A further example of the production of nanotubes by heat treatment is the formation of double-walled nanotubes from tubes filled with fullerenes, as discussed in Chapter 10 (p. 257).

Finally in this section it should be mentioned that multiwalled carbon nanotubes, albeit with rather unusual structures, can sometimes be found in commercial samples of

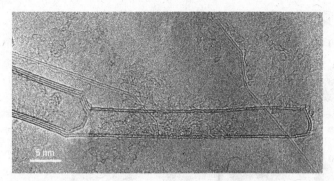

Fig. 2.8   Nanotube structure found in a commercial graphite.

synthetic graphite (2.56). An example is shown in Fig. 2.8. Since synthetic graphite is produced from solid precursors (typically petroleum coke) by high-temperature heat treatment, this provides further evidence for a solid-state growth mechanism.

2.4   Production of single-walled nanotubes by arc-evaporation

The discovery of single-walled carbon nanotubes, like so many in fullerene science, was serendipitous. In early 1993, several groups reported that foreign materials could be encapsulated inside carbon nanoparticles or nanotubes by carrying out arc-evaporation using modified electrodes. Rodney Ruoff's group in the USA (2.57) and Yahachi Saito's group in Japan (2.58) prepared encapsulated crystals of LaC_2 by employing electrodes impregnated with La, while Supapan Seraphin and colleagues reported that YC_2 could be introduced into nanotubes by using electrodes containing Y (2.59). This work opened the way to a whole new field based on the use of nanoparticles and nanotubes as 'molecular containers', as described in Chapter 10, but it also led indirectly to a quite different discovery, with even more important implications.

Donald Bethune and his colleagues of the IBM Almaden Research Center in San Jose, California were particularly interested in the papers of Ruoff and others. This group were working on magnetic materials for applications in information storage, and believed that ferromagnetic transition metal crystallites encapsulated in carbon shells might be of great value in this area. Bethune therefore set out to try some arc-evaporation experiments using electrodes impregnated with the ferromagnetic transition metals Fe, Co and Ni and an atmosphere of He (100–500 torr). But the result of this experiment was not at all what he expected. To begin with, the soot produced by arc-evaporation was quite unlike the normal material produced by the arc-evaporation of pure graphite. Sheets of soot hung like cobwebs from the chamber walls, while the material deposited on the walls themselves had a rubbery texture, and could be peeled away in strips. When Bethune and his colleague Robert Beyers examined this strange new material using high-resolution electron microscopy they were astonished to find that it contained multitudes of nanotubes with single-atomic-layer walls. These ultra-fine tubes were entangled with amorphous soot and particles of metal or metal carbide, holding the material together in a way

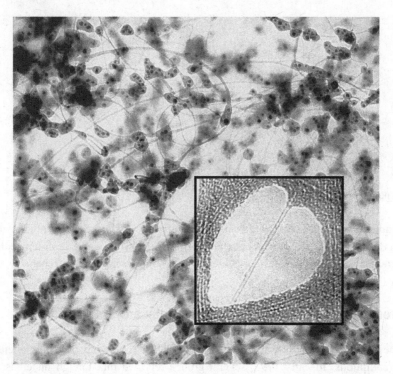

Fig. 2.9   Images from the work of Bethune *et al.* (2.60) showing single-walled carbon nanotubes produced by co-vaporization of graphite and cobalt. The tubes have diameters of approximately 1.2 nm.

that would account for its strange texture. This work was written up for *Nature* and appeared in June 1993 (2.60). Micrographs taken from their paper are shown in Fig. 2.9.

Independently of the American group, Sumio Iijima and Toshinari Ichihashi of the NEC laboratories in Japan were also experimenting with arc-evaporation using modified electrodes. In addition, they were interested in the effect of varying the atmosphere inside the arc-evaporation chamber. Like Bethune and colleagues, they discovered that certain conditions produced a quite different type of soot from that normally formed by arc-evaporation. For this work, the Japanese researchers impregnated their electrodes with Fe, and the atmosphere in the arc-evaporation chamber was a mixture of methane and argon rather than pure He. When examined by high-resolution electron microscopy, the arc-evaporated material was found to contain extremely fine single-walled nanotubes running like threads between clusters containing amorphous carbon and metal particles (2.61).

As noted in Chapter 1, single-walled nanotubes differ from multiwalled tubes produced by conventional arc-evaporation in having a very narrow range of diameters. In the case of the multiwalled tubes, the inner diameter can range from *c.* 1.5 to *c.* 15.0 nm, and the outer diameter from *c.* 2.5 to *c.* 30 nm. The single-layer tubes, on the other hand, all have extremely narrow diameters. In the material produced by Bethune and colleagues, the tubes had diameters of 1.2 (±0.1) nm, while Iijima and Ichihashi found that the tube diameters ranged from about 0.7 to 1.6 nm with the average being approximately 1.05 nm. Like tubes produced by conventional arc-evaporation, all the single-layer tubes appeared to be capped, and there appeared to be no evidence that catalytic metal particles were present at the ends of the tubes.

Following these initial studies, a great deal of work has now been carried out on optimizing the arc synthesis of single-walled tubes (e.g. 2.62–2.68). Helium at around 500–800 torr appears to be the most favourable atmosphere for SWNT production, and Fe, Co and Ni, or mixtures such as Ni/Y (2.63) are the most commonly used 'promoters' (we avoid the term 'catalysts' for these additives, as the process of nanotube formation is not catalytic). It has been shown that the addition of sulphur to the Co in the anode (either as elemental S, or as CoS), resulted in a much wider range of nanotube diameters than obtained from Co alone. Thus, single-walled nanotubes with diameters ranging from 1 to 6 nm were produced when sulphur was present in the cathode, compared to *c.* 1–2 nm for pure Co (2.64). It was subsequently shown that bismuth and lead could similarly promote the formation of large-diameter tubes (2.65). In 1997 a French group showed that high yields of single-walled nanotubes could be achieved with arc-evaporation (2.66). Their method was similar to the original technique of Bethune and colleagues, but with a slightly different reactor geometry. Also, the promoter used was a Ni/Y mixture rather than the Co generally favoured by the Bethune group. The highest concentration of SWNTs was found to form in a 'collar' around the cathodic deposit, which made up approximately 20% of the total mass of evaporated material. Overall, the yield of tubes was estimated to be 70–90%. Examination of the 'collar' material by high-resolution electron microscopy showed many bundles of tubes, with diameters around 1.4 nm. In 2007, Marc Monthioux and colleagues showed that significantly improved yields could be achieved by using anodes made from either diamond powder or small grain graphite, rather than the usual large-grain graphite (2.68).

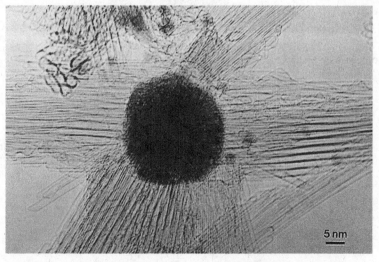

5 nm

Fig. 2.10   Single-layer nanotubes growing radially on a lanthanum particle (2.72).

Although the iron group metals, with or without additives, are the most commonly used promoters, other metals including Rh, Pd and Pt (2.69, 2.70) and rare earth metals (2.71–2.73) can also be used. When produced using rare earth metals, the tubes tend to be rather short, and are often found growing radially from the metal particles. An example, taken from the work of Saito and colleagues (2.72), is shown in Fig. 2.10. Unlike the iron group metals, the rare earth elements are not known as catalysts for the production of multiwalled nanotubes, so the formation of tubes on these elements is rather surprising, and the fact that the tubes grow on relatively large particles suggests that the mechanism may be different. It is worth noting that the radial growth of multiwall tubes from iron group metal particles was observed many years ago by Baker and others (e.g. 2.74).

2.5   Production of single-walled nanotubes by laser vaporization

As discussed in the previous chapter, C_{60} was first produced at Rice University in 1985 as a result of a programme of experiments on the vaporization of graphite using a Nd:YAG laser. In 1995 Smalley's group reported the laser vaporization synthesis of single-walled nanotubes (2.75). Subsequent refinements of the method led to enhanced yields of single-walled tubes, which tended to form large bundles or 'ropes' (2.76). The apparatus used by the Rice team illustrated schematically in Fig. 2.11. The furnace is heated to a temperature of approximately 1200 °C and an inert gas (typically argon) flows through

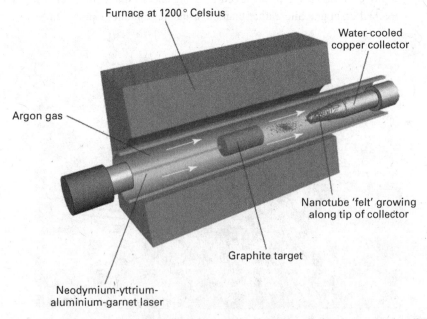

Furnace at 1200° Celsius

Water-cooled copper collector

Argon gas

Nanotube 'felt' growing along tip of collector

Graphite target

Neodymium-yttrium-aluminium-garnet laser

Fig. 2.11   Oven laser-vaporization apparatus for the synthesis of single-walled carbon nanotubes (2.77).

the 5 cm diameter tube at a constant pressure of 500 torr. A cylindrical graphite target doped with small amounts of catalyst metal (typically 0.5–1.0% each of Co and Ni) is mounted at the centre of the furnace. Vaporization of the target is performed by a Nd: YAG laser. In the refined process (2.76), a double laser pulse was used to provide a more even vaporization of the target. This method was capable of producing up to 1 g per day of SWNTs, and the Rice group began selling samples commercially. The availability of these high-quality samples of SWNTs gave an important boost to nanotube research, and some important results were achieved using these samples.

The laser vaporization method has been taken up by several other groups to make single-walled nanotubes, although the high cost of the powerful lasers required has perhaps prevented it becoming more widely used. In 1999 Iijima and colleagues produced SWNTs by irradiating a graphite–Co/Ni target with a 1 kW CO_2 laser (2.78). It was found that nanotubes were produced even at room temperature, although the yield increased significantly when the oven temperature was increased to 1100–1200 °C. The effect of oven temperature on nanotube yield has been studied by a number of groups, as discussed in the next section. Several groups have explored the possibility of scaling up the laser vaporization process. In 2002, Peter Eklund of the Pennsylvania State University and colleagues used a 1 kW free electron laser at the Jefferson Lab in Virginia to produce SWNTs with production rates as high as $1.5 \, \mathrm{g \, h^{-1}}$ (2.79). It is interesting to note that Hongjie Dai's team have reported that the laser vaporization method appears to preferentially produce metallic SWNTs (2.80) – see also p. 65. This potentially important finding does not seem to have been confirmed. Useful reviews of the production of single-walled nanotubes by laser vaporization have been given by Sivaram Arepalli (2.81) and by Christopher Kingston and Benoit Simard (2.82).

2.6   Growth mechanisms of SWNTs in the arc and laser methods

There are good reasons for assuming that the mechanisms of single-walled nanotube formation in the arc-evaporation and laser-vaporization processes are broadly similar. Both use similar starting materials, namely a graphite–metal mixture, and both involve the vaporization of this mixture followed by condensation in an inert atmosphere. Moreover, the nanotube-containing soot produced by both methods is identical in appearance, containing bundles of SWNTs together with disordered carbon and metal particles. Therefore, in the discussion that follows, we assume that mechanisms proposed for one process are applicable to both.

Although many different models have been mooted for the growth of SWNTs by the arc or laser methods, it is generally accepted that the mechanism probably involves 'root growth' rather than 'tip growth'. In other words, the tubes grow away from the metal particles, with carbon being continuously supplied to the base. This is supported by the fact that metal particles are not found at the tips of SWNTs produced by arc-evaporation or laser-vaporization, as would be the case if tip growth had occurred. Also, most of the particles observed in the SWNT-containing soot have diameters much larger than those of the individual tubes. The mechanisms which have been put forward for nanotube growth in the arc-evaporation and laser-vaporization methods will now be considered in detail.

2.6.1   Vapour–liquid–solid models

Much the most popular mechanism for the growth of single-walled carbon nanotubes in the arc and laser methods is the vapour–liquid–solid (VLS) model. This kind of mechanism was first put forward in the early 1960s to describe the growth of whiskers of Si, Ge and other materials (2.83). It was adapted by Tibbetts in 1984 (2.84) to explain the catalytic growth of multiwalled carbon nanotubes (see also next chapter, p. 53) and was applied by Saito in 1995 to the growth of SWNTs in the arc (2.85).

The model assumes that the first stage of nanotube formation involves the co-condensation of carbon and metal atoms from the vapour phase to form a liquid metal carbide particle. When the particles are supersaturated, solid phase nanotubes begin to grow, as illustrated in Fig. 2.12. The driving force for the diffusion of carbon through the particles is either a temperature gradient or a concentration gradient. A number of detailed modelling studies, have been carried out based on the application of this mechanism to SWNT growth (e.g. 2.86–2.90). An illustration of the VLS growth of single-walled carbon nanotubes, from the work of Annick Loiseau and colleagues (2.88), is shown in Fig. 2.13. The first stage in this process is the formation of a liquid

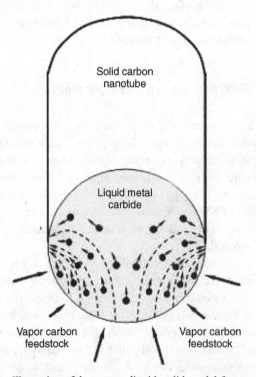

Solid carbon
nanotube

Liquid metal
carbide

Vapor carbon
feedstock

Vapor carbon
feedstock

Fig. 2.12   Illustration of the vapour–liquid–solid model for nanotube growth (2.86).

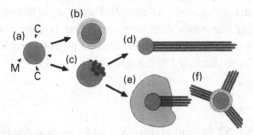

Fig. 2.13   A scenario, based on the VLS mechanism, for nucleation and growth of SWNTs, from the work of Annick Loiseau and colleagues (2.88).

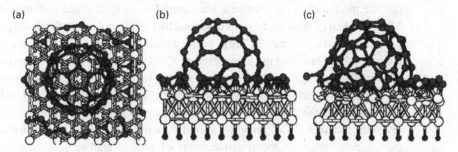

Fig. 2.14   Simulation of the root growth mechanism for SWNTs extruded from large metal nanoparticles by Loiseau *et al.* (2.89).

nanoparticle of metal supersaturated with carbon (Fig. 2.13a). On cooling, carbon begins to precipitate out of the solution, and can either form a graphitic coating on the particle surface (Fig. 2.13b) or can form seeds for the nucleation of single-walled nanotubes (Fig. 2.13c). Nanotube growth then proceeds through further incorporation of carbon atoms at the root (Fig. 2.13d). Figures 2.13(d) and (f) show situations where growth has been perturbed for some reason, resulting in the formation of short tubes and amorphous or graphitic C on the particle surfaces.

Loiseau and co-workers used molecular dynamics simulations to model some of these processes. They firstly modelled the diffusion–segregation process occurring at the surface of the catalytic particle. The starting point was a cluster containing 51 Co atoms and 102 C atoms, which was first heated to 1727 °C, leading to random positions of Co and C within the cluster, and then cooled to 1227 °C, causing C atoms to segregate to the surface. The formation of a hexagon connected with two pentagons on the surface was observed, which the authors considered to be a possible first stage of the nucleation of a nanotube. A second simulation was then carried out which modelled the growth of a tube from an initial seed. This is illustrated in Fig. 2.14. The starting point was a small (6, 6) nanotube portion capped by a fullerene hemisphere, which was is placed on a slab of HCP Co, with 20 additional isolated carbon atoms on the particle surface. The system was then heated to 1227 °C, and diffusion of C atoms to the tube base, and incorporation into the nanotube structure, were observed.

Although Annick Loiseau and colleagues have modelled some of the key processes in the VLS growth of single-walled nanotubes, their treatment cannot be considered to be a complete simulation. The VLS model is also considered to be a plausible mechanism for the growth of SWNTs by CVD, as discussed in the next chapter (p. 65).

2.6.2   Solid-state models

All the theoretical studies of SWNT formation discussed so far assume that the process involves a transformation of vapour phase carbon to solid carbon tubes, induced by the metal particles. Studies of SWNT production using the laser vaporization method published in 2001 and 2002 suggested that this model might not be correct, and that in fact the mechanism may involve a transformation of solid phase carbon (2.43, 2.91–2.95). The first of these studies was described by David Geohegan of Oak Ridge National Laboratory and colleagues in 2001 (2.91), whose work will now be summarized. This group had been studying the preparation of SWNTs by Nd:YAG laser vaporization of a graphite/Ni–Co target (2.92, 2.93). Their studies suggested that nanotube growth did not occur during the early stages of the process when carbon was in the vapour phase, but at a later stage when the 'feedstock' would be aggregated clusters and nanoparticles. In order to test the idea that SWNT growth is a solid-state transformation, they carried out further experiments involving the heat treatment of nanoparticulate soot containing short (~50 nm long) nanotube 'seeds'. This 'seeded' soot was produced by carrying out laser vaporization for shorter periods at a lower temperature than that used to produce full-length nanotubes. The soot collected from the laser vaporization apparatus was placed inside a graphite crucible under argon, and heated by a CO_2 laser to temperatures up to 1600 °C. It was found that these heat treatments could produce micron length SWNTs, with optimum growth occurring at temperatures in the range 1000–1300 °C.

Geohegan and colleagues put forward the following growth mechanism for SWNTs by laser vaporization. The Nd:YAG laser pulse initially produces an atomic-molecular vapour containing both carbon species and Ni/Co atoms. This evaporated material remains in the vapour phase for approximately 100 μs. The plasma then cools rapidly, and the carbon condenses and forms clusters ~200 μs after ablation (the metal atoms condense much later, at about 2 ms). The size of the carbon particles within the plume at these times does not exceed 20 nm at temperatures around 1100 °C. Geohegan et al. estimate the onset of SWNT growth to occur at 2 ms after ablation. By this time, both the carbon and the metal atoms are in a condensed form, so nanotube growth is largely a solid-state process.

Studies similar to those of Geohegan and colleagues were carried out at about the same time by two other groups. Andre Gorbunov of Dresden University of Technology and co-workers prepared soot using laser vaporization, but at a temperature too low to induce nanotube formation (2.94). This soot was then annealed at 1200 °C in an Ar atmosphere. This resulted in the formation of large numbers of single-walled nanotubes. On the basis of this observation they put forward a growth model that involved the conversion of solid disordered carbon into nanotubes via liquid phase metal particles. The mechanism is clearly very similar to the vapour–liquid–solid model, but with a solid carbon source, and

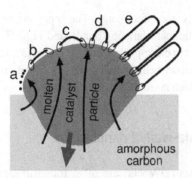

Fig. 2.15   The solid–liquid–solid mechanism for the growth of single-walled carbon nanotubes from the work of Gorbunov *et al.* (2.94).

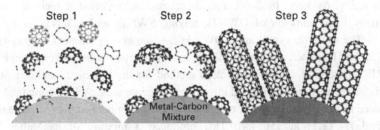

Fig. 2.16   An illustration of a model for SWNT growth proposed by Kataura and colleagues (2.96).

was named the solid–liquid–solid (SLS) mechanism. The process is illustrated in Fig. 2.15. The first stage involves a molten catalyst nanoparticle penetrating a disordered carbon aggregate, dissolving it and precipitating carbon atoms at the opposite surface. These atoms then form a graphene sheet, whose orientation parallel to the supersaturated metal–carbon melt is not energetically favourable. Any local defect of this graphene sheet will therefore result in its buckling and the formation of a SWNT nucleus.

The third study which independently demonstrated solid phase growth of single-walled nanotubes was described by Hiromichi Kataura, then at Tokyo Metropolitan University, and colleagues (2.95). Here, soot was obtained by laser ablation of Ni–Co–graphite composite targets at temperatures in the range 25–700 °C. The soot was then heated to 1200 °C in Ar. It was found that the soot which had been prepared using temperatures above about 550 °C yielded single-walled nanotubes after the 1200 °C treatment, but the soot prepared at lower temperatures did not. This is a very significant finding, for reasons we will return to below. On the basis of these observations, and of previous studies (2.96), Kataura and colleagues posited a model for SWNT growth, which is generally similar to the Geohegan mechanism, but which emphasizes the key role played by fullerene-like carbon fragments in nucleating growth. The model is illustrated in Fig. 2.16. In the first phase, which occurs at a very early stage (μs) and at very high temperatures (2000–3000 °C), small carbon clusters nucleate. These have fullerene-like structures, rich in pentagonal rings. At this stage the metal atoms are still in the gas phase. As the system cools, metal atoms condense, forming particles or droplets, which become supersaturated with carbon, at around the eutectic temperature. The particles then become covered with fullerene-like carbon fragments. The 'open edges' of these fragments tend to stick to the particles to eliminate dangling bonds, and TEM observations suggest that in some cases the fragments form close-packed arrays. The fragments then act as precursors for SWNT growth, with carbon being supplied by precipitation from the particles or from the disordered carbon which surrounds the particles.

2.7   Arc-evaporation synthesis of double-walled nanotubes

Although double-walled carbon nanotubes (DWNTs) are often found in samples prepared by arc-evaporation (an example was included in Iijima's classic 1991 paper – see Fig. 5.15), specific methods of preparing DWNTs in high yield were not developed until about ten years later. In 2000, French researchers reported a method that yielded relatively large numbers of DWNTs, among SWNTs and MWNTs, using catalysis (2.97). The catalytic synthesis of DWNTs is discussed in the next chapter (p. 68). In 2001, John Hutchison and Jeremy Sloan, in collaboration with workers from Russia and the USA described an arc-evaporation method for the selective synthesis of DWNTs (2.98). Arc discharge was carried out using a graphite anode which contained a catalyst prepared from a mixture of Ni, Co, Fe and S. The atmosphere employed was a mixture of Ar and H_2 (1:1) at 350 torr. This produced a mixture of nanotubes in which double-walled tubes were the dominant type. The outer diameter of the DWNTs generally fell into the range 1.9–5 nm, and the tubes often formed bundles. Figure 2.17

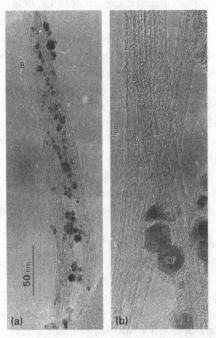

(a)   (b)

Fig. 2.17   Double-walled nanotubes produced by arc-evaporation (2.98).

shows some of the material produced by Hutchison and colleagues. It can be seen that the DWNTs are accompanied by rather large catalyst particles covered with graphitic carbon.

It is not entirely clear why the method of Hutchison et al. favours DWNT production, although the authors themselves believe that the use of the Ar/H_2 mixture is a key factor. The temperature of the arc in this atmosphere is likely to be rather lower than when He is used (approximately 2400–2600 K, compared with 3600–3900 K). This lower tempera-ture, coupled with the presence of H_2 means that carbon may be vaporized in the form of hydrocarbon fragments, radicals and molecules, rather than as pure carbon species, and this may favour the formation of DWNTs. On the other hand, a team from Sony in Japan showed that DWNTs could be grown by arc-discharge in an H_2 free atmosphere (2.99). The influence of different atmospheres and catalysts on the formation of DWNTs in the arc has been examined by Chinese workers (2.100).

2.8   Discussion

Despite the huge volume of research on nanotube production since the publication of Iijima's paper in 1991, the arc-evaporation method remains the best method for the synthesis of high-quality multi walled tubes (for single-walled nanotubes the situation is rather different, as discussed below). The arc technique, however, suffers from a number of disadvantages. Firstly, it is labour intensive and requires some skill to achieve a satisfactory level of reproducibility. Secondly, the yield is rather low, since most of the evaporated carbon is deposited on the walls of the vessel rather than on the cathode, and the nanotubes are 'contaminated' with nanoparticles and other graphitic debris. Thirdly, it is a 'batch' rather than a continuous process, and it does not easily lend itself to scale-up. Clearly there is a need for an alternative method for producing the best quality MWNTs. Some researchers have argued that catalytic methods can produce tubes of a comparable degree of perfection to those made by arc-evaporation. However, the mechanical properties of catalytically-produced MWNTs have been shown to be sub-stantially poorer than those of arc-grown tubes (see p. 182), suggesting a much less perfect structure. It is possible that high-temperature heat treatment methods of the kind described in Section 2.3 might provide a scaleable method for making high-quality MWNTs, and this is an area which would benefit from more research.

Turning now to single-walled nanotubes, it appears that the quality of the tubes is less dependent on the production method. Single-walled tubes produced by the arc and laser techniques are undoubtedly highly perfect, as shown by their excellent mechanical properties (see p. 188), but there is evidence that catalytically-produced SWNTs are equally perfect or more so, as discussed in the next chapter (p. 71). Catalytic methods are therefore increasingly becoming the favoured methods for producing single-walled tubes, owing to the relatively simple apparatus required and scaleability.

As far as growth mechanisms in the arc and laser methods are concerned, these remain obscure. In the case of multiwalled nanotubes, many different growth models have been put forward, as we saw in Section 2.2, but there is no consensus on which one is correct.

It is not even clear whether growth is a vapour, liquid or solid phase phenomenon. For single-walled tubes produced by arc or laser vaporization, two types of growth model have been posited – the vapour–liquid–solid mechanism and the solid-state mechanism. Of these, the former is more widely accepted, and has been the subject of far more theoretical modelling. However, there appears to be experimental support for a solid-state mechanism, and some modelling studies of this type of growth would be of great value. The lack of understanding of the growth of both MWNTs and SWNTs by arc/laser vaporization is a serious impediment to progress in producing tubes with a defined structure or to developing methods for the mass production of high-quality tubes.

References

(2.1) S. Iijima, 'Helical microtubules of graphitic carbon', *Nature*, **354**, 56 (1991).

(2.2) T. W. Ebbesen and P. M. Ajayan, 'Large-scale synthesis of carbon nanotubes', *Nature*, **358**, 220 (1992).

(2.3) T. W. Ebbesen, 'Carbon nanotubes', *Ann. Rev. Mater. Sci.*, **24**, 235 (1994).

(2.4) T. W. Ebbesen, *Carbon Nanotubes: Preparation and Properties*, ed. T. W. Ebbesen, CRC Press, Boca Raton, 1997, p. 139.

(2.5) Y. Saito, K. Nishikubo, K. Kawabata *et al.*, 'Carbon nanocapsules and single-layered nanotubes produced with platinum-group metals (Ru, Rh, Pd, Os, Ir, Pt) by arc-discharge', *J. Appl. Phys.* **80**, 3062 (1996).

(2.6) C. T. Kingston and B. Simard, 'Fabrication of carbon nanotubes', *Anal. Lett.*, **36**, 3119 (2003).

(2.7) T. W. Ebbesen, H. Hiura, J. Fujita *et al.*, 'Patterns in the bulk growth of carbon nanotubes', *Chem. Phys. Lett.*, **209**, 83 (1993).

(2.8) G. H. Taylor, J. D. Fitzgerald, L. Pang *et al.*, 'Cathode deposits in fullerene formation – microstructural evidence for independent pathways of pyrolytic carbon and nanobody formation', *J. Cryst. Growth*, **135**, 157 (1994).

(2.9) X. Zhao, M. Ohkohchi, M. Wang *et al.*, 'Preparation of high-grade carbon nanotubes by hydrogen arc discharge', *Carbon*, **35**, 775 (1994).

(2.10) X. K. Wang, X. W. Lin, V. P. Dravid *et al.*, 'Carbon nanotubes synthesized in a hydrogen arc discharge', *Appl. Phys. Lett.*, **66**, 2430 (1995).

(2.11) L. C. Qin, X. L. Zhao, K. Hirahara *et al.*, 'The smallest carbon nanotube', *Nature*, **408**, 50 (2000).

(2.12) S. Cui, P. Scharff, C. Siegmund *et al.*, 'Investigation on preparation of multiwalled carbon nanotubes by DC arc discharge under N_2 atmosphere', *Carbon*, **42**, 931 (2004).

(2.13) H. Yokomichi, M. Matoba, H. Sakima, *et al.*, 'Synthesis of carbon nanotubes by arc discharge in CF_4 gas atmosphere', *Jpn. J. Appl. Phys.*, **37**, 6492 (1998).

(2.14) K. Shimotani, K. Anazawa, H. Watanabe *et al.*, 'New synthesis of multi-walled carbon nanotubes using an arc discharge technique under organic molecular atmospheres', *Appl. Phys. A*, **73**, 451 (2001).

(2.15) N. Wang, Z. K. Tang, G. D. Li *et al.*, 'Single-walled 4 Å carbon nanotube arrays', *Nature*, **408**, 50 (2000).

(2.16) M. Ishigami, J. Cumings, A. Zettl *et al.*, 'A simple method for the continuous production of carbon nanotubes', *Chem. Phys. Lett.*, **319**, 457 (2000).

(2.17) H. W. Zhu, X. S. Li, B. Jiang *et al.*, 'Formation of carbon nanotubes in water by the electric-arc technique', *Chem. Phys. Lett.*, **366**, 664 (2002).

(2.18) X. S. Li, H. W. Zhu, B. Jiang *et al.*, 'High-yield synthesis of multi-walled carbon nanotubes by water-protected arc discharge method', *Carbon*, **41**, 1664 (2003).

(2.19) N. Sano, M. Naito, M. Chhowalla *et al.*, 'Pressure effects on nanotubes formation using the submerged arc in water method', *Chem. Phys. Lett.*, **378**, 29 (2003).

(2.20) I. Alexandrou, H. Wang, N. Sano *et al.*, 'Structure of carbon onions and nanotubes formed by arc in liquids', *J. Chem. Phys.*, **120**, 1055 (2004).

(2.21) N. Sano, H. Wang, M. Chhowalla *et al.*, 'Synthesis of carbon "onions" in water', *Nature*, **414**, 506 (2001).

(2.22) H. Wang, M. Chhowalla, N. Sano *et al.*, 'Large-scale synthesis of single-walled carbon nanohorns by submerged arc', *Nanotechnology*, **15**, 546 (2004).

(2.23) D. Bera, E. Brinley, S. C. Kuiry *et al.*, 'Optoelectronically automated system for carbon nanotubes synthesis via arc-discharge in solution', *Rev. Sci. Instrum.*, **76**, 033903 (2005).

(2.24) D. Bera, G. Johnston, H. Heinrich *et al.*, 'A parametric study on the synthesis of carbon nanotubes through arc-discharge in water', *Nanotechnology*, **17**, 1722 (2006).

(2.25) K. Anazawa, K. Shimotani, C. Manabe *et al.*, 'High-purity carbon nanotubes synthesis method by an arc discharging in magnetic field', *Appl. Phys. Lett.*, **81**, 739 (2002).

(2.26) J. S. Qiu, Y. Zhou, L. N. Wang *et al.*, 'Formation of carbon nanotubes and encapsulated nanoparticles from coals with moderate ash contents', *Carbon*, **36**, 465 (1998).

(2.27) J. S. Qiu, Y. F. Li, Y. P. Wang *et al.*, 'High-purity single-wall carbon nanotubes synthesized from coal by arc discharge', *Carbon*, **41**, 2170 (2003).

(2.28) J. S. Qiu, Y. F. Li, Y. P. Wang *et al.*, 'Production of carbon nanotubes from coal', *Fuel Proc. Technol.*, **85**, 1663 (2004).

(2.29) Z. Y. Wang, Z. B. Zhao and J. S. Qiu, 'Synthesis of branched carbon nanotubes from coal', *Carbon*, **44**, 1321 (2006).

(2.30) R. E. Smalley, 'From dopyballs to nanowires', *Mater. Sci. Eng. B*, **19**, 1 (1993).

(2.31) D. T. Colbert, J. Zhang, S. M. McClure *et al.*, 'Growth and sintering of fullerene nanotubes', *Science*, **266**, 1218 (1994).

(2.32) A. Maiti, C. J. Brabec, C. M. Roland *et al.*, 'Growth energetics of carbon nanotubes', *Phys. Rev. Lett.*, **73**, 2468 (1994).

(2.33) L. Lou, P. Nordlander and R. E. Smalley, 'Fullerene nanotubes in electric fields', *Phys. Rev. B*, **52**, 1429 (1995).

(2.34) D. T. Colbert and R. E. Smalley, 'Electric effects in nanotube growth', *Carbon*, **33**, 921 (1995).

(2.35) J. C. Charlier, A. De Vita, X. Blase *et al.*, 'Microscopic growth mechanisms for carbon nanotubes', *Science*, **275**, 646 (1997).

(2.36) Y.-K. Kwon, Y. H. Lee, S.-G. Kim *et al.*, 'Morphology and stability of growing multiwall carbon nanotubes', *Phys. Rev. Lett.*, **79**, 2065 (1997).

(2.37) Y. Saito, T. Yoshikawa, M. Inagaki *et al.*, 'Growth and structure of graphitic tubules and polyhedral particles in arc-discharge', *Chem. Phys. Lett.*, **204**, 277 (1993).

(2.38) E. G. Gamaly and T. W. Ebbesen, 'Mechanism of carbon nanotube formation in the arc discharge', *Phys. Rev. B*, **52**, 2083 (1995).

(2.39) O. A. Louchev, 'Transport-kinetical phenomena in nanotube growth', *J. Cryst. Growth*, **237**, 65 (2002).

(2.40) O. A. Louchev, Y. Sato and H. Kanda, 'Morphological stabilization, destabilization, and open-end closure during carbon nanotube growth mediated by surface diffusion', *Phys. Rev. E*, **66**, 011601 (2002).

(2.41) Y. W. Liu, L. Wang and H. Zhang, 'A possible mechanism of uncatalyzed growth of carbon nanotubes', *Chem. Phys. Lett.*, **427**, 142 (2006).

(2.42) W. A. De Heer, P. Poncharal, C. Berger *et al.*, 'Liquid carbon, carbon-glass beads, and the crystallization of carbon nanotubes', *Science*, **307**, 907 (2005).

(2.43) P. J. F. Harris, 'Solid state growth mechanisms for carbon nanotubes', *Carbon*, **45**, 229 (2007).

(2.44) P. J. F. Harris, S. C. Tsang, J. B. Claridge *et al.*, 'High-resolution electron microscopy studies of a microporous carbon produced by arc-evaporation', *J. Chem. Soc., Faraday Trans.*, **90**, 2799 (1994).

(2.45) D. Zhou and L. Chow, 'Complex structure of carbon nanotubes and their implications for formation mechanism', *J. Appl. Phys.* **93**, 9972 (2003).

(2.46) J. Y. Huang, S. Chen, Z. F. Ren *et al.*, 'Real-time observation of tubule formation from amorphous carbon nanowires under high-bias Joule heating', *Nano Lett.* **6**, 1699 (2006).

(2.47) S. C. Tsang, P. J. F. Harris, J. B. Claridge *et al.*, 'A microporous carbon produced by arc-evaporation', *Chem. Commun.*, 1519 (1993).

(2.48) L. A. Bursill and L. N. Bourgeois, 'Image-analysis of a negatively curved graphitic sheet model for amorphous-carbon', *Mod. Phys. Lett. B*, **9**, 1461 (1995).

(2.49) W. A. de Heer and D. Ugarte, 'Carbon onions produced by heat-treatment of carbon soot and their relation to the 217.5 nm interstellar absorption feature', *Chem. Phys. Lett.*, **207**, 480 (1993).

(2.50) D. Ugarte, 'High-temperature behaviour of "fullerene black"' *Carbon*, **32**, 1245 (1994).

(2.51) A. A. Setlur, S. P. Doherty, J. Y. Dai *et al.*, 'A promising pathway to make multiwalled carbon nanotubes', *Appl. Phys. Lett.*, **76**, 3008 (2000).

(2.52) S. P. Doherty and R. P. H. Chang, 'Synthesis of multiwalled carbon nanotubes from carbon black', *Appl. Phys. Lett.*, **81**, 2466 (2002).

(2.53) D. B. Buchholz, S. P. Doherty and R. P. H. Chang, 'Mechanism for the growth of multi-walled carbon-nanotubes from carbon black', *Carbon*, **41**, 1625 (2003).

(2.54) S. P. Doherty, D. B. Buchholz and R. P. H. Chang, 'Semi-continuous production of multi-walled carbon nanotubes using magnetic field assisted arc furnace', *Carbon*, **44**, 1511 (2006).

(2.55) L. T. Chadderton and Y. Chen, 'Nanotube growth by surface diffusion', *Phys. Lett. A*, **263**, 401 (1999).

(2.56) P. J. F. Harris, 'Ultrathin graphitic structures and carbon nanotubes in a purified synthetic graphite', *J. Phys. - Cond. Matter*, **21**, 355009 (2009).

(2.57) R. S. Ruoff, D. C. Lorents, B. Chan *et al.*, 'Single crystal metals encapsulated in carbon nanoparticles', *Science*, **259**, 346 (1993).

(2.58) M. Tomita, Y. Saito and T. Hayashi, 'LaC_2 encapsulated in graphite nanoparticle', *Jap. J. Appl. Phys.*, **32**, L280 (1993).

(2.59) S. Seraphin, D. Zhou, J. Jiao *et al.*, 'Yttrium carbide in nanotubes', *Nature*, **362**, 503 (1993).

(2.60) D. S. Bethune, C. H. Kiang, M. S. de Vries *et al.*, 'Cobalt-catalysed growth of carbon nanotubes with single-atomic-layer walls', *Nature*, **363**, 605 (1993).

(2.61) S. Iijima and T. Ichihashi, 'Single-shell carbon nanotubes of 1-nm diameter', *Nature*, **363**, 603 (1993).

(2.62) B. Hornbostel, M. Haluska, J. Cech *et al.*, *Carbon Nanotubes, Proceedings of the NATO Advanced Study Institute*, ed. V. N. Popov and P. Lambin, Springer-Verlag, Berlin, 2006, p. 1.

(2.63) M. E. Itkis, D. E. Perea, S. Niyogi *et al.*, 'Optimization of the Ni–Y catalyst composition in bulk electric arc synthesis of single-walled carbon nanotubes by use of near-infrared spectroscopy', *J. Phys. Chem. B*, **108**, 12770 (2004).

(2.64) C. H. Kiang, W. A. Goddard III, R. Beyers *et al.*, 'Carbon nanotubes with single-layer walls', *Carbon*, **33**, 903 (1995).

(2.65) C. H. Kiang, P. H. M. van Loosdrecht, R. Beyers *et al.*, 'Novel structures from arc-vaporized carbon and metals: single-layer nanotubes and metallofullerenes', *Surf. Rev. Lett.*, **3**, 765 (1996).

(2.66) C. Journet, W. K. Maser, P. Bernier *et al.*, 'Large-scale production of single-walled nanotube nanotubes by the electric-arc technique', *Nature*, **388**, 756 (1997).

(2.67) M. Takizawa, S. Bandow, M. Yudasaka *et al.*, 'Change of tube diameter distribution of single-wall carbon nanotubes induced by changing the bimetallic ratio of Ni and Y catalysts', *Chem. Phys. Lett.*, **326**, 351 (2000).

(2.68) A. Mansour, M. Razafinimanana, M. Monthioux *et al.*, 'A significant improvement of both yield and purity during SWCNT synthesis via the electric arc process', *Carbon*, **45**, 1651 (2007).

(2.69) Y. Saito, K. Nishikubo, K. Kawabata *et al.*, 'Carbon nanocapsules and single-layered nanotubes produced with platinum-group metals (Ru, Rh, Pd, Os, Ir, Pt) by arc discharge', *J. Appl. Phys.*, **80**, 3062 (1996).

(2.70) Z. H. Li, M. Wang and Y. Saito, 'Effect of Rh and Pt on growth and structure of single-walled carbon nanotubes', *Inorg. Mater.*, **42**, 605 (2006).

(2.71) S. Subramoney, R. S. Ruoff, D. C. Lorents *et al.*, 'Radial single-layer nanotubes', *Nature*, **366**, 637 (1993).

(2.72) Y. Saito, M. Okuda, M. Tomita *et al.*, 'Extrusion of single-wall carbon nanotubes via formation of small particles condensed near an arc evaporation source', *Chem. Phys. Lett.*, **236**, 419 (1995).

(2.73) D. Zhou, S. Seraphin and S. Wang, 'Single-walled carbon nanotubes growing radially from YC_2 particles', *Appl. Phys. Lett.*, **65**, 1593 (1994).

(2.74) R. T. K. Baker and R. J. Waite, 'Formation of carbonaceous deposits from catalysed decomposition of acetylene', *J. Catalysis*, **37**, 101 (1975).

(2.75) T. Guo, P. Nikolaev, A. Thess *et al.*, 'Catalytic growth of single-walled nanotubes by laser vaporization', *Chem. Phys. Lett.*, **243**, 49 (1995).

(2.76) A. Thess, R. Lee, P. Nikolaev *et al.*, 'Crystalline ropes of metallic carbon nanotubes', *Science*, **273**, 483 (1996).

(2.77) B. I. Yakobson and R. E. Smalley, 'Fullerene nanotubes: $C_{1,000,000}$ and beyond', *Amer. Sci.*, **85**, 324 (1997).

(2.78) F. Kokai, K. Takahashi, M. Yudasaka *et al.*, 'Growth dynamics of single-wall carbon nanotubes synthesized by CO_2 laser vaporization', *J. Phys. Chem. B*, **103**, 4346 (1999).

(2.79) P. C. Eklund, B. K. Pradhan, U. J. Kim *et al.*, 'Large-scale production of single-walled carbon nanotubes using ultrafast pulses from a free electron laser', *Nano Lett.*, **2**, 561 (2002).

(2.80) Y. M. Li, D. Mann, M. Rolandi *et al.*, 'Preferential growth of semi-conducting single-walled carbon nanotubes by a plasma enhanced CVD method', *Nano Lett.*, **4**, 317 (2004).

(2.81) S. Arepalli, 'Laser ablation process for single-walled carbon nanotube production', *J. Nanosci. Nanotech.*, **4**, 317 (2004).

(2.82) C. T. Kingston and B. Simard, 'Recent advances in laser synthesis of single-walled carbon nanotubes', *J. Nanosci. Nanotech.*, **6**, 1225 (2006).

(2.83) R. S. Wagner and W. C. Ellis, 'Vapor-liquid-solid mechanism of single crystal growth', *Appl. Phys. Lett.*, **4**, 89 (1964).

(2.84) G. G. Tibbetts, 'Why are carbon filaments tubular?', *J. Cryst. Growth*, **66**, 632 (1984).

(2.85) Y. Saito, 'Nanoparticles and filled nanocapsules', *Carbon*, **33**, 979 (1995).

(2.86) K. Bolton, F. Ding and A. Rosén, 'Atomistic simulations of catalyzed carbon nanotube growth', *J. Nanosci. Nanotech.*, **6**, 1211 (2006).

(2.87) Y. Shibuta and S. Maruyama, 'Molecular dynamics simulation of generation process of SWNTs', *Physica B*, **323**, 187 (2002).

(2.88) J. Gavillet, A. Loiseau, C. Journet *et al.*, 'Root-growth mechanism for single-wall carbon nanotubes', *Phys. Rev. Lett.*, **87**, 275504 (2001).

(2.89) J. Gavillet, A. Loiseau, F. Ducastelle *et al.*, 'Microscopic mechanisms for the catalyst assisted growth of single-wall carbon nanotubes', *Carbon*, **40**, 1649 (2002).

(2.90) C. L. Luo, H. W. Yu, Y. Q. Zhang *et al.*, 'Simulations of nucleation of single-walled carbon nanotubes', *Physica Status Solidi A*, **204**, 555 (2007).

(2.91) D. B. Geohegan, H. Schittenhelm, X. Fan *et al.*, 'Condensed phase growth of single-wall carbon nanotubes from laser annealed nanoparticulates', *Appl. Phys. Lett.*, **78**, 3307 (2001).

(2.92) A. A. Puretzky, D. B. Geohegan, X. Fan *et al.*, '*In situ* imaging and spectroscopy of single-wall carbon nanotube synthesis by laser vaporization', *Appl. Phys. Lett.*, **76**, 182 (2000).

(2.93) A. A. Puretzky, H. Schittenhelm, X. Fan *et al.*, 'Investigations of single-wall carbon nanotube growth by time-restricted laser vaporization', *Phys. Rev. B*, **65**, 245425 (2002).

(2.94) A. Gorbunov, O. Jost, W. Pompe *et al.*, 'Solid–liquid-solid growth mechanism of single-wall carbon nanotubes', *Carbon*, **40**, 113 (2002).

(2.95) R. Sen, S. Suzuki, H. Kataura *et al.*, 'Growth of single-walled carbon nanotubes from the condensed phase', *Chem. Phys. Lett.*, **349**, 383 (2001).

(2.96) H. Kataura, Y. Kumazawa, Y. Maniwa *et al.*, 'Diameter control of single-walled carbon nanotubes', *Carbon*, **38**, 1691 (2000).

(2.97) E. Flahaut, A. Peigney, C. Laurent *et al.*, 'Synthesis of single-walled carbon nanotube–Co–MgO composite powders and extraction of the nanotubes', *J. Mater. Chem.*, **10**, 249 (2000).

(2.98) J. L Hutchison, N. A. Kiselev, E. P Krinichnaya, *et al.*, 'Double-walled carbon nanotubes fabricated by a hydrogen arc discharge method', *Carbon*, **39**, 761 (2001).

(2.99) H. Huang, H. Kajiura, S. Tsutsui *et al.*, 'High-quality double-walled carbon nanotube super bundles grown in a hydrogen-free atmosphere', *J. Phys. Chem. B*, **107**, 8794 (2003).

(2.100) Z. H. Li, M. Wang, B. Yang *et al.*, 'The influence of different atmosphere gases on the growth and structure of double-walled carbon nanotubes', *Inorg. Mater.*, **43**, 475 (2007).

3 Synthesis II: catalytic chemical vapour deposition and related methods

The preparation of carbon nanotubes by catalysis has a number of potential advantages over the arc and laser methods discussed in the previous chapter. In particular, catalysis (or chemical vapour deposition, as the process is often called) is much more amenable to scale-up than arc- or laser-evaporation, and many successful processes for the large-scale catalytic synthesis of both SWNTs and MWNTs have been developed. Catalytic techniques also enable nanotube synthesis to be achieved under relatively mild conditions, giving more control over the growth process. Thus, it is possible, using catalytic methods, to grow arrays of aligned nanotubes on substrates. Such arrays are showing great promise as field-emission displays. It may also be possible to construct nano-electronic circuits by using catalysis to grow defined networks of nanotubes. It is widely believed that the main disadvantage of catalytic methods is that nanotubes produced in this way are structurally inferior to those made by the high-temperature arc and laser techniques. While this may still be true for multiwalled tubes, there is evidence that catalytically-produced SWNTs can have a high degree of structural perfection, as noted at the end of the last chapter. This may suggest that a common growth mechanism is involved.

The long history of filamentous carbon production by catalysis was outlined in Chapter 1. The present chapter begins with a brief summary of the work carried out in the 1970s and 1980s by Baker, Endo and others, but concentrates mainly on post-1991 research. Methods for growing aligned MWNTs on substrates, and for producing nanotube yarns are described. The structure and possible growth mechanisms of catalytically produced MWNTs are then discussed. The next section of the chapter covers the rapidly growing subject of the catalytic synthesis of single-walled carbon nanotubes. The experimental conditions required to produce SWNTs are described, and specialized topics including the preparation of SWNT strands and the directed growth of SWNTs discussed. Finally, brief accounts are given of the catalytic synthesis of double-walled nanotubes, and of the production of MWNTs by electrochemistry and by heat treatment of metal-doped carbon. No attempt is made in this chapter to give detailed descriptions of the experimental set-ups or the precise conditions used in the catalytic synthesis of nanotubes. Methods used to prepare the catalysts are also not covered. For such details, the reader should consult the many excellent reviews available (3.1–3.7).

3.1   Catalytic synthesis of multiwalled nanotubes: pre-1991 work

In the early 1970s, notable work on the formation of filamentous carbon was carried out by Terry Baker's group at Harwell and by Tom Baird, John Fryer and co-workers at the University of Glasgow (e.g. 3.8–3.13). Carbon formation both from the disproportionation of carbon monoxide and from the decomposition of hydrocarbons was investigated. Considering CO disproportionation first, this reaction can be represented by the Boudouard equilibrium:

$$2CO_{(g)} < ==== > C_{(s)} + CO_{2(g)}$$

The maximum rates of carbon deposition from this reaction were found to occur at temperatures of around 550 °C in the presence of metal particles of the iron subgroup. Transmission electron microscope studies of filaments produced in this way showed that they were often helical or twisted, and could be either hollow, i.e. tubular, or solid. The filaments had diameters ranging from 10 nm to 0.5 μm and could be up to 10 μm in length. It was not clear from Baker's work, or from the other early studies, whether the active catalytic site for filament production was the metal or the carbide.

Turning now to the catalysed decomposition of hydrocarbons, Baker *et al.* found that once again Fe, Co and Ni were the most active catalysts. The rate of filament growth from a number of unsaturated hydrocarbons was measured (3.8). It was found that the growth rates increased in the same order as the exothermicity of the hydrocarbon decomposition, with acetylene producing the most rapid growth. On the basis of these observations, a mechanism was proposed for the filament growth process. This mechanism assumes that a temperature gradient develops on the catalyst particle due to the exothermic character of the hydrocarbon decomposition reaction, and that this promotes diffusion of carbon through the particle, to be precipitated at the trailing end in the form of a filament. A problem with this mechanism is that hydrocarbons which undergo endothermic decompositions, such as paraffins, should not produce carbon filaments, and it is known that filaments can be grown from CH_4. Theories of the catalytic growth of MWNTs will be dealt with in detail in Section 3.3. The morphologies of the filaments observed in these early studies varied quite widely, but were generally very disordered. A typical TEM micrograph is shown in Fig. 3.1(a). As noted above, helically coiled filaments were commonly seen, as in Fig. 3.1(b) (3.9). Baker has given comprehensive reviews of this early work (3.14, 3.15).

Morinobu Endo carried out his doctoral studies with Agnes Oberlin in France, and began working on the catalytic growth of carbon filaments in the early 1970s (3.16, 3.17). Much of his early work involved the controlled decomposition of benzene on a catalytic substrate. Thus, in a typical experiment, high-purity hydrogen would be passed through benzene, and the resulting mixture would then flow across a catalytically-treated substrate held in a furnace at an initial temperature of approximately 1000 °C. This initially produced hollow nanofibres approximately 10 nm in diameter. These could subsequently be thickened by raising the temperature to promote direct decomposition of the benzene.

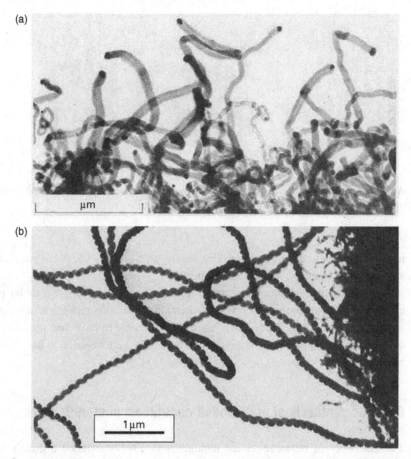

Fig. 3.1   Early images of catalytically-produced carbon nanofibres. (a) Fibres produced by decomposition of butadiene over Ni, (b) helically coiled fibres, produced by Co-catalysed decomposition of acetylene (3.9).

In this way, fibres with diameters up to 10 μm and lengths up to about 25 cm could be produced (3.18).

In the late 1980s, Endo's team, working in collaboration with Mildred Dresselhaus and colleagues of MIT introduced an important innovation in nanofibre production, the 'volume seeding' or 'floating catalyst' technique (3.19). In this method catalyst particles are introduced directly into the feedstock so that fibres can grow in the three-dimensional space of the reactor, rather than just on a two-dimensional surface. This opened the way for the *continuous* production of the filaments. The 'vapour grown carbon fibres' (VGCF) tended to have rather large diameters compared with today's carbon nanotubes – typically between 100 nm and 1μm. Like the filaments produced by Baker and his colleagues, these benzene-produced fibres were hollow, with a small catalytic particle remaining at the tip, but they were generally quite straight rather than curled or helical. The degree of graphitization in the catalytically-grown fibres is rather

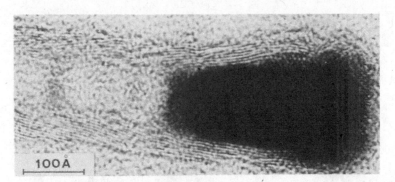

Fig. 3.2   A high-resolution TEM micrograph of a 'herringbone' carbon nanofibre, from the work
of Audier *et al.* (3.21).

low, but the fibres can be graphitized by heating at ~3000 °C. The fibres produced in this
way can have excellent mechanical properties.

As well as tubular structures, other forms of carbon nanofibre can be produced
catalytically. Quite commonly observed are fibres with a 'herringbone' appearance, as
shown in Fig. 3.2 (e.g. 3.20–3.23). These fibres appear to have few useful properties,
compared with nanotubes, although at one time they were believed to have exceptional
hydrogen uptake capacities (see Chapter 10 p. 263).

3.2   Catalytic synthesis of multiwalled nanotubes: post-1991 work

The discovery of fullerene-related nanotubes in 1991 stimulated a greatly increased
demand for samples of nanotubes. Since MWNTs could be made far more cheaply by
catalysis than by arc-evaporation, many commercial companies began to supply these
materials (one of the most prominent of these, Hyperion, had actually been in existence
since 1982). The methods used by these companies to produce bulk samples of MWNTs
remain closely-guarded secrets, but are probably broadly similar to the methods developed by Morinobu Endo and his colleagues.

The huge volume of research on the catalytic production of multiwalled nanotubes
since 1991 has resulted in significant improvements in quality. This can be seen by
comparing Fig. 3.3, taken from a 2004 review by Ken Teo and colleagues of the
Cambridge Engineering Department (3.2), with the earlier image shown in Fig. 3.1(a).
The tubes shown in Fig. 3.3 were prepared using ferrocene as both the carbon source and
the catalyst. The use of metal complexes in this way to produce multiwalled carbon
nanotubes has become increasingly popular. It appears that Endo and co-workers experimented with this approach in the late 1980s (3.18), but the use of metal complexes was
not taken up widely until the mid 1990s. In 1994, Gary Tibbetts *et al.* reported the
gas-phase synthesis of carbon fibres in flowing mixtures of CH_4 or hexane with orga-
nometallics including ferrocene (3.24) and iron pentacarbonyl (3.25). C. N. R. Rao and
colleagues have also described the production of nanotubes by the pyrolysis of pure

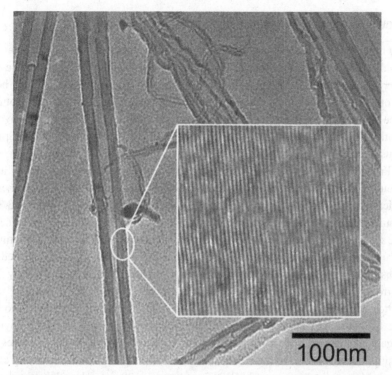

Fig. 3.3 Multiwalled tubes grown using a ferrocene-toluene mixture at 760 °C with the inset showing a high degree of graphitization (3.2).

metallocenes (3.26) and of metallocene–hydrocarbon mixtures. The advantages of using organometallic compounds are that the carbon source and catalyst are in the same phase and that there is no catalyst support to remove at the end of the reaction. Metal complexes are used in the production of aligned arrays of MWNTs, and of nanotube yarns, as discussed below. Complexes are also used to make SWNTs (see p. 55). A review of the organometallic precursor route to nanotubes, concentrating on MWNTs, has been given by Govindaraj and Rao (3.27).

In 2006, Christian Deck and Kenneth Vecchio from the University of California, San Diego, described a detailed study of the catalytic growth of MWNTs using a wide variety of transition metals catalysts (3.28). They found that Fe, Co and Ni were the only active catalysts, with Cr, Mn, Zn, Cd, Ti, Zr, La, Cu, V and Gd showing no activity. They suggested that the key to catalytic activity is the solubility of carbon in the metals. Successful catalysts had carbon solubility limits of 0.5–1.5 wt% carbon, while unsuccessful catalysts had either nearly zero carbon solubility, or formed intermediate carbides, making it difficult for the diffusion required for graphite precipitation to occur.

It should be mentioned that sophisticated catalyst formulations are not necessary in order to make multiwalled carbon nanotubes. Among the materials from which MWNTs have been made are volcanic ash and cat litter (3.29, 3.30).

3.2.1   Growth of aligned MWNTs on substrates

An important growth area in nanotube science since the late 1990s has been the catalytic synthesis of aligned nanotubes on substrates. The primary application of these nanotube arrays has been as field emission sources (see p. 170), but they have also been used in other applications including photonics (3.31) and in the spinning of nanotube yarns (see Section 4.3.3). A further novel application for the arrays is as dry adhesives, which work in the same way as gecko foot-hairs (3.32), and could be used in 'Spiderman' suits (3.33).

Probably the first demonstration of aligned growth was given in 1996 by a Chinese group who used a technique in which Fe nanoparticles were embedded in mesoporous silica and then used to catalyse the decomposition acetylene at 700 °C (3.34). This resulted in the formation of straight MWNTs growing in a direction perpendicular to the surface of the silica. It was reported that nanotube arrays of several square millimetres in area could be grown in this way, and that the arrays could be readily detached from the substrate. A slightly different method was described a short time later by Mauricio Terrones and colleagues (3.35). These researchers used laser etching to prepare a patterned cobalt catalyst on a silica substrate, which was then heat treated to break up the Co films into discrete particles. The patterned catalyst was used to pyrolyse an organic compound at 950 °C, resulting in the formation of aligned nanotubes growing approximately parallel to the substrate. In 1999, workers from the University of Toronto reported the growth of aligned MWNT arrays up to 100 μm in length by the pyrolysis of C_2H_2 on Co particles within a 'nanochannel' alumina template (3.36). At about the same time, Hongjie Dai and colleagues from Stanford grew aligned arrays of nanotubes on both porous silicon and plain silicon substrates, and demonstrated their field emission properties (3.37).

A very popular approach to growing carbon nanotubes on substrates involves the use of a plasma. The technique of plasma enhanced chemical vapour deposition (PECVD) was first used for nanotubes by Zhifeng Ren and his team from the State University of New York in 1998 (3.38). In this work, the tubes were grown on nickel particles deposited onto glass. A DC plasma was used with acetylene as the carbon source. Ammonia was introduced into the reaction chamber, and appeared to have an important role as a catalyst as well as acting as a dilution gas. Aligned arrays of tubes were formed over several square centimetres. An SEM micrograph showing the excellent alignment achieved is shown in Fig. 3.4. Plasma enhanced chemical vapour deposition had been widely used for many years as a method for coating glass plates and other substrates for applications in flat panel displays, solar cells or other devices (3.39). It is widely used for the deposition of diamond-like carbon films. A plasma is an excited/ionized gas, and the processing plasmas, usually known as 'cold' plasmas, are generated using DC, RF or microwave excitation. When used for fabricating diamond-like carbon, the role of the plasma is to activate carbon-containing precursor molecules in the gas phase, in order to allow deposition to occur at a lower temperature. However, it is not entirely clear whether this is always the case for PECVD growth of nanotubes. The alignment that occurs during plasma enhanced CVD is believed to be due to the presence of the electric field.

8μm 300(

Fig. 3.4   A scanning electron micrograph of aligned nanotubes grown on glass using catalytic methods (3.38).

The effect of the electric field was demonstrated by Otto Zhou of the University of North Carolina at Chapel Hill and colleagues (3.40). In this study, aligned nanotubes were grown on various substrates under PECVD conditions. The experiment was then repeated under identical conditions, but with the plasma source turned off. This resulted in the growth of 'curly' nanotubes, with no obvious alignment. Other groups have similarly demonstrated the aligning effect of the plasma (e.g. 3.41).

Various methods have been used to apply catalytic nanoparticles to substrates. A number of groups have used solutions containing salts of the catalytic metals, which are deposited onto the substrate, dried and then reduced to leave the particles on the surface. In other studies, including the original work by Ren and colleagues, physical techniques have been used. These include ion beam sputtering (e.g. 3.41) and electron gun evaporation (e.g. 3.42). The advantage of physical deposition techniques is that catalytic particles can be deposited in defined patterns using e-beam lithography. This is important, since the best field emission properties are achieved with tubes which are spaced out rather than closely packed, as discussed on p. 171. An illustration of the degree of control that can be achieved using lithographic methods is given in Fig. 3.5, taken from the work of Bill Milne and his team at the Engineering Department at Cambridge (3.43–3.46). In this work, arrays of Ni nanodots with diameters ranging from 100 up to 800 nm were deposited onto a substrate. For Ni particles smaller than 300 nm, it was found that each particle generally produced a single nanotube. Excellent control over both the height and diameter of individual tubes grown in this way has also been demonstrated by Milne's group and others. A wide variety of patterns have been grown – an example is shown in Fig. 3.6. Interesting work on the plasma enhanced CVD growth of nanotubes

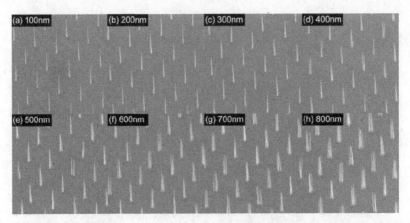

Fig. 3.5   Arrays of MWNTs produced on Ni dots of different diameters, from the work of Milne and colleagues (3.44).

Fig. 3.6   Patterned growth of MWNTs on a substrate (3.43).

on surfaces has been carried out by Ravi Silva and colleagues from the University of Surrey (3.47). These workers showed that tubes could be grown with the substrate at room temperature. This was achieved by using plasma energy rather than thermal energy to decompose the CH_4 carbon source. This approach enables plastic and even biological samples to be used as substrates. Useful reviews of the growth of carbon nanotubes and nanofibres by plasma enhanced chemical vapour deposition have been given by the Cambridge group (3.44, 3.46).

As mentioned above, aligned MWNT bundles can also be produced by the decomposition of organometallic precursors which are injected into the reactor together with the carbon source (e.g. 3.48, 3.49). Here, the alignment seems to be simply a result of self-assembly promoted by van der Waals interactions. An advantage of this approach over the use of a patterned catalyst is that aligned tubes can be grown parallel to a substrate as well as in a perpendicular direction.

3.2.2 Direct spinning of nanotube yarns

The production of continuous nanotube fibres can been achieved through post-synthesis methods, as discussed in the next chapter (p. 89). However, there would be significant advantages if a direct method for producing such fibres could be developed. Alan Windle and his co-workers in Cambridge's Materials Department described such a method in 2004 (3.50). The technique, which was partly inspired by polymer technology, involved spinning the fibres from the CVD synthesis zone of a furnace onto a rod or spindle. The starting material was ethanol containing 0.23–2.3 wt% ferrocene, and between 1.0 and 4.0 wt% thiophene. This solution was injected from the top of the furnace into a hydrogen carrier gas. Multiwalled nanotubes formed in the furnace's hot zone at a temperature of 1100–1180 °C. The tubes formed an aerogel, which could then be wound onto a rotating rod, as shown in Fig. 3.7. Single-walled nanotubes could also be formed by adjusting the thiophene concentration, the hydrogen flow rate and the temperature. The mechanical

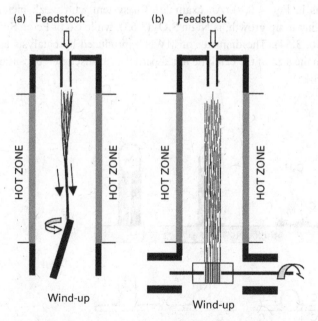

Fig. 3.7 The apparatus used by Windle and colleagues for the direct spinning of carbon nanotube fibres (3.50). (a) An arrangement with an offset rotating spindle, (b) spindle rotating normal to the furnace.

properties of the fibres produced in this way were initially unimpressive, but subsequent improvements to the process have enabled the production of fibres with high stiffness and toughness (3.51).

3.3 Growth mechanisms of catalytically produced MWNTs

Baker and his colleagues obtained detailed insights into the mechanism of filament growth from the skilful application of controlled atmosphere electron microscopy (CAEM) from about 1972 onwards (e.g. 3.9). This work demonstrated directly for the first time that filament growth involved the deposition of carbon behind an advancing metal particle, with the forward face remaining apparently clean. The CAEM technique also enabled the kinetics of the process to be determined directly, showing that the activation energy for filament growth was about the same as the activation energy for bulk carbon diffusion in nickel. This result led Baker *et al.* to propose mechanisms for both the tip and base growth (sometimes called root growth) of carbon filaments. These are illustrated in Fig. 3.8. Tip growth involves the decomposition of the carbon-containing gas on the 'front' surface of the metal particle, producing carbon, which then dissolves in the metal, as shown in Fig. 3.8(a). The dissolved carbon then diffuses through the particle, to be deposited on the trailing face, forming the filament. They proposed that temperature and concentration gradients were the main driving forces for this process. If there is a strong interaction between the metal particles and the support material, the particle remains anchored to the surface and base growth occurs, as in Fig. 3.8(b). An example of a system with weak metal-support interaction, resulting in tip growth, is Ni on SiO_2 (3.53), while Co or Fe on SiO_2 favour base growth (3.36, 3.54). The diameter of MWNTs produced by catalysis appears to depend largely on the size of the catalytic metal particles, as discussed by Susan Sinnott and colleagues (3.55).

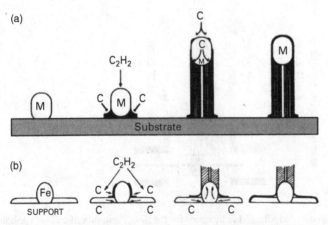

Fig. 3.8 An illustration of the tip growth mechanism (upper diagram) and the base growth mechanism (lower diagram) of nanotube growth (3.14, 3.52).

An interesting discussion of the growth mechanisms of catalytically produced MWNTs was given by Gary Tibbetts in 1984 (3.56). In this paper Tibbetts suggested that the catalytic particles might be liquid during the growth of the tubes, so that the mechanism could be described as vapour–liquid–solid (VLS). Mechanisms of this kind had been previously posited to describe the growth of whiskers of Si, Ge and many other materials (3.57), as mentioned in the previous chapter. An apparent problem with the model when applied to the growth of carbon nanotubes is that the temperatures used to produce MWNTs are usually well below the melting points of the catalytic metals. Thus, the catalytic synthesis of MWNTs might typically be carried out at 500–700 °C, while the melting point of bulk Fe is 1536 °C, and that of Ni is 1453 °C. However, it has been known for many years that the melting temperature of nanoparticles can be significantly depressed, compared to that of the bulk (3.58, 3.59). Could the catalytic particles used to produce MWNTs be in a liquid state? This actually seems rather unlikely. Moisala *et al.* give a formula for the melting temperature of metals as a function of particle diameter (3.1). This shows that the melting temperature for a 10 nm Fe particle (which is a typical size used in MWNT synthesis) would be about 940 °C, while that of a Ni particle of the same size would be ~1180 °C. These temperatures are higher than those normally used to produce MWNTs. Another factor to take into account is that the iron group metals form eutectic mixtures with carbon, i.e. mixtures with melting points lower than those of the individual components. The Fe–C eutectic point lies at 1130 °C, while for Ni, the eutectic occurs at 1326 °C. In neither case is this effect likely to be sufficient to result in particle melting at the synthesis temperatures of MWNTs. Evidence that the catalytic particles which produce MWNTs are indeed solid comes from a consideration of the growth rates of nanotubes. It was pointed out by Baker (3.10) that the activation energy of MWNT growth correlates closely with that for the diffusion rate of carbon through the solid metals. Diffusion of carbon through a liquid metal would occur much more rapidly. In the case of SWNTs, on the other hand, where higher temperatures and smaller catalytic particles are used, the possibility that the particles are in a liquid state needs to be considered (see p. 65).

Another area of controversy is the chemical state of the catalyst, i.e. are the active particles metal or metal carbide? There are many conflicting reports, with some groups finding evidence that carbides are the active phase (e.g. 3.17, 3.60), and others arguing that the particles remain metallic (3.61, 3.62). It may be that different phases are active under different conditions.

So far, it has been assumed that the catalytic growth of MWNTs involves the diffusion of carbon through a solid or liquid particle. An alternative mechanism, based on surface diffusion of carbon around the metal particle was put forward by Tom Baird and colleagues in 1974 (3.13) and elaborated by Oberlin, Endo and Koyama in 1976 (3.17). The model is illustrated in Fig. 3.9. Apparent support for this mechanism came in a study by Danish researchers in 2004 using controlled atmosphere transmission electron microscopy (3.63, 3.64). As noted above, some of the earliest insights into the growth mechanism of catalytically produced MWNTs were obtained using this technique. The Danish workers were able to use the much higher resolution available with modern TEMs to achieve atomic-scale images of the catalytic reaction between CH_4 and

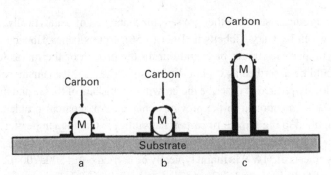

Fig. 3.9   An illustration of the growth mechanism based on the surface diffusion of carbon around the metal particle (3.52).

supported nickel nanoparticles at about 500 °C. Nanotubes were found to grow by the tip growth mechanism, with growth apparently promoted by abrupt shape changes in the catalyst particle itself, from spherical to elongated and back. In its elongated form the particle served as a template, assisting the alignment of the graphene layers into a tubular form as carbon atoms diffused across its surface. A detailed analysis of the images revealed that the nucleation and growth of these graphene layers occurred at single-atom step-edges that developed and disappeared continuously. This study provides quite compelling evidence for the surface diffusion mechanism, but it is possible that bulk diffusion may be important under different conditions.

The role of the catalyst in the growth of aligned nanotubes on a substrate has been quite widely investigated. In general, it seems that the mechanism involves base-growth, whether the catalytic particles are pre-deposited on the substrate or produced *in situ* by decomposition of organometallic precursors. In some cases, however, catalytic particles are seen at the tips of the nanotubes, showing that tip growth has occurred (3.65). Liming Dai and his colleagues studied the mechanism that operates when the catalyst particles are formed by decomposition of organometallic precursors (3.49). They suggested the following mechanism for the process. When injected into the reactor, the Fe-containing precursor decomposes to produce Fe particles, surrounded by carbon, on the substrate surface. These metal particles then grow until they reach an optimal size for carbon nanotube nucleation, when nanotube growth begins. Once again, however, there is evidence that tip growth can occur under similar conditions (3.66).

If the fundamental processes involved in the catalytic formation of MWNTs from gaseous carbon sources remain unclear, the mechanisms whereby helical morphologies can form are also poorly understood. Severin Amelinckx and colleagues at the University of Antwerp have discussed the growth of these tubes in terms of a locus of active sites around the periphery of the catalytic particle, and growth velocity vectors (3.67, 3.68). In the simplest case the locus of active sites is circular and the extrusion velocity constant, producing a straight tube propagating at a constant rate. If the catalytic activity varies around the circle, such that the velocity vectors terminate in a plane which is not parallel with the locus of active sites, a curved tube results. In practice, of course, the locus of active sites may not be circular, and this introduces another level of complexity. Amelinckx *et al.* consider the case of an elliptical locus of active sites. In this case, they show that a catalytic activity which varies around the ellipse will produce helical growth. A detailed discussion of their model is given in ref. (3.67).

Several groups have discussed the growth mechanism of bamboo-like structures (e.g. 3.69–3.72). The most direct study of the process was reported in 2007 by a group from the National University of Singapore (3.72). These workers used *in situ* TEM to make real-time observations of the growth of bamboo tubes on a Ni–MgO catalyst by the catalytic decomposition of C_2H_2 at 650 °C. A series of their micrographs is shown in Fig. 3.10. The nucleation of an 'internal cap' begins in micrograph (c), and the new graphene layers grow around the trailing edge of the particle. As growth continues, the Ni particle becomes elongated until eventually it is ejected from the tube leaving a fully formed internal cap. It was also found that the catalyst particle remained as metallic Ni during the growth process.

The growth mechanism of herringbone fibres, and the reasons why these structures sometimes form instead of tubes has also been widely discussed (e.g. 3.73–3.75). A notable feature of the catalytic particles found at the end of the herringbone fibres is that they tend to be strongly faceted, and this seems to provide a clue to the growth mechanism. It is well established from fundamental studies of catalysis that metal particles can undergo restructuring as a result of adsorption. Baker and his colleagues have suggested that herringbone fibres form when the carbon-containing gases cause reconstruction of the metal particles, to produce faces on which decomposition of the reactant gases can occur, and other faces on which graphitic carbon is precipitated. The formation of herringbone structures is apparently favoured by alloy catalysts (3.11, 3.73), although Pd has also been used alone under certain growth conditions to yield similar structures (3.74). It has been suggested (3.75) that hydrogen plays a significant role in the formation of nanofibres, as the presence of hydrogen in abundance can terminate the large number of dangling bonds at the edges of the stacked graphite platelets. Without hydrogen termination, the more stable form of the carbon filament would be a closed tube with no dangling bonds.

3.4 Catalytic synthesis of single-walled nanotubes

3.4.1 Conditions required to produce SWNTs

In order to produce single-walled rather than multiwalled nanotubes by catalysis, a number of factors need to be carefully controlled. These include temperature, feedstock and the nature and size of the catalytic particles. The temperatures used for synthesizing SWNTs are higher than those used for MWNTs: typically 900–1200 °C. This can cause problems when certain feedstocks are used. At temperatures higher than about 900 °C, the rate of pyrolysis of many hydrocarbons becomes very high, resulting in the formation of large amounts of amorphous carbon. For this reason, many groups have chosen to use CO or CH_4 as the carbon source, since these both have relatively high thermal stability. The addition of H_2 or benzene to the CH_4 flow has been shown to enhance the SWNT

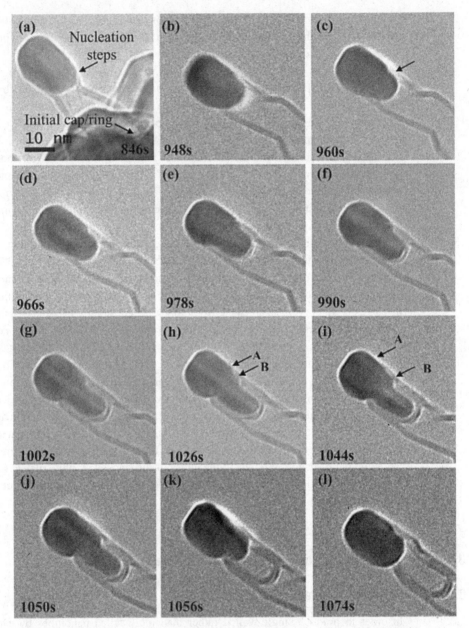

Fig. 3.10   Sequence of *in situ* TEM images showing the growth of a bamboo-like carbon nanotube catalysed by a Ni particle at 650 °C (3.72).

yield (3.76). Smalley's group also produced SWNTs in high yield using ethylene (3.77). An interesting aspect of this study was that SWNT formation was apparently favoured by a limited carbon supply, and this enabled relatively low temperatures to be used (700–850 °C). Other hydrocarbons have been used, including hexane (e.g. 3.78) and benzene (e.g. 3.79), but in such cases H_2 is normally added to the feedstock to break

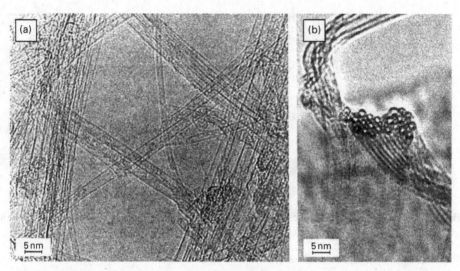

Fig. 3.11   TEM images of typical catalytically-produced single-walled carbon nanotubes, from the work of Colomer *et al.* (3.80).

down the excess carbon. Some micrographs of typical CVD-grown single-walled nano-tubes are shown in Fig. 3.11.

The first successful catalytic synthesis of SWNTs, by the Rice group in 1996 (3.81), employed CO as the feedstock. This was passed over a catalyst containing Mo particles a few nanometres in diameter, at a temperature of 1200 °C, resulting in the formation of SWNTs in the size range 1–5 nm. The best catalyst for SWNT synthesis depends to some extent on the feedstock. The most commonly used metals have been Fe, Ni, Co and Mo, or some combination of these, but oxides have also been used. In the first example of SWNT synthesis from CH_4, in 1998, Hongjie Dai's group used a supported Fe_2O_3 catalyst (3.82). In subsequent work, this group found that the optimum catalyst for SWNT synthesis from CH_4 consisted of Fe/Mo bimetallic species supported on silica-alumina (3.83). More recently it has been shown that SWNTs can be produced from metals not known to be catalysts for carbon formation, including Au, Ag and Cu (3.84–3.86). This came as a surprise, since it has been shown experimentally that Cu cannot catalyse the growth of multiwalled nanotubes (3.28), while theoretical work has suggested that Au should be incapable of catalysing SWNT formation (3.87). These findings have implications for understanding the growth mechanisms of SWNTs by CVD, as discussed later. As well as supported catalysts, 'floating catalysts' have been successfully employed. Mildred Dresselhaus and co-workers were probably the first to use this technique to make SWNTs, using ferrocene as the catalyst precursor and benzene as the feedstock (3.79). The high-pressure CO disproportionation (HiPco) process for high-volume SWNT pro-duction utilizes $Fe(CO)_5$ as a precursor for Fe clusters, as discussed in the next section.

In addition to temperature, feedstock and the type of catalyst, the other major factor determining whether SWNTs are produced is the size of the catalyst particles. Work by Jie Liu and his group from Duke University suggested that catalyst particles must be smaller than about 8.5 nm in order to promote SWNT growth (3.88). Hongjie Dai's group

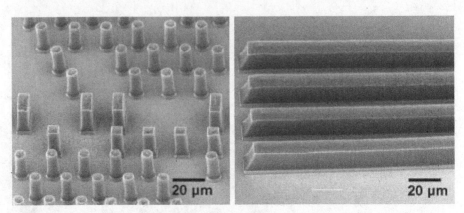

Fig. 3.12   Single-walled nanotubes grown vertically on a SiO_2/Si wafer by O_2-assisted PECVD. (a) Square and circular towers, (b) lines (3.93).

observed similar trends: nanoparticles less than 1.8 nm in size were the most active in producing single walled tubes, while nanoparticles with diameters above ~7 nm did not show any SWNT growth (3.89). Workers from Harvard prepared monodispersed nanoclusters of iron with diameters of 3, 9 and 13 nm (3.90). After exposure to ethylene, single walled and double walled tubes were nucleated from the 3 and 9 nm diameter nanoclusters, whereas only multiwalled tubes were observed from the 13 nm nanoclusters. Further evidence for the importance of catalyst particle size in determining the type of nanotubes produced has come form work by Emmanuel Flahaut and colleagues from the University of Toulouse, discussed in Section 3.6 below. Despite these observations, it is also apparently possible for SWNTs to grow from larger catalyst particles (3.91, 3.92). Similar behaviour is seen in the arc-discharge synthesis of SWNTs (see p. 29).

A few groups have looked at the application of PECVD to SWNT synthesis. This process has been widely used in growing aligned MWNTs on substrates discussed in Section 3.2.1 above, but in general has proved more difficult for SWNT growth. However, in 2005, Dai's group showed that PECVD could be used to grow vertical arrays of single-walled carbon nanotubes on wafers of SiO_2/Si, using an Fe catalyst (3.93). Nanotube growth was found to be boosted by the addition of 1% O_2 to the reaction mixture. The role of the O_2 was apparently to remove reactive hydrogen radicals, and to provide a carbon-rich and hydrogen-deficient condition conducive to SWNT growth. Some micrographs of the SWNT arrays are shown in Fig. 3.12. In another interesting study, Dai's group claimed that PECVD synthesis can result in the preferential formation of semiconducting SWNTs (3.94). This work, and other studies aimed at preparing SWNTs with defined structures is discussed in Section 3.4.5 below.

3.4.2   Large-scale catalytic synthesis of SWNTs

Making SWNTs catalytically clearly has great potential for scale-up, and many groups have used this approach to develop bulk synthesis processes. An early example was the already-mentioned work of Dresselhaus and colleagues, who produced large numbers of SWNTs from benzene, using ferrocene as the catalyst precursor. The addition of thiophene was found to be effective in increasing the yield of nanotubes (3.79). The work of Dai *et al.* on SWNT synthesis from CH_4 with a supported Fe/Mo bimetallic catalyst was also mentioned in the previous section (3.83). In this way, gram quantities of SWNTs could be synthesized in ~0.5 h. Jean-François Colomer of the University of Antwerp and co-workers reported the large-scale production of SWNTs using a catalyst comprising metal particles supported on MgO (3.80). Following the reaction, the MgO support could be easily removed by an acid treatment. Cheol Lee, of Hanyang University, Korea and colleagues (3.95) described a method for the large-scale production of SWNTs from ethylene using a Fe–Mo/MgO catalyst at 800 °C. The tubes were produced in high yield, and were relatively free of amorphous carbon, as can be seen in Fig. 3.13. As in the Colomer work, the MgO support material could be readily removed by mild acid treatment, giving high-purity SWNTs.

Some of the most successful methods for the large-scale catalytic synthesis of SWNTs have employed CO as the feedstock. Daniel Resasco and co-workers at the University of Oklahoma showed that increased yields of SWNTs could be achieved by using Mo oxides combined with Co as catalysts for CO disproportionation (3.96–3.98). They believe that interactions between Mo oxides and Co stabilize the Co catalyst against aggregation through high-temperature sintering. The Resasco group have developed a method for high-volume production of SWNTs using this catalyst, which they named the CoMoCAT process (3.98).

Perhaps the most important method for the large-scale catalytic synthesis of SWNTs is the high-pressure CO disproportionation (HiPco) process introduced by the Smalley group in 1999 (3.99). This process is based on the decomposition of $Fe(CO)_5$ to form Fe clusters for the catalytic production of SWNTs from CO at about 1000 °C. As noted earlier, the injection of organometallic compounds as catalyst precursors directly into the reactant feedstock greatly facilitates the continuous production of nanotubes. A problem in the early work, however, was that the HiPco nanotubes contained comparatively large amounts of Fe, typically 14% by weight. Smalley's group (3.100) and others (3.101) have reported purification protocols, which can reduce the Fe content to 3 wt% or lower. The purification of nanotubes is discussed in detail in the next chapter. At least two large-scale HiPco reactors are currently operational: one at Rice University and one at a spin-off company, Carbon Nanotechnologies, Inc. A number of reviews of the process have been given (3.4, 3.102, 3.103).

3.4.3 Preparation of SWNT strands

Strands or fibres of aligned SWNTs can be prepared by the processing of ready-made material, as discussed in the next chapter (p. 95). The direct synthesis of macroscopic strands of SWNTs using catalysis was first reported by Pulickel Ajayan and colleagues in 2002 (3.78). The carbon source in this work was liquid n-hexane, into which thiophene and ferrocene were dissolved. This was then sprayed into a H_2 stream, and fed into the top of a vertical furnace. It was found that this method could result in the formation of nanotube strands, up to 20 cm in length, which could be collected at the bottom of the furnace.

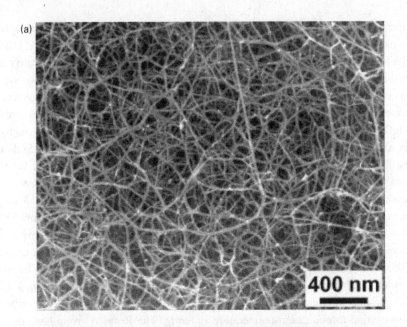

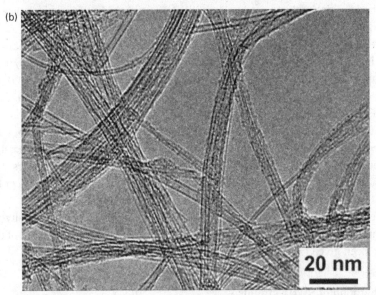

Fig. 3.13   Images of SWNTs prepared by the Fe–Mo/MgO catalysed decomposition of ethylene, from
the work of Lee *et al*. (3.95). (a) Low-magnification SEM images, (b) TEM image.

Figure 3.14 shows some typical electron micrographs of the SWNT strands prepared
by Ajayan *et al*. The strand shown in Fig. 3.14(a) is approximately 70 μm in diameter;
some other strands were as large as 500 μm (0.5 mm) in diameter. The strands appeared to
be made up from smaller ropes of SWNTs, which could be peeled apart, as in Fig. 3.14(b).
High-resolution TEM (Fig. 3.14c) confirmed that the ropes contained aligned SWNTs.
Subsequent work showed that most tubes had a diameter of 1.1 nm, and the 'lattice

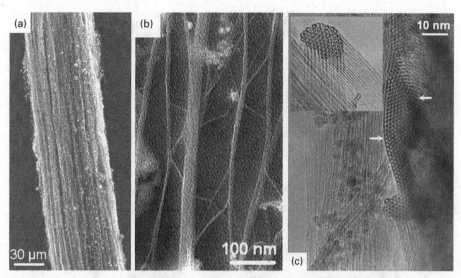

Fig. 3.14   SWNT strands prepared by Ajayan and colleagues (3.78). (a) Low-magnification SEM image of a strand. (b) High-resolution SEM of an array of SWNT ropes peeled from a strand. (c) HRTEM image of a top view of a SWNT rope. The inset shows the cross-sectional view of polycrystalline bundle.

constant' was 1.42 nm (3.104, 3.105). Tensile tests were carried out on some of the ropes teased away from the larger strands, and produced values of the Young's modulus in the range 49–77 GPa. These figures are similar to those achieved for SWNT/polyvinyl alcohol composite fibres prepared by Baughman and colleagues (see Chapter 9, p. 234). Researchers from South Korea have also explored the synthesis of long strands of SWNTs using catalysis (3.106).

3.4.4   Directed growth of SWNTs

Producing defined arrays of single-walled nanotubes on a substrate would be of great interest in the development of nanoscale devices, as well as in other areas. There are two ways in which this could be achived: by directed *in situ* growth of nanotubes, or by post-synthesis arrangement of the tubes. This section is concerned with the guided growth of SWNTs on surfaces. The alternative approach is discussed in the next chapter.

The earliest attempt to achieve controlled growth of SWNTs was described by Hongjie Dai and his colleagues in 1998 (3.107). This group used silicon wafers patterned with µm-scale islands of an iron compound to catalyse the formation of single-walled nanotubes. The resulting tubes grew out from the catalytic regions, in some cases forming bridges between adjacent islands. However, although this work was impressive, it did not represent *directed* SWNT growth. True directed growth of SWNTs, using an electric field, was demonstrated by Dai's group in 2001 (3.108). Their method involved depositing a poly-silicon film on a quartz wafer and then using photolithography and plasma

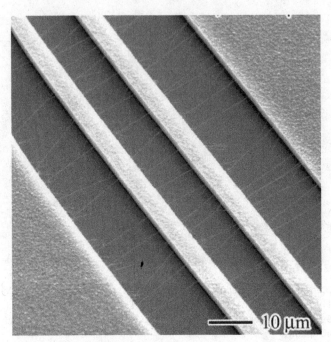

Fig. 3.15   SEM micrograph showing electric field-directed growth of SWNTs between Si strips, from the work of Dai *et al.* (3.108).

etching to form parallel trenches in the film. A catalyst consisting of metal particles supported in mesoporous alumina was then deposited onto the poly-Si structures. Some outer poly-Si strips acted as electrodes. Note that in this arrangement the nanotubes are growing in free space, rather than being supported on the substrate. If the nanotubes were pinned on a surface, strong van der Waals interactions would reduce the field-directing effect. In an initial control experiment, nanotube growth was carried out with no applied electric field. As expected, the resulting growth exhibited no preferred orientation. However, when a voltage was applied to the outer poly-Si electrodes, aligned growth was observed across the trenches, as shown in Fig. 3.15. Both DC and AC fields produced alignment.

A short time later, Ernesto Joselevich, of the Weizmann Institute of Science, Israel, and Charles Lieber also reported the directed growth of SWNTs using an electric field (3.109). In this work, iron oxide nanoparticles were deposited on a silicon chip between a pair of microfabricated electrodes, 25 μm apart. The SWNTs were grown at 800 °C in a flow of 0.2% C_2H_4, 40% H_2 and 60% Ar. Applying an electric field of 100 V across the microelectrodes resulted in directed growth of the tubes perpendicular to the electrodes. They speculated that field-directed growth could discriminate between metallic and semiconducting nanotubes during their formation, although this has not yet been proved experimentally.

Attempting to grow SWNTs in two dimensions using electric fields has proved more difficult. Initial experiments by Dai's group involved depositing four Mo electrodes in a cross arrangement on an Si substrate (3.110). Catalyst islands were then deposited onto the electrodes, and SWNTs were grown at 800–900 °C using a CH_4/C_2H_4 mixture as the carbon source. It was found that nanotubes assembled in the regions of the most intense electric field and tended to follow the local field lines. It was also observed that the tubes grew from negative toward positive electrodes.

The application of electric fields is not the only way to achieved directed growth of SWNTs on a substrate. An alternative approach is to use the crystallography of the substrate material to define the growth direction. The first demonstration of this was given by Jie Liu and colleagues in 2000 (3.111). The SWNTs were grown by catalytic CVD of CH_4 on silicon surfaces terminated with hydrogen, native oxide, and an ultra-thin aluminium layer. On Si (100)-based surfaces, the nanotubes grew in two perpendicular directions, while on Si (111)-based surfaces they grew in three preferred directions separated by 60°. Ernesto Joselevich and his colleagues have studied the growth of SWNTs on sapphire wafers with a high concentration of atomic steps (3.112, 3.113). They showed that SWNT growth occurred preferentially along the atomic steps. In subsequent work they deliberately created faceted nanosteps on the surface of sapphire crystals by annealing, and once again demonstrated preferential growth along the steps. This kind of intentional modification of surfaces to enhance epitaxy is known as 'graphoe-pitaxy', and may have considerable potential in the guided growth of SWNTs on surfaces.

A still further way to achieve oriented growth of SWNTs was described by Liu's team in 2003 (3.114). It was shown that the growth direction of tubes on a substrate could be determined simply by the flow direction of the feedstock gas. Thus, when Fe/Mo catalyst nanoparticles on a SiO_2/Si support were rapidly heated in the presence of a CO/H_2 gas mixture, SWNTs were found to grow almost exclusively in the gas flow direction. These tubes could be grown to lengths of several mm. In subsequent work (3.115), this method was used to produce aligned growth in two dimensions. This was achieved by firstly growing tubes along one direction, then rotating the sample and growing a second set of tubes in the new direction. Examples of nanotube arrays produced in this way are shown in Fig. 3.16.

Finally in this section, another fascinating piece of work by Dai's group should be mentioned (3.116). Although this did not involve directed growth as such, it probably represents the first successful integration of nanotube transistors into a silicon circuit. The starting point was a silicon metal oxide semiconductor (MOS) chip, onto which were deposited catalyst islands. By using these islands to grow single-walled tubes, it was possible to produce nanotube transistors connected to the NMOS circuit. Thousands of these connections could be made on a 1 cm^2 chip, and the path that led to an individual nanotube could be isolated. In this way it could be determined whether the tube was a semiconductor and/or a metal. This would seem to be an important step towards the utilization of carbon nanotubes in integrated circuits.

3.4.5 Synthesis of SWNTs with defined structures

It was noted in Section 3.4.1 that the sizes of catalytic particles can determine whether single-walled or multiwalled nanotubes are formed. It is much less clear whether catalytic synthesis can be used to produce SWNTs with defined structures. Certainly no

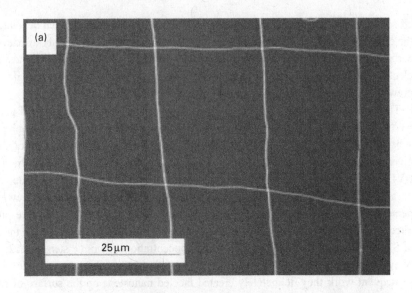

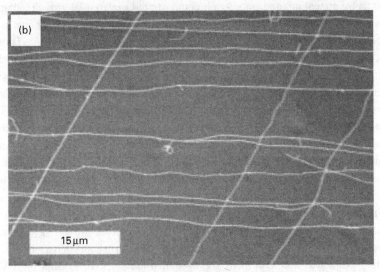

Fig. 3.16   SEM images of networks of crossed SWNTs on a SiO₂/Si surface fabricated by Liu *et al.* (3.115). (a) A crossed-network SWNT array with a 90° included angle; (b) an array with a 60° included angle.

procedures yet exist that are guaranteed to produce tubes with a known structure. However, a few reports have been published which suggest that this might be possible. One of the first of these was published by Jean-François Colomer and his colleagues in 2002 (3.117). These workers used transmission electron diffraction to study SWNT bundles produced by arc-discharge, laser ablation and catalytic CVD. Diffraction patterns of arc-discharge-produced bundles showed a diffuse pattern, characteristic of a random chirality dispersion within the bundle. Diffraction from the laser-ablation-produced bundles showed evidence of a rather narrower range of chiralities, while bundles produced by CVD appeared to exhibit just one or two tube helicities.

Daniel Resasco and co-workers have also reported that SWNTs with certain chiralities can be formed preferentially by thermal CVD (3.118). They used fluorescence spectroscopy to analyse SWNTs prepared by the CoMoCAT method. As discussed in Chapter 7 (p. 190), spectrofluorimetric measurements on semiconducting SWNTs in solution can distinguish between tubes with different structures. Resasco and colleagues found evidence that tubes with the (6, 5) and (7, 5) structures made up more than 50% of the semiconducting SWNTs produced by the CoMoCAT process. In subsequent work it was shown that the (n, m) distribution could be controlled by varying the gaseous feed composition, the reaction temperature, and the type of catalyst support used (3.119). The reasons why these near-armchair structures appear to be favoured is not well understood.

In 2004, Hongjie Dai and co-workers reported that plasma enhanced CVD could result in the preferential growth of semiconducting SWNTs (3.94). This preferential formation was particularly marked in the smallest tubes (3.120), and may indicate that semiconducting SWNTs have higher stabilities than metallic ones. As already noted (p. 31), Dai *et al.* carried out some control experiments which indicated that laser vaporization preferentially produces metallic SWNTs.

3.5 Growth mechanisms of catalytically produced SWNTs

Single-walled tubes produced by CVD are generally similar in appearance to those made by the arc and laser methods. They typically have diameters of the order of 1.2 nm, and often form close-packed bundles (see Fig. 3.13b). This suggests that many tubes can grow from a single metal particle, and this has been directly observed in some cases (3.121). In other studies, small metal particles have been seen at the tips of individual CVD-grown single-walled tubes (3.89, 3.122). Unlike multiwalled carbon nanotubes produced catalytically, CVD-grown single-walled tubes seem to be highly perfect, with few defects. The growth mechanisms of catalytically produced SWNTs have been widely discussed, and a number of useful reviews are available (3.3–3.7, 3.123). Virtually all the discussions have assumed that CVD growth of SWNTs involves a VLS-type process, and this type of mechanism will now be considered.

3.5.1 Vapour–liquid–solid mechanisms

The VLS model has been widely discussed with reference to the arc and laser methods of SWNT synthesis (see previous chapter, p. 32). A number of groups have also developed models of VLS growth that relate specifically to catalytic synthesis (e.g. 3.123–3.128). Kim Bolton of the University of Gothenburg and colleagues have simulated the CVD growth of SWNTs on floating Fe particles at temperatures in the range 600–1600 K (3.123, 3.125). They assume a vapour–liquid–solid mechanism in which Fe particles become saturated with carbon and then carbon strings, polygons and small graphitic

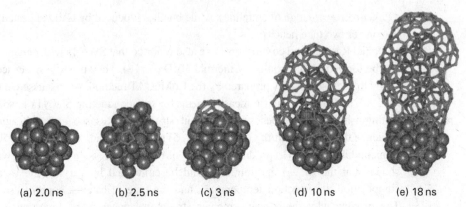

(a) 2.0 ns   (b) 2.5 ns   (c) 3 ns   (d) 10 ns   (e) 18 ns

Fig. 3.17   A simulation of SWNT growth at 900 K by Bolton *et al*. The cluster contains 50 Fe atoms, and one carbon atom is added to the central part of the cluster every 40 ps. Iron atoms are represented as balls and carbon atoms as a stick-like structure (3.125).

islands nucleate on the cluster surface. Then, if the temperature is sufficiently high, a graphitic island can lift off the particle to form a cap which can potentially grow into a tube. They found that the critical temperature for cap nucleation is 800 K. Below this temperature, graphene sheets encapsulate the particles, whereas above 1600 K, carbon is deposited as a soot-like structure rather than as nanotubes. These results broadly agree with experimental observations. One of their simulations, corresponding to a temperature of 900 K, is shown in Fig. 3.17.

As noted previously, a basic assumption of the VLS mechanism is that carbon diffuses through a liquid metal particle, driven by either a temperature gradient or a concentration gradient. Bolton and colleagues have pointed out that, for very small catalytic particles, it is unlikely that any temperature gradient could exist (3.126). However, they have shown that a carbon concentration gradient can provide a sufficient driving force for diffusion. On the other hand, the idea that diffusion through the metal particle occurs at all during nanotube growth has been questioned by Jean-Yves Raty and colleagues (3.87). These researchers modelled the CVD growth of SWNTs on Fe nanoparticles using *ab initio* molecular dynamics. In contrast to the work of Bolton *et al.*, they found that the carbon atoms did not dissolve in the Fe, but instead diffused onto the surface of the metal and then formed an sp^2 sheet, with the form of a nanotube cap, as shown in Fig. 3.18. Single C atoms could then diffuse to the root and become incorporated into the growing tube. In this work approximately half of the nanoparticle surface atoms were passivated with H to mimic the presence of a supporting surface.

3.5.2   A solid-state mechanism for CVD growth?

As outlined in the previous chapter (p. 34), several groups have proposed that the production of single-walled nanotubes by laser vaporization might involve a solid-state mechanism in which the metal particles convert solid, disordered carbon into nanotubes. The suggestion that the CVD synthesis of SWNTs might also involve a solid-state

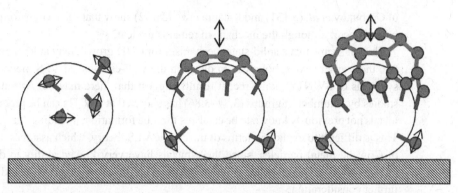

Fig. 3.18 A simulation of SWNT growth on an Fe catalyst, from the work of Raty *et al.* (3.87). (a) Diffusion of single C atoms on the surface of the catalyst. (b) Formation of a graphene sheet on the catalyst surface with edge atoms covalently bonded to the metal. (c) Root incorporation of diffusing single C atoms (or dimers).

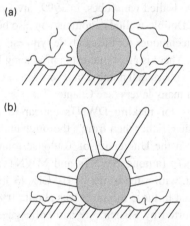

Fig. 3.19 Schematic illustration of the 'solid-state' mechanism for the growth of single-walled carbon nanotubes by catalytic CVD. (a) Deposition of carbon fragments on surfaces of the catalyst, (b) transformation of disordered carbon into nanotubes.

transformation has not been previously raised, so it is worth considering whether such a mechanism would be feasible. A possible scenario is given in Fig. 3.19. The first stage in the process (Fig. 3.19a) is the condensation of curved, fullerene-related, carbonaceous fragments on the surfaces of the catalyst. Literature reports suggest that significant amounts of disordered carbon could deposit on the catalyst surfaces under the conditions used for CVD. For example Bai *et al.* have shown that methane readily decomposes to carbon over alumina at 850 °C (3.129). It cannot be said with certainty whether this carbon would have a fullerene-related structure, but there are many studies which show that C formed by condensation from the vapour consists of assemblies of small curved sheets (3.130), and this curvature may indicate the presence of fullerene-like elements. The second stage in the process is the conversion of the rather disordered carbon clusters into single-walled tubes, promoted by the metal particles (Fig. 3.19b). The experiments of Gorbunov *et al.* (3.131) and Kataura *et al.* (3.132) show that such a transformation can indeed occur, although the mechanism remains unclear.

Thus, it seems that a solid-state mechanism for CVD growth may at least be possible. Are there any reasons, however, to question the well-established VLS model for CVD synthesis of SWNTs? Some recent results suggest that there may be. As noted above, studies by a number of groups (3.84–3.86) has shown that SWNTs can be produced from metals not previously known to be catalysts for tube formation, including Au, Ag and Cu. This is difficult to explain in terms of the classic VLS theory, which assumes dissolution of C in the metal particles, since these metals have very limited ability to dissolve C (3.133, 3.134). Therefore the possibility of a solid-state mechanism should be given further consideration.

3.6   Catalytic synthesis of double-walled nanotubes

The first controlled production of double-walled nanotubes, in 1999, involved thermal treatments of C_{60} 'peapods' (see p. 257). Double-walled nanotubes can also be produced by arc-evaporation (p. 36). A number of techniques for the catalytic synthesis of DWNTs in high yield are also now also available. There is great interest in producing DWNTs in this way, since CVD-produced MWNTs with a small number of layers can have excellent mechanical properties, unlike those with many layers (see Chapter 7).

In most cases the catalytic methods for making DWNTs appear to have been developed using a trial-and-error approach, rather than from a thorough understanding of the processes involved. Workers from the University of Toulouse found in 2000 that relatively large numbers of DWNTs (among SWNTs and MWNTs) could be produced using an $Mg_{1-x}Co_xO$ catalyst, with a feedstock of H_2–CH_4 (3.135, 3.136). The highest fraction of DWNTs observed in these mixtures was about 50%. A some-what higher yield of DWNTs was reported by Sishen Xie and colleagues from the Chinese Academy of Science, Beijing in 2002 (3.137). Their method involved pyr-olysing C_2H_2 on a floating iron catalyst at 900–1100 °C. An interesting aspect of this work was that the growth of DWNTs was strongly promoted by the presence of sulphur in the reaction mixture, although the reasons for this are unclear. A short time later, another Chinese group, led by Hui-Ming Cheng (3.138) synthesized DWNTs, in good yield, using the floating catalyst method with ferrocene as catalyst precursor and CH_4 as carbon source. Again sulphur, introduced into the reactor in the form of thiophene, was found to promote nanotube formation. In subsequent work, they were able to obtain aligned DWNT ropes with a narrow diameter distributions using a similar technique (3.139).

Cheol Lee's group at Hanyang University, Korea (3.140) described in 2003 the production of high-quality DWNTs in good yield by the catalytic decomposition of alcohol over an Fe–Mo/Al_2O_3 catalyst at 800 °C. Using an ethylene feedstock under similar conditions produced mainly SWNTs (see p. 59 above), for reasons that are not well understood. The DWNTs synthesized by Lee and colleagues had outer diameters in the range of 1.52–3.54 nm, with a slightly larger interlayer distance than usually observed for MWNTs (approximately 0.38 nm). At about the same time the Toulouse group reported the synthesis of gram-scale quantities of nanotubes, 77% of which were DWNTs, using a $Mg_{1-x}Co_xO$ catalyst containing additions of Mo oxide (3.141). The average diameter was about 2 nm. In later work, this group showed that preparing the Mg–Co–Mo–O catalyst in different ways could result in different distributions of nanotubes (3.142). Thus, one preparation could result in a sample containing more than 90% double- and triple-walled nanotubes, while another could give a sample containing almost 80% double-walled tubes. Since the chemical compositions of the catalysts were identical, the authors believe that the differing products resulted from differences in the morphology and particle sizes of the catalysts.

Numerous other recipes for preparing double-walled carbon nanotubes are now available (e.g. 3.143–3.146). These include a method for growing vertically aligned DWNT arrays (3.146). Perhaps the most effective method for large-scale DWNT production was described by a multinational team from Shinshu University, Japan, IPICYT, Mexico and MIT (3.147). Their method involved using an Mo 'conditioning catalyst' at one end of the reaction furnace and an Fe catalyst in the middle of the furnace. The conditioning catalyst promoted the growth of DWNTs, rather than SWNTs, possibly by increasing the amount of active carbon species. The DWNTs produced in this way were then purified in a two-step process. An HCl treatment removed the iron catalyst and supporting material, while oxidation in air removed amorphous carbon and chemically active single-walled carbon nanotubes. A filtration step then produced sheets of DWNT 'buckypaper', which contained more than 95% double-walled tubes. Images of the product are shown in Fig. 3.20 (see Section 4.4.3 for a fuller discussion of buckypaper).

(a)

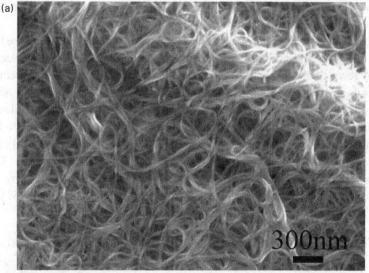

300nm

Fig. 3.20 Images of DWNT 'buckypaper', from the work of Endo and colleagues (3.147). (a) SEM micrograph showing bulk structure, (b) HRTEM image showing a cluster of tubes.

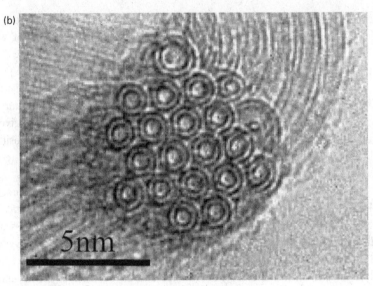

Fig. 3.20   (cont.)

3.7   Electrochemical synthesis of multiwalled nanotubes

An electrochemical method for the synthesis of multiwalled nanotubes was described by the Sussex group (with the present author) in the mid 1990s (3.148, 3.149). This involved the electrolysis of molten lithium chloride using a graphite cell in which the anode was a graphite crucible and the cathode a graphite rod immersed in the melt. A current of about 30 A was passed through the cell for 1 minute, after which the electrolyte was allowed to cool, and then added to water to dissolve the lithium chloride and react with the lithium metal. The mixture was left for 4 hours, and toluene was then added to the aqueous suspension and the whole agitated for several minutes.

This treatment resulted in most of the solid material passing into the toluene layer, which was then separated from the aqueous layer by decanting. Transmission electron microscopy revealed that the material contained large numbers of rather imperfect multiwalled nanotubes which were similar in appearance to catalytically-formed tubes. Nanoparticles were also found, and both tubes and nanoparticles often contained encapsulated material, presumably lithium chloride or oxide. Thus, the technique could prove to be a useful method of preparing filled carbon nanostructures. There has been relatively little subsequent work on the electrochemical synthesis of nanotubes, although Ian Kinloch and co-workers have examined the effect of electrolysis duration, current density and voltage on nanotube yield, and the feasibility of scaling-up the process (3.150, 3.151).

3.8   Synthesis of MWNTs by heat treatment of metal-doped carbon

We saw in the previous chapter (p. 24) that multiwalled nanotubes can be produced by the high-temperature heat treatment of disordered carbons, and that single-walled tubes can also form as a result of a solid-state transformation. In recent work, Chinese workers have shown that multiwalled tubes can be made by heating a glassy carbon containing Fe nanoparticles at temperatures of 800–1000 °C (3.152). Clearly the mechanism must involve a solid phase transformation, which may be similar to that of the electrochemical synthesis described above. A group from the Czech Republic have also reported a solid phase synthesis of MWNTs from an Fe–C 'nanopowder' (3.153). Further work in this area would be welcome.

3.9 Discussion

As we have seen in this chapter, amazing progress has been made in the catalytic synthesis of nanotubes since the early 1990s. Processes for the bulk production of multiwalled nanotubes have advanced to the point where several companies claim to be making more than 100 tons per year. Single-walled tubes can also now be produced in bulk using CVD, although production volumes are inevitably rather lower, and techniques for making double-walled nanotubes catalytically have been developed. Remarkable advances have also been made in the growth of nanotubes in a controlled manner on substrates. In the case of MWNTs, this has generally involved the vertical growth of tubes on surfaces, producing large arrays which have applications in areas such as field emission. The directed growth of SWNTs, on the other hand, has mainly involved the formation of defined patterns on substrates, with the aim of constructing nanoscale electronic devices. Another area where important progress has been made, notably by the Windle group in Cambridge, is the continuous spinning of nanotube yarns, which may have important applications as components of composite materials.

Are there any disadvantages to the catalytic synthesis of nanotubes? In the case of multiwalled tubes, the answer is yes, as the quality of CVD-produced multiwalled nanotubes is still inferior to those made by arc-evaporation. Generally speaking, catalytically-made MWNTs contain far more defects than those produced in the arc, and this is reflected in their mechanical properties. It might be thought that the properties of CVD-MWNTs could be improved by high-temperature annealing, but this does not seem to be the case. As discussed in Chapter 7 (p. 187), the mechanical properties of CVD tubes are not significantly improved even by heating at temperatures up to 2400 °C. Interestingly, however, catalytically-produced SWNTs, or MWNTs with a small number of layers, can have excellent mechanical properties, suggesting that they contain many fewer defects than many-walled tubes (see pp. 187–188). Further evidence for the high quality of catalytically-produced SWNTs comes from work by Philip Collins and colleagues (3.154). These workers showed that the most chemically reactive sites on SWNT surfaces could be selectively decorated by electrochemically deposited nickel particles. Using this method they found that the best quality SWNTs contained just one defect per 4 μm on average, with most defects being found in areas of SWNT curvature. They pointed out that this defect density compares favourably to high-quality silicon single-crystals. Collins *et al.* used their technique to compare SWNTs prepared by different processes and found, perhaps surprisingly, that tubes made by arc-evaporation contained rather more defects than these made catalytically.

The growth mechanisms of catalytically-made nanotubes are little better understood than those of arc- or laser- produced tubes. In the case of MWNT growth, the generally accepted view is that the carbon-containing molecules decompose on one side of the metal particle, and the tube grows on the opposite side. However, there is uncertainty about whether the carbon diffuses through the particle or around the edge, and about the physical and chemical state of the catalytic particle. For single-walled tubes, the vapour–liquid–solid mechanism is quite widely thought to be correct, but this has not been definitely established. Other possible mechanisms exist, such as the solid-state model discussed in Section 3.5.2.

References

(3.1) A. Moisala, A. G. Nasibulin and E. I. Kauppinen, 'The role of metal nanoparticles in the catalytic production of single-walled carbon nanotubes – a review', *J. Phys. Cond. Matter*, **15**, S3011 (2003).

(3.2) K. B. K. Teo, C. Singh, M. Chhowalla *et al.*, 'Catalytic synthesis of carbon nanotubes and nanofibres', in *Encyclopedia of Nanoscience and Nanotechnology*, vol. 1, ed. H. S. Nalwa, 2004, p. 665.

(3.3) J. B. Nagy, G. Bister, A. Fonseca *et al.*, 'On the growth mechanism of single-walled carbon nanotubes by catalytic carbon vapor deposition on supported metal catalysts', *J. Nanosci. Nanotech.*, **4**, 326 (2004).

(3.4) P. Nikolaev, 'Gas-phase production of single-walled carbon nanotubes from carbon monoxide: a review of the HiPco process', *J. Nanosci. Nanotech.*, **4**, 307 (2004).

(3.5) A. C. Dupuis, 'The catalyst in the CCVD of carbon nanotubes – a review', *Prog. Mater. Sci.*, **50**, 929 (2005).

(3.6) M. L. Terranova, V. Sessa and M. Rossi, 'The world of carbon nanotubes: an overview of CVD growth methodologies', *Chem. Vapor Deposition*, **12**, 315 (2006).

(3.7) E. Lamouroux, P. Serp and P. Kalck, 'Catalytic routes towards single wall carbon nanotubes', *Catalysis Rev. Sci. Eng.*, **49**, 341 (2007).

(3.8) R. T. K. Baker, M. A. Barber, P. S. Harris *et al.*, 'Nucleation and growth of carbon deposits from the nickel catalyzed decomposition of acetylene', *J. Catalysis*, **26**, 51 (1972).

(3.9) R. T. K. Baker, P. S. Harris, R. B. Thomas *et al.*, 'Formation of filamentous carbon from iron, cobalt and chromium catalysed decomposition of acetylene', *J. Catalysis*, **30**, 86 (1973).

(3.10) R. T. K. Baker, G. R. Gadsby and S. Terry, 'Formation of carbon filaments from catalysed decomposition of hydrocarbons', *Carbon*, **13**, 245 (1975).

(3.11) R. T. K. Baker and R. J. Waite, 'Formation of carbonaceous deposits from the platinum–iron catalysed decomposition of acetylene', *J. Catalysis*, **37**, 101 (1975).

(3.12) T. Baird, J. R. Fryer and B. Grant, 'Structure of fibrous carbon', *Nature*, **233**, 329 (1971).

(3.13) T. Baird, J. R. Fryer and B. Grant, 'Carbon formation on iron and nickel foils by hydrocarbon pyrolysis – reactions at 700 °C', *Carbon*, **12**, 591 (1974).

(3.14) R. T. K. Baker and P. S. Harris, 'The formation of filamentous carbon', *Chem. Phys. Carbon*, **14**, 83 (1978).

(3.15) R. T. K. Baker, 'Catalytic growth of carbon filaments', *Carbon*, **27**, 315 (1989).

(3.16) T. Koyama, M. Endo and Y. Onuma, 'Carbon fibers obtained by thermal decomposition of vaporized hydrocarbon', *Jpn. J. Appl. Phys.*, **11**, 445 (1972).

(3.17) A. Oberlin, M. Endo and T. Koyama, 'Filamentous growth of carbon through benzene decomposition', *J. Cryst. Growth*, **32**, 335 (1976).

(3.18) M. Endo, 'Grow carbon fibres in the vapour phase', *Chemtech*, **18**, 568 (1988) (September issue).

(3.19) M. S. Dresselhaus, G. Dresselhaus, K. Sugihara *et al.*, *Graphite Fibers and Filaments*, Springer-Verlag, Berlin, 1988.

(3.20) E. Boellaard, P. K. Debokx, A. J. H. M. Kock *et al.*, 'The formation of filamentous carbon on iron and nickel catalysts. 3. Morphology', *J. Catalysis*, **96**, 481 (1985).

(3.21) M. Audier, A. Oberlin and M. Coulon, 'Crystallographic orientations of catalytic particles in filamentous carbon, case of simple conical particles', *J. Cryst. Growth*, **55**, 549 (1981).

(3.22) M. Audier, A. Oberlin and M. Coulon, 'Study of biconic micro-crystals in the middle of carbon tubes obtained by catalytic disproportionation of CO', *J. Cryst. Growth*, **57**, 524 (1982).

(3.23) M. Audier, A. Oberlin, M. Oberlin *et al.*, 'Morphology and crystalline order in catalytic carbons', *Carbon*, **19**, 217 (1981).

(3.24) G. G. Tibbetts, D. W. Gorkiewicz and R. L. Alig, 'A new reactor for growing carbon-fibers from liquid-phase and vapor-phase hydrocarbons', *Carbon*, **31**, 809 (1993).

(3.25) G. G. Tibbetts, C. A. Bernardo, D. W. Gorkiewicz *et al.*, 'Role of sulfur in the production of carbon-fibers in the vapor-phase', *Carbon*, **32**, 569 (1994).

(3.26) R. Sen, A. Govindaraj, C. N. R. Rao, 'Metal-filled and hollow carbon nanotubes obtained by the decomposition of metal-containing free precursor molecules', *Chem. Mater.*, **9**, 2078 (1997).

(3.27) A. Govindaraj and C. N. R. Rao, 'Organometallic precursor route to carbon nanotubes', *Pure Appl. Chem.*, **74**, 1571 (2002).

(3.28) C. P. Deck and K. Vecchio, 'Prediction of carbon nanotube growth success by the analysis of carbon-catalyst binary phase diagrams', *Carbon*, **44**, 267 (2006).

(3.29) D. S. Su and X. W. Chen, 'Natural lavas as catalysts for efficient production of carbon nanotubes and nanofibers', *Angew. Chem. Int. Ed.*, **46**, 1823 (2007).

(3.30) D. S. Su, 'Efficient fabrications of nanocarbon using Mount Etna lavas and cat sand as catalyst and support', paper presented at NanoteC07, University of Sussex, 2007.

(3.31) K. Kempa, B. Kimball, J. Rybczynski *et al.*, 'Photonic crystals based on periodic arrays of aligned carbon nanotubes', *Nano Lett.*, **3**, 13 (2003).

(3.32) B. Yurdumakan, N. R. Raravikar, P. M. Ajayan *et al.*, 'Synthetic gecko foot-hairs from multiwalled carbon nanotubes', *Chem. Commun.*, 3799 (2005).

(3.33) N. M. Pugno, 'Towards a Spiderman suit: large invisible cables and self-cleaning releasable superadhesive materials', *J. Phys. Cond. Matter*, **19**, 395001 (2007).

(3.34) W. Z. Li, S. Xie, L. X. Qian *et al.*, 'Large-scale synthesis of aligned carbon nanotubes', *Science*, **274**, 1701 (1996).

(3.35) M. Terrones, N. Grobert, J. Olivares *et al.*, 'Controlled production of aligned-nanotube bundles', *Nature*, **388**, 52 (1997).

(3.36) J. Li, C. Papadopoulos, J. M. Xu *et al.*, 'Highly-ordered carbon nanotube arrays for electronics applications', *Appl. Phys. Lett.*, **75**, 367 (1999).

(3.37) S. S. Fan, M. G. Chapline, N. R. Franklin *et al.*, 'Self-oriented regular arrays of carbon nanotubes and their field emission properties', *Science*, **283**, 512 (1999).

(3.38) Z. F. Ren, Z. P. Huang, J. W. Xu *et al.*, 'Synthesis of large arrays of well-aligned carbon nanotubes on glass', *Science*, **282**, 1105 (1998).

(3.39) J. Perrin, J. Schmitt, C. Hollenstein *et al.*, 'The physics of plasma-enhanced chemical vapour deposition for large-area coating', *Plasma Phys. Control. Fusion*, **42**, B353 (2000).

(3.40) C. Bower, W. Zhu, S. H. Jin *et al.*, 'Plasma-induced alignment of carbon nanotubes', *Appl. Phys. Lett.*, **77**, 830 (2000).

(3.41) L. Delzeit, I. McAninch, B. A. Cruden *et al.*, 'Growth of multiwall carbon nanotubes in an inductively coupled plasma reactor', *Appl. Phys. Lett.*, **91**, 6027 (2002).

(3.42) V. I. Merkulov, D. H. Lowndes, Y. Y. Wei *et al.*, 'Patterned growth of individual and multiple vertically aligned carbon nanofibers', *Appl. Phys. Lett.*, **76**, 3555 (2000).

(3.43) K. B. K. Teo, M. Chhowalla, G. A. J. Amaratunga *et al.*, 'Uniform patterned growth of carbon nanotubes without surface carbon', *Appl. Phys. Lett.*, **79**, 1534 (2001).

(3.44) W. I. Milne, K. B. K. Teo, G. A. J. Amaratunga *et al.*, 'Carbon nanotubes as field emission sources', *J. Mater. Chem.*, **14**, 933 (2004).

(3.45) M. S. Bell, K. B. K. Teo, R. G. Lacerda *et al.*, 'Carbon nanotubes by plasma-enhanced chemical vapour deposition', *Pure Appl. Chem.*, **78**, 1117 (2006).

(3.46) M. S. Bell, K. B. K. Teo and W. I. Milne, 'Factors determining properties of multi-walled carbon nanotubes/fibres deposited by PECVD', *J. Phys. D*, **40**, 2285 (2007).

(3.47) B. O. Boskovic, V. Stolojan, R. U. A. Khan *et al.*, 'Large-area synthesis of carbon nanofibres at room temperature', *Nat. Mater.*, **1**, 165 (2002).

(3.48) Z. W. Pan, S. S. Xie, B. H. Chang *et al.*, 'Direct growth of aligned open carbon nanotubes by chemical vapor deposition', *Chem. Phys. Lett.*, **299**, 97 (1999).

(3.49) D. C. Li, L. M. Dai, S. M. Huang *et al.*, 'Structure and growth of aligned carbon nanotube films by pyrolysis', *Chem. Phys. Lett.*, **316**, 349 (2000).

(3.50) Y. L. Li, I. A. Kinloch and A. H. Windle, 'Direct spinning of carbon nanotube fibers from chemical vapor deposition synthesis', *Science*, **304**, 276 (2004).

(3.51) K. Koziol, J. Vilatela, A. Mòisala *et al.*, 'High-performance carbon nanotube fiber', *Science*, **318**, 1892 (2007).

(3.52) M. Terrones, 'Carbon nanotubes: synthesis and properties, electronic devices and other emerging applications', *Int. Mater. Rev.*, **49**, 325 (2004).

(3.53) M. Chhowalla, K. B. K. Teo, C. Ducati *et al.*, 'Growth process conditions of vertically aligned carbon nanotubes using plasma enhanced chemical vapor deposition', *J. Appl. Phys.*, **90**, 5308 (2001).

(3.54) C. Bower, O. Zhou, W. Zhu *et al.*, 'Nucleation and growth of carbon nanotubes by microwave plasma chemical vapor deposition', *Appl. Phys. Lett.*, **77**, 2767 (2000).

(3.55) S. B. Sinnott, R. Andrews, D. Qian *et al.*, 'Model of carbon nanotube growth through chemical vapor deposition', *Chem. Phys. Lett.*, **315**, 25 (1999).

(3.56) G. G. Tibbetts, 'Why are carbon filaments tubular?', *J. Cryst. Growth*, **66**, 632 (1984).

(3.57) R. S. Wagner and W. C. Ellis, 'The vapor–liquid–solid mechanism of crystal growth and its application to silicon', *Trans. Metallurg. Soc. AIME*, **233**, 1053 (1965).

(3.58) P. Buffat and J. P. Borel, 'Size effect on the melting temperature of gold particles,' *Phys. Rev. A*, **13**, 2287 (1976).

(3.59) Q. S. Mei and K. Lu, 'Melting and superheating of crystalline solids: from bulk to nanocrystals', *Progr. Mater. Sci.*, **52**, 1175 (2007).

(3.60) M. Audier, P. Bowen and W. Jones, 'Electron-microscopic and Mössbauer study of the iron carbides θ-Fe$_3$C and χ-Fe$_5$C$_2$ formed during the disproportionation of CO', *J. Cryst. Growth*, **64**, 291 (1983).

(3.61) R. T. K. Baker, J. R. Alonzo, J. A. Dumesic *et al.*, 'Effect of the surface-state of iron on filamentous carbon formation', *J. Catalysis*, **77**, 74 (1982).

(3.62) K. L. Yang and R. T. Yang, 'The accelerating and retarding effects of hydrogen on carbon deposition on metal-surfaces', *Carbon*, **24**, 687 (1986).

(3.63) S. Helveg, C. Lopez-Cartes, J. Sehested *et al.*, 'Atomic-scale imaging of carbon nanofibre growth', *Nature*, **427**, 426 (2004).

(3.64) P. M. Ajayan, 'How does a nanofibre grow?', *Nature*, **427**, 402 (2004).

(3.65) M. Meyyappan, L. Delzeit, A. Cassell *et al.*, 'Carbon nanotube growth by PECVD: a review', *Plasma Sources Sci. Technol.*, **12**, 205 (2003).

(3.66) X. Xhang, A. Cao, B. Wei *et al.*, 'Rapid growth of well-aligned carbon nanotube arrays', *Chem. Phys. Lett.*, **362**, 285 (2002).

(3.67) S. Amelinckx, X. B. Zhang, D. Bernaerts *et al.*, 'A formation mechanism for catalytically grown helix-shaped graphite nanotubes', *Science*, **265**, 635 (1994).

(3.68) D. Bernaerts, X. B. Zhang, X. F. Zhang *et al.*, 'Electron microscopy study of coiled carbon tubules', *Phil. Mag. A*, **71**, 605 (1995).

(3.69) C. J. Lee and J. Park, 'Growth model of bamboo-shaped carbon nanotubes by thermal chemical vapor deposition', *Appl. Phys. Lett.*, **77**, 3397 (2000).

(3.70) J. L. Chen, Y. D. Li, Y. M. Ma *et al.*, 'Formation of bamboo-shaped carbon filaments and dependence of their morphology on catalyst composition and reaction conditions', *Carbon*, **39**, 1467 (2001).

(3.71) L. T. Chadderton and Y. Chen, 'A model for the growth of bamboo and skeletal nanotubes: catalytic capillarity', *J. Cryst. Growth*, **240**, 164 (2002).

(3.72) M. Lin, J. P. Y. Tan, C. B. Boothroyd *et al.*, 'Dynamical observation of bamboo-like carbon nanotube growth', *Nano Lett.*, **7**, 2234 (2007).

(3.73) M. S. Kim, N. M. Rodriguez and R. T. K. Baker, 'The role of interfacial phenomena in the structure of carbon deposits', *J. Catalysis*, **134**, 253 (1992).

(3.74) H. Terrones, T. Hayashi, M. Munoz-Navia *et al.*, 'Graphitic cones in palladium catalysed carbon nanofibres', *Chem. Phys. Lett.*, **343**, 241 (2001).

(3.75) P. E. Nolan, D. C. Lynch and A. H. Cutler, 'Carbon deposition and hydrocarbon formation on group VIII metal catalysts', *J. Phys. Chem. B*, **102**, 4165 (1998).

(3.76) N. R. Franklin and H. J. Dai, 'An enhanced chemical vapor deposition method to extensive single-walled nanotube networks with directionality', *Adv. Mater.*, **12**, 890, (2000).

(3.77) J. H. Hafner, M. J. Bronikowski, B. R. Azamian *et al.*, 'Catalytic growth of single-wall carbon nanotubes from metal particles', *Chem. Phys. Lett.*, **296**, 195 (1998).

(3.78) H. W. Zhu, C. L. Xu, D. H. Wu *et al.*, 'Direct synthesis of long single-walled carbon nanotube strands', *Science*, **296**, 884 (2002).

(3.79) H. M. Cheng, F. Li, G. Su *et al.*, 'Large-scale and low-cost synthesis of single-walled carbon nanotubes by the catalytic pyrolysis of hydrocarbons', *Appl. Phys. Lett.*, **72**, 3282 (1998).

(3.80) J. F. Colomer, C. Stephan, S. Lefrant *et al.*, 'Large-scale synthesis of single-wall carbon nanotubes by catalytic chemical vapor deposition (CCVD) method', *Chem. Phys. Lett.*, **317**, 83 (2000).

(3.81) H. J. Dai, A. G. Rinzler, P. Nikolaev *et al.*, 'Single-wall nanotubes produced by metal-catalyzed disproportionation of carbon monoxide', *Chem. Phys. Lett.*, **260**, 471 (1996).

(3.82) J. Kong, A. M. Cassell and H. J. Dai, 'Chemical vapor deposition of methane for single-walled carbon nanotubes', *Chem. Phys. Lett.*, **292**, 567 (1998).

(3.83) A. M. Cassell, J. A. Raymakers, J. Kong *et al.*, 'Large scale CVD synthesis of single-walled carbon nanotubes', *J. Phys. Chem. B*, **103**, 6484 (1999).

(3.84) D. Takagi, Y. Homma, H. Hibino *et al.*, 'Single-walled carbon nanotube growth from highly activated metal nanoparticles', *Nano Lett.*, **6**, 2642 (2006).

(3.85) W. W. Zhou, Z. Y. Han, J. Y. Wang *et al.*, 'Copper catalyzing growth of single-walled carbon nanotubes on substrates', *Nano Lett.*, **6**, 2987 (2006).

(3.86) S. Bhaviripudi, E. Mile, S. A. Steiner *et al.*, 'CVD synthesis of single-walled carbon nanotubes from gold nanoparticle catalysts', *J. Amer. Chem. Soc.*, **129**, 1516 (2007).

(3.87) J. Y. Raty, F. Gygi and G. Galli, 'Growth of carbon nanotubes on metal nanoparticles: a microscopic mechanism from ab initio molecular dynamics simulations', *Phys. Rev. Lett.*, **95**, 096103 (2005).

(3.88) Y. Li, J. Liu, Y. Q. Wang *et al.*, 'Preparation of monodispersed Fe–Mo nanoparticles as the catalyst for CVD synthesis of carbon nanotubes', *Chem. Mater.*, **13**, 1008 (2001).

(3.89) Y. M. Li, W. Kim, Y. G. Zhang *et al.*, 'Growth of single-walled carbon nanotubes from discrete catalytic nanoparticles of various sizes', *J. Phys. Chem. B*, **105**, 11 424 (2001).

(3.90) C. L. Cheung, A. Kurtz, H. Park *et al.*, 'Diameter-controlled synthesis of carbon nanotubes', *J. Phys. Chem. B*, **106**, 2429 (2002).

(3.91) J. F. Colomer, G. Bister, I. Willems *et al.*, 'Synthesis of single-wall carbon nanotubes by catalytic decomposition of hydrocarbons', *Chem. Commun.*, 1343 (1999).

(3.92) H. Ago, S. Ohshima, K. Uchida *et al.*, 'Gas-phase synthesis of single-wall carbon nanotubes from colloidal solution of metal nanoparticles', *J. Phys. Chem. B*, **105**, 10 453 (2001).

(3.93) G. Y. Zhang, D. Mann, L. Zhang *et al.*, 'Ultra-high-yield growth of vertical single-walled carbon nanotubes: hidden roles of hydrogen and oxygen', *Proc. Nat. Acad. Sci. USA*, **102**, 16 141 (2005).

(3.94) Y. M. Li, D. Mann, M. Rolandi *et al.*, 'Preferential growth of semi-conducting single-walled carbon nanotubes by a plasma enhanced CVD method', *Nano Lett.*, **4**, 317 (2004).

(3.95) S. C. Lyu, B. C. Liu, S. H. Lee *et al.*, 'Large-scale synthesis of high-quality single-walled carbon nanotubes by catalytic decomposition of ethylene', *J. Phys. Chem. B*, **108**, 1613 (2004).

(3.96) J. E. Herrera, L. Balzano, A. Borgna *et al.*, 'Relationship between the structure/composition of Co–Mo catalysts and their ability to produce single-walled carbon nanotubes by CO disproportionation', *J. Catalysis*, **204**, 129 (2001).

(3.97) W. E. Alvarez, B. Kitiyanan, A. Borgna *et al.*, 'Synergism of Co and Mo in the catalytic production of single-wall carbon nanotubes by decomposition of CO', *Carbon*, **39**, 547 (2001).

(3.98) D. E. Resasco, W. E. Alvarez, F. Pompeo *et al.*, 'A scalable process for production of single-walled carbon nanotubes (SWNTs) by catalytic disproportionation of CO on a solid catalyst', *J. Nanoparticle Res.*, **4**, 131 (2002).

(3.99) P. Nikolaev, M. J. Bronikowski, R. K. Bradley *et al.*, 'Gas-phase catalytic growth of single-walled carbon nanotubes from carbon monoxide', *Chem. Phys. Lett.*, **313**, 91 (1999).

(3.100) I. W. Chiang, B. E. Brinson, A. Y. Huang *et al.*, 'Purification and characterization of single-wall carbon nanotubes (SWNTs) obtained from the gas-phase decomposition of CO (HiPco process)', *J. Phys. Chem. B*, **105**, 8297 (2001).

(3.101) C. M. Yang, K. Kaneko, M. Yudasaka *et al.*, 'Effect of purification on pore structure of HiPco single-walled carbon nanotube aggregates', *Nano Lett.*, **2**, 385 (2002).

(3.102) M. J. Bronikowski, P. A. Willis, C. T. Colbert *et al.*, 'Gas-phase production of carbon single-walled nanotubes from carbon monoxide via the HiPco process: a parametric study', *J. Vac. Sci. Tech. A*, **19**, 1800 (2001).

(3.103) R. L. Carver, H. Peng, A. K. Sadana *et al.*, 'A model for nucleation and growth of single wall carbon nanotubes via the HiPco process; a catalyst concentration study', *J. Nanosci. Nanotech.*, **5**, 1035 (2005).

(3.104) B. Q. Wei, R. Vajtai, Y. Y. Choi *et al.*, 'Structural characterizations of long single-walled carbon nanotube strands', *Nano Lett.*, **2**, 1105 (2002).

(3.105) R. Vajtai, B. Q. Wei, Y. J. Jung *et al.*, 'Building and testing organized architectures of carbon nanotubes', *IEEE Trans. Nanotechnol.*, **2**, 355 (2003).

(3.106) N. Van Quy, Y. S. Cho, G. S. Choi *et al.*, 'Synthesis of a long strand of single-wall carbon nanotubes', *Nanotechnology*, **16**, 386 (2005).

(3.107) J. Kong, H. T. Soh, A. M. Cassell *et al.*, 'Synthesis of individual single-walled carbon nanotubes on patterned silicon wafers', *Nature*, **395**, 878 (1998).

(3.108) Y. G. Zhang, A. L. Chang, J. Cao *et al.*, 'Electric-field-directed growth of aligned single-walled carbon nanotubes', *Appl. Phys. Lett.*, **79**, 3155 (2001).

(3.109) E. Joselevich and C. M Lieber, 'Vectorial growth of metallic and semiconducting single-wall carbon nanotubes', *Nano Lett.*, **2**, 1137 (2002).

(3.110) A. Nojeh, A. Ural, R. F. Pease *et al.*, 'Electric-field-directed growth of carbon nanotubes in two dimensions', *J. Vac. Sci. Tech. B*, **22**, 3421 (2004).

(3.111) M. Su, Y. Li, B. Maynor *et al.*, 'Lattice-oriented growth of single-walled carbon nanotubes', *J. Phys. Chem. B*, **104**, 6505 (2000).

(3.112) A. Ismach, L. Segev, E. Wachtel *et al.*, 'Atomic-step-templated formation of single wall carbon nanotube patterns', *Angew. Chem. Int. Ed.*, **43**, 6140 (2004).

(3.113) A. Ismach, D. Kantorovich and E. Joselevich, 'Carbon nanotube graphoepitaxy: highly oriented growth by faceted nanosteps', *J. Amer. Chem. Soc.*, **127**, 11 554 (2005).

(3.114) S. M. Huang, X. Y. Cai and J. Liu, 'Growth of millimeter-long and horizontally aligned single-walled carbon nanotubes on flat substrates', *J. Amer. Chem. Soc.*, **125**, 5636 (2003).

(3.115) S. M. Huang, B. Maynor, X. Y. Cai *et al.*, 'Ultralong, well-aligned single-walled carbon nanotube architectures on surfaces', *Adv. Mater.*, **15**, 1651 (2003).

(3.116) Y. C. Tseng, P. Q. Xuan, A. Javey *et al.*, 'Monolithic integration of carbon nanotube devices with silicon MOS technology', *Nano Lett.*, **4**, 123 (2004).

(3.117) J. F. Colomer, L. Henrard, P. Lambin *et al.*, 'Electron diffraction and microscopy of single-wall carbon nanotube bundles produced by different methods', *Eur. Phys. J. D*, **27**, 111 (2002).

(3.118) S. M. Bachilo, L. Balzano, J. E. Herrera *et al.*, 'Narrow (n, m)-distribution of single-walled carbon nanotubes grown using a solid supported catalyst', *J. Amer. Chem. Soc.*, **125**, 11 186 (2003).

(3.119) G. Lolli, L. A. Zhang, L. Balzano *et al.*, 'Tailoring (n, m) structure of single-walled carbon nanotubes by modifying reaction conditions and the nature of the support of CoMo catalysts', *J. Phys. Chem. B*, **110**, 2108 (2006).

(3.120) Y. Li, S. Peng, D. Mann *et al.*, 'On the origin of preferential growth of semiconducting single-walled carbon nanotubes', *J. Phys. Chem. B*, **109**, 6968 (2005).

(3.121) W. C. Ren, F. Li, S. Bai *et al.*, 'The effect of sulfur on the structure of carbon nanotubes produced by a floating catalyst method', *J. Nanosci. Nanotech.*, **6**, 1339 (2006).

(3.122) H. W. Zhu, K. Suenaga, A. Hashimoto *et al.*, 'Atomic-resolution imaging of the nucleation points of single-walled carbon nanotubes', *Small*, **1**, 1180 (2005).

(3.123) K. Bolton, F. Ding and A. Rosén, 'Atomistic simulations of catalyzed carbon nanotube growth', *J. Nanosci. Nanotech.*, **6**, 1211 (2006).

(3.124) Y. Shibuta and S. Maruyama, 'Molecular dynamics simulation of formation process of single-walled carbon nanotubes by CCVD', *Chem. Phys. Lett.*, **382**, 381 (2003).

(3.125) F. Ding, K. Bolton and A. Rosén, 'Nucleation and growth of single-walled carbon nanotubes: a molecular dynamics study', *J. Phys. Chem. B*, **108**, 17 369 (2004).

(3.126) F. Ding, K. Bolton and A. Rosén, 'Molecular dynamics study of SWNT growth on catalyst particles without temperature gradients', *Comput. Mater. Sci.*, **35**, 243 (2006).

(3.127) Y. Shibuta and J. A. Elliott, 'A molecular dynamics study of the carbon-catalyst interaction energy for multi-scale modelling of single wall carbon nanotube growth', *Chem. Phys. Lett.*, **427**, 365 (2006).

(3.128) H. Duan, F. Ding, A. Rosén *et al.*, 'Initial growth of single-walled carbon nanotubes on supported iron clusters: a molecular dynamics study', *Eur. Phys. J. D*, **43**, 185 (2007).

(3.129) Z. Q. Bai, H. Chen, B. Q. Li *et al.*, 'Catalytic decomposition of methane over activated carbon', *J. Anal. Appl. Pyrolysis*, **73**, 335 (2005).

(3.130) R. Vander Wal, A. J. Tomasek, M. I. Pamphlet *et al.*, 'Analysis of HRTEM images for carbon nanostructure quantification', *J. Nanoparticle Res.*, **6**, 555 (2004).

(3.131) A. Gorbunov, O. Jost, W. Pompe *et al.*, 'Solid–liquid–solid growth mechanism of single-wall carbon nanotubes', *Carbon*, **40**, 113 (2002).

(3.132) R. Sen, S. Suzuki, H. Kataura *et al.*, 'Growth of single-walled carbon nanotubes from the condensed phase', *Chem. Phys. Lett.*, **349**, 383 (2001).

(3.133) G. Mathieu, S. Guiot and J. Cabane, 'Solubility of carbon in silver, copper and gold', *Scripta Metallurgica*, **7**, 421 (1973).

(3.134) G. A. López and E. J. Mittemeijer, 'The solubility of C in solid Cu', *Scripta Materialia*, **51**, 1 (2004).

(3.135) E. Flahaut, A. Peigney, C. Laurent *et al.*, 'Synthesis of single-walled carbon nanotube–Co–MgO composite powders and extraction of the nanotubes', *J. Mater. Chem.*, **10**, 249 (2000).

(3.136) R. R. Bacsa, C. Laurent, A. Peigney *et al.*, 'High specific surface area carbon nanotubes from catalytic chemical vapor deposition process', *Chem. Phys. Lett.*, **323**, 556 (2000).

(3.137) L. J. Ci, Z. L. Rao, Z. P. Zhou *et al.*, 'Double wall carbon nanotubes promoted by sulfur in a floating iron catalyst CVD system', *Chem. Phys. Lett.*, **359**, 63 (2002).

(3.138) W. C. Ren, F. Li, J. Chen *et al.*, 'Morphology, diameter distribution and Raman scattering measurements of double-walled carbon nanotubes synthesized by catalytic decomposition of methane', *Chem. Phys. Lett.*, **359**, 196 (2002).

(3.139) W. C. Ren and H. M. Cheng, 'Aligned double-walled carbon nanotube long ropes with a narrow diameter distribution', *J. Phys. Chem. B*, **109**, 7169 (2005).

(3.140) S. C. Lyu, B. C. Liu, C. J. Lee *et al.*, 'High-quality double-walled carbon nanotubes produced by catalytic decomposition of benzene', *Chem. Mater.*, **15**, 3951 (2003).

(3.141) E. Flahaut, R. Bacsa, A. Peigney *et al.*, 'Gram-scale CCVD synthesis of double-walled carbon nanotubes', *Chem. Commun.*, 1442 (2003).

(3.142) E. Flahaut, C. Laurent and A. Peigney, 'Catalytic CVD synthesis of double and triple-walled carbon nanotubes by the control of the catalyst preparation', *Carbon*, **43**, 375 (2005).

(3.143) T. Hiraoka, T. Kawakubo, J. Kimura *et al.*, 'Selective synthesis of double-wall carbon nanotubes by CCVD of acetylene using zeolite supports', *Chem. Phys. Lett.*, **382**, 679 (2003).

(3.144) J. Q. Wei, B. Jiang, D. H. Wu *et al.*, 'Large-scale synthesis of long double-walled carbon nanotubes', *J. Phys. Chem. B*, **108**, 8844 (2004).

(3.145) A. Gruneis, M. H. Rummeli, C. Kramberger *et al.*, 'High quality double wall carbon nanotubes with a defined diameter distribution by chemical vapor deposition from alcohol', *Carbon*, **44**, 3177 (2006).

(3.146) L. Ci, R. Vajtai and P. M. Ajayan, 'Vertically aligned large-diameter double-walled carbon nanotube arrays having ultralow density', *J. Phys. Chem. C*, **111**, 9077 (2007).

(3.147) M. Endo, H. Muramatsu, T. Hayashi *et al.*, '"Buckypaper" from coaxial nanotubes', *Nature*, **433**, 476 (2005).

(3.148) W. K. Hsu, J. P. Hare, M. Terrones *et al.*, 'Condensed phase nanotubes', *Nature*, **377**, 687 (1995).

(3.149) W. K. Hsu, M. Terrones, J. P. Hare *et al.*, 'Electrolytic formation of carbon nanostructures', *Chem. Phys. Lett.*, **262**, 161 (1996).

(3.150) A. T. Dimitrov, G. Z. Chen, I. A. Kinloch *et al.*, 'A feasibility study of scaling-up the electrolytic production of carbon nanotubes in molten salts', *Electrochimica Acta*, **48**, 91 (2002).

(3.151) I. A. Kinloch, G. Z. Chen, J. Howes *et al.*, 'Electrolytic, TEM and Raman studies on the production of carbon nanotubes in molten NaCl', *Carbon*, **41**, 1127 (2003).

(3.152) G. X. Du, C. Song, J. H. Zhao *et al.*, 'Solid phase transformation of glass-like carbon nanoparticles into nanotubes and the related mechanism', *Carbon*, **46**, 92 (2008).

(3.153) B. David, N. Pizurova, O. Schneeweiss *et al.*, 'Multi-walled carbon nanotubes formed by condensed-phase conversions of Fe–C-based nanopowder in vacuum', *Czech. J. Phys.*, **56**, E51 (2006).

(3.154) Y. W. Fan, B. R. Goldsmith and P. G. Collins, 'Identifying and counting point defects in carbon nanotubes', *Nat. Mater.*, **4**, 906 (2005).

4   Purification and processing

None of the techniques discussed in the previous two chapters produce pure carbon nanotubes. When multiwalled nanotubes are produced by arc-evaporation, for example, they are always accompanied by nanoparticles and disordered carbon. When produced catalytically, nanotubes are contaminated with catalyst particles and support material. In order to make use of the nanotubes, therefore, it is often necessary to purify them. This chapter begins with a summary of the chemical and physical methods available for purifying nanotubes. Techniques for processing nanotubes are then reviewed. Procedures for aligning tubes and for forming them into fibres and sheets are covered, and methods for sorting them by length and structure outlined. Ways of improving the solubility of nanotubes by functionalization are covered later in the book (Chapter 8), while the incorporation of nanotubes into composite materials is discussed in Chapter 9.

4.1   Purification of multiwalled tubes

4.1.1   MWNTs produced by arc-evaporation

The first successful technique for purifying multiwalled tubes produced by arc-evaporation was described by Thomas Ebbesen and co-workers in 1994 (4.1). Following the demonstration that nanotube caps could be selectively attacked by oxidizing gases (see Chapter 10), these researchers realized that nanoparticles, with their defect-rich structures might be oxidized much more readily than the relatively perfect nanotubes. Therefore they subjected raw nanotube samples to a range of oxidizing treatments, in the hope that the nanoparticles would be preferentially oxidized away. They found that a significant relative enrichment of nanotubes could be achieved in this way, but only at the expense of losing of a major proportion of the original sample. Thus, in order to remove all the nanoparticles, it was necessary to oxidize more than 99% of the raw sample. When 95% of the original material was oxidized, about 10–20% of the remaining sample consisted of nanoparticles, while an oxidation of 85% resulted in no enrichment at all. These results suggest that the reactivities of nanotubes and nanoparticles towards oxidation are very similar, so that only a very narrow 'window' exists between the selective removal of nanoparticles and complete oxidation of the sample.

A different and slightly less destructive approach was introduced by a Japanese group at about the same time (4.2). This technique made use of the fact that nanoparticles and other graphitic contaminants have relatively 'open' structures, and can therefore be more readily intercalated with a variety of materials than can closed nanotubes. By intercalating with copper chloride, and then reducing this to metallic copper, the Japanese group were able to preferentially oxidize the nanoparticles away, using copper as an oxidation catalyst. A similar purification technique, which involves intercalation with bromine followed by oxidation, was described by the Oxford group (4.3). Another Japanese team combined wet grinding, hydrothermal treatment, and oxidation to purify arc-grown MWNTs (4.4). In this way, 16 mg of purified MWNTs were obtained from 850 mg of inner core material from the cathode.

Other workers have employed solubilization methods to achieve purification of arc-synthesized MWNTs (see Chapter 8, p. 211, for a discussion of solubilization by surfactants). Thus, a group from the Ecole Polytechnique Fédérale de Lausanne in Switzerland used sodium dodecyl sulphate (SDS) to produce a stable suspension of nanotubes and nanoparticles in water, then allowed the nanotubes to flocculate, leaving the nanoparticles in suspension (4.5). The sediment could then be removed, and further flocculation procedures carried out. This not only enabled the nanoparticles to be removed, but also resulted in some degree of length separation of the tubes. Chromatographic methods can also be used for the purification and size selection of arc-produced MWNTs, as shown by Georg Duesberg and colleagues (4.6). In recent years, interest in purifying multiwalled nanotubes produced by arc-evaporation has dwindled, as the focus has shifted to catalytically-produced tubes, and at present there are no completely satisfactory ways of achieving pure arc-grown MWNTs.

4.1.2 Catalytically-produced MWNTs

When produced catalytically, MWNT samples inevitably contain residual metal catalyst particles, and support material, if used. These impurities can be more harmful than the carbon contaminants that accompany arc-produced tubes. However, removing this unwanted material can be achieved rather more easily than is the case for arc-grown tubes, as shown by a number of groups (e.g. 4.7–4.11). It seems that the most successful methods involve high-temperature annealing. Rodney Andrews and colleagues from the University of Kentucky described the purification of catalytically-produced MWNTs produced by annealing at 'graphitization' temperatures (1600–3000 °C) (4.7). The effect of this treatment was not only to remove the catalyst impurities but also to improve the structural perfection of the tubes, as shown in Fig. 4.1.

Acid treatments, often combined with heat treatments, have also been used to purify CVD-produced MWNTs. Workers from Hunan University, China, used this approach to purify MWNTs produced with Ni–Mg–O catalysts (4.8). Treatment with concentrated HNO_3 and HCl followed by oxidation in air at 510 °C reportedly resulted in 96% pure tubes. A disadvantage of acid treatments is that they may damage the structure of the nanotubes, while heat treatments tend to improve them. Fei Wei and co-workers from Tsing Hua University compared vacuum annealing and acid methods (4.9, 4.10), and

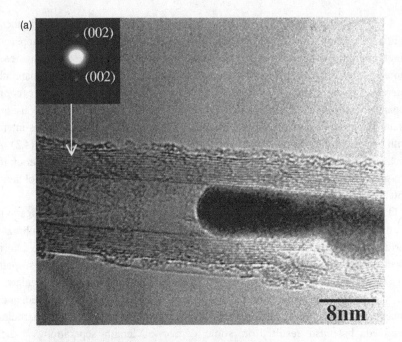

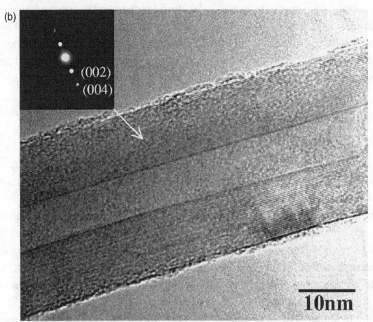

Fig. 4.1   Micrographs showing (a) as-prepared MWNT and (b) MWNT following heat treatment at 3000 °C. The insets show diffraction patterns illustrating the improvement in structural perfection of the tubes (4.7).

found annealing to be more effective, with the capability of producing 99.9% purity tubes. Hui-Ming Cheng and colleagues described an effective purification method involving high-temperature annealing (2600 °C) followed by an extraction treatment with a dispersing agent (4.11). The annealing process evaporated the metal particles, while the extraction treatment removed carbon nanoparticles.

4.2   Purification of single-walled tubes

Purifying single-walled nanotubes presents an even greater challenge than for MWNTs. Early attempts to apply harsh oxidation methods, like those used to purify MWNTs, proved unsuccessful with SWNTs, so it was necessary to develop more sophisticated approaches. As with MWNTs, slightly different methods are needed depending on how the tubes were prepared. When produced by the arc-evaporation and laser vaporization methods, SWNTs are typically accompanied by large amounts of amorphous carbon and metal particles, often themselves coated with carbon. These coated particles can be particularly difficult to remove. Catalytically-produced SWNT samples also contain residual catalyst material, but tend to have less amorphous carbon. A huge effort has been put into developing techniques for purifying SWNTs, and some of this work will now be summarized. For more detailed discussions, a number of excellent reviews have been published (4.12–4.14).

4.2.1   Acid treatment and oxidation

Of the many recipes which have been published for the purification of single-walled nanotubes, the majority involve acid treatments and/or gas phase oxidation. Rather than attempt to review all of the available methods, this section will describe two very effective techniques for the purification of SWNTs which were described by Ivana Chiang and co-workers at Rice University (4.15–4.16). Firstly, a little background is given.

When oxidation methods were applied to SWNT samples, it was found that the tubes were destroyed along with the contaminating material. The reasons for this were not immediately clear. It is now appreciated, however, that the problem lies with the metal catalyst particles present in the soot. In the presence of oxidizing gases, the metal particles catalyse low-temperature oxidation of carbons indiscriminately, destroying the SWNTs. Methods are therefore needed to remove the metal particles before applying more rigorous oxidation treatments. Frequently the method used for this involves an acid reflux, as first used by Andrew Rinzler et al. in a 1998 paper (4.17) describing the production of 'buckypaper' (see p. 96 below).

The first of the Chiang papers described the purification of SWNTs produced by the laser vaporization method (4.15). In this case, the as-received soot had already been subjected to nitric acid treatment, which would have dissolved away most of the metal particles. The soot was supplied as a suspension in toluene. This suspension was filtered and washed with methanol to remove additional soluble residue left from the initial

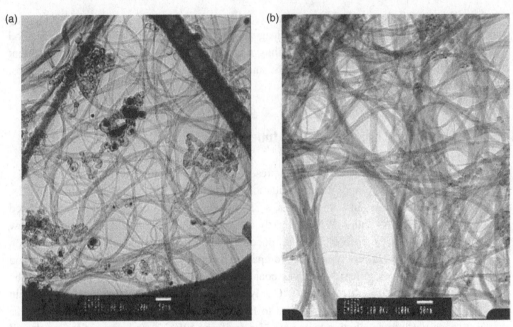

Fig. 4.2   Micrographs illustrating the purification of laser-grown single-walled nanotubes, from the work of Chiang *et al.* (4.15). (a) Untreated soot, (b) soot after the purification treatment.

nitric acid cleaning treatment. This left a black puffy paper ('buckypaper'), which was refluxed in water. This was followed by two oxidations, first at 300 °C, then at 500 °C, with an HCl extraction step after each oxidation. The authors claimed that this could produce samples containing 99.9% SWNTs. Two images illustrating the purification process are shown in Fig. 4.2. In a subsequent paper they described the purification of SWNTs produced by the HiPco process (4.16). As noted in Section 3.4.2, HiPco nanotubes contain comparatively large amounts of Fe, typically 14% by weight. Here, the first step was a low-temperature (225 °C), wet Ar/O_2 oxidation treatment. This was found to facilitate the removal of carbon coated metal particles by breaking open the carbon shell and converting the metal to the oxide or hydroxide. Stirring the material in concentrated HCl following this treatment resulted in dissolution of the iron. After filtering off the acid and drying, the oxidation and acid extraction cycle was repeated once more at 325 °C, followed by oxidative heating at 425 °C and annealing in Ar at 800 °C. The authors do not seem to give a figure for the SWNT purity following this treatment, but they state that the final Fe content is approximately 0.02%.

When using acid and oxidative treatments for purification, it is important to recognize that these treatments can result in the formation of carboxylic acid and other groups at the tube ends and possibly, at defects on the sidewalls (see p. 205). It is possible that the introduction of these groups could affect the tubes' properties. For a detailed survey of acid and oxidative treatments for the purification of SWNTs, the review by Stanislaus Wong and colleagues can be recommended (4.13).

4.2.2 Functionalization

The chemical functionalization of carbon nanotubes is discussed in detail in Chapter 8. A few researchers have explored the idea of using functionalization as a way of purifying single-walled tubes. It has been argued that organic functionalization, as opposed to treatment with acids or gas phase oxidants, renders SWNTs easier to handle and therefore better for potential practical uses (4.13). In 2001, Maurizio Prato of the University of Trieste and co-workers described the functionalization of SWNTs with azomethine ylides (see p. 208). They showed that this treatment could be used to purify SWNTs produced by the HiPco process (4.18). It was found that the functionalization greatly increased the solubility of SWNTs in organic solvents, while leaving the metal particles insoluble. The metal particles could therefore readily be separated from the nanotubes. To remove the remaining amorphous carbon and nanoparticles, a slow precipitation process was used which involved the addition of diethyl ether to a chloroform solution of SWNTs. This was found to precipitate most of the contaminating carbon, leaving the nanotubes in solution. After filtering off the tubes, the final step of the purification process was the removal of functional groups by thermal treatment at 350 °C followed by annealing to 900 °C. Transmission electron microscopy showed that the resulting nanotubes were largely free of impurities.

As discussed in Section 8.1.2, Sarbajit Banerjee and Stanislaus Wong have used ozonolysis to functionalize SWNT sidewalls (4.19). It was also found that ozonolysis had the effect of purifying the nanotube samples. The reason for this seems to be that, upon ozonolysis, amorphous carbon and nanoparticles become heavily functionalized with oxygenated groups and thereby have increased solubility in the polar solvents that were used to wash the samples.

4.2.3 Physical techniques

Physical techniques that have been used for the purification of single-walled nanotubes include filtration, chromatography, centrifugation and laser treatment. The potential advantages of physical treatments are that they are, in principle, less destructive than chemical methods and there is less danger of chemical modification of the tubes. Smalley and co-workers were probably the first to use microfiltration to purify SWNT samples. In 1997 they described a technique involving the use of a cationic surfactant to suspend the nanotubes and accompanying material in solution, and then trapping the tubes on a membrane filter (4.20). However, multiple filtration was required, with sample resuspension after each filtration, in order to achieve a significant level of purification, making the procedure slow and inefficient. An improved method was described (4.21) in which ultrasonication was used to keep the material suspended during the filtration, thus enabling large amounts of sample to be filtered continuously. In this way, up to 150 mg of soot could be purified in 3–6 hours, with the resulting material containing more than 90% of SWNTs. The oxidation–filtration method of Chiang et al. was described above.

Size exclusion chromatography has been used to purify single-walled tubes, as well as to separate small quantities of SWNTs into fractions with a small length and diameter distribution. Duesberg et al. have described a method similar to that used for

MWNTs (see above), which proved effective for SWNTs (4.22). Robert Haddon's group at Kentucky have described a chromatographic purification of soluble single-walled tubes (4.23).

Haddon's group has also shown that centrifugation can be effective in removing both amorphous carbon and carbon nanoparticles from nitric acid-treated SWNT soot (4.23, 4.24). Low-speed centrifugation had the effect of preferentially suspending the amorphous carbon, leaving the SWNTs in the sediment. By contrast, high-speed centrifugation of well-dispersed preparations is effective in sedimenting carbon nanoparticles, while leaving the SWNTs suspended in aqueous media.

The use of laser treatments to purify SWNTs was demonstrated by John Lehman of the USA's National Institute of Standards and Technology in 2007 (4.25). Samples produced by arc-discharge and CVD were exposed to a 248 nm excimer laser, and improvements in purity were shown by Raman measurements. The technique seemed to be particularly effective at removing amorphous carbon and nano-crystalline graphite.

4.2.4   Assessing purity

Perhaps the most useful method of assessing the purity of samples of single-walled nanotubes is Raman spectroscopy. The application of this technique to nanotubes is discussed in detail in Chapter 7 (p. 192). Of particular interest in relation to purity is the so-called D line, at around $1340 \ cm^{-1}$, which is assigned to disordered graphitic material. The ratio of the D band intensity to the intensity of the G band at $\sim 1582 \ cm^{-1}$ (I_D/I_G) provides a good estimation of sample purity. However, Raman spectroscopy does not supply information about the amount of metal impurities.

Near-infrared (NIR) spectroscopy is another technique that has been used to evaluate the purity of bulk quantities of single-walled carbon nanotubes (e.g. 4.26–4.28). The purity can be assessed by determining the integrated intensity of the S_{22} transitions compared with the S_{22} intensity of a reference sample. Thermogravimetric analysis (TGA) has proved useful in determining the amount of metal catalyst particles in samples of SWNTs (4.15, 4.16, 4.29), but is of less value in easily distinguishing between different carbonaceous species. Of course, the most direct way to evaluate purity is transmission electron microscopy, but obtaining a representative sample is not always easy since the raw material can be highly inhomogeneous.

In a comparative study, Haddon and colleagues concluded that solution-phase near-infrared spectroscopy and solution-phase Raman spectroscopy are the best techniques for determining the purity of samples of bulk single-walled nanotubes (4.30).

4.3   Processing of multiwalled nanotubes

4.3.1   Multiwalled nanotube suspensions and assemblies of pure MWNTs

A great deal of work has been carried out on the acid treatment of nanotubes, both in connection with purification (see above) and as a method for opening the tubes, as

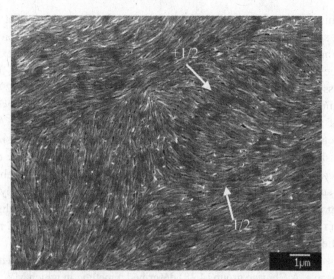

Fig. 4.3   Scanning electron micrograph of dried nematic MWNT film showing a pair of disclinations, from the work of Song and Windle (4.34).

discussed in Chapter 10. The effect of acid treatment is often to introduce oxygen containing surface groups, including phenolic, carboxylic and lactonic groups. Alan Windle and colleagues have shown that these have the effect of stabilizing dispersions of nanotubes at much higher concentrations than are possible with the raw material (4.31). The properties of nanotube dispersions prepared in this way have been investigated, and analogies drawn between the dispersions and polymer solutions (4.32). Optical bireflection revealed that aqueous dispersions of multiwalled tubes underwent a transition above a certain concentration (~4.3 vol.%), in which the arrangement changed from isotropic to liquid crystalline (4.33, 4.34). The liquid crystal phase exhibited a Schlieren texture which is typical of a nematic liquid crystal phase, i.e. a phase in which the components have long-range orientational order. This kind of organization is commonly seen in nanoscale rigid rod systems, such as the tobacco mosaic virus. The evaporation of the solvent from the dispersions gave a solid sample with a very similar microstructure to the original dispersions, and this solid material could be studied in the scanning electron microscope. A typical SEM image is shown in Fig. 4.3. Here, disclinations which are typical of nematic liquid crystals can be seen. Windle *et al.* point out that the existence of these phases suggest that carbon nanotubes could be processed using methods analogous to those used for rigid chain polymers such as the aramids. In practice, this has proved difficult, due to the strong interactions between tubes and their tendency to stick together in bundles or ropes.

4.3.2   Alignment and arrangement of MWNTs

Methods for the catalytic growth of aligned MWNTs on substrates are well established (see Section 3.2.1), but there is often a requirement for the post-synthesis alignment of nanotubes. This can be achived in a variety of ways. It has been shown that nanotube 'yarns' can be spun from aligned arrays of tubes, as discussed in the next section. Another way of aligning nanotubes is to incorporate them into a matrix and then extrude the matrix in some way, so that the tubes become aligned along the direction of flow. This kind of process is discussed in the chapter on nanotube composites (see p. 229).

One of the first methods used to align multiwalled tubes was dielectrophoresis, i.e. the application of an electric field to a sample held between electrodes. Seiji Akita and colleagues from Osaka demonstrated significant alignment of MWNTs suspended in isopropanol using this method (4.35, 4.36). When AC fields were used, the degree of alignment increased with increasing frequency of the applied field, an observation confirmed in studies of single-walled tubes (see Section 4.4.1). More recently, Rodney Ruoff's group at Northwestern University used an array of electrodes to align multi-walled tubes in solution (4.37). It was found that using combined AC and DC fields (i.e. a biased AC field) gave the best results.

Functionalization has been used to facilitate the arrangement of multiwalled nanotubes into aligned arrays. A Chinese group (4.38) described a method in which MWNTs were shortened using an oxidation–sonication treatment and then functionalized with acyl chloride in thionyl chloride ($SOCl_2$). Multilayer polyelectrolyte films were then deposited onto the substrates and the modified substrates were dipped into a tetrahydrofuran suspension of the functionalized tubes. This resulted in the formation of perpendicularly aligned arrays of MWNTs, as shown in Fig. 4.4. Whether this approach will prove more effective than the direct growth of MWNTs on substrates remains to be seen.

A novel approach to the assembly of multiwalled tubes was described by Pulickel Ajayan and Ravi Kane and their colleagues in 2004 (4.39). The starting point for this

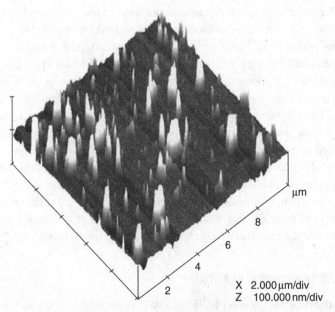

X   2.000 µm/div
Z   100.000 nm/div

Fig. 4.4   An AFM image of self-assembled MWNTs on a substrate (4.38).

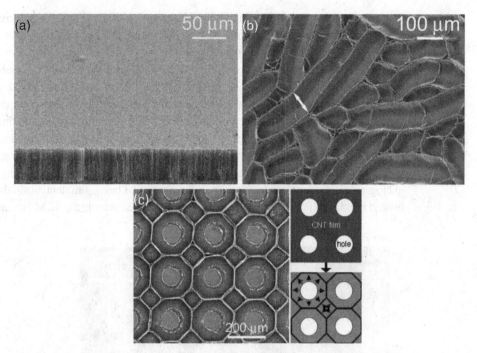

Fig. 4.5 The formation of cellular patterns by the evaporation of liquids from vertically aligned multiwalled carbon nanotube films (4.39). (a) A SEM image of original array, (b) an image showing the cellular structures formed by the evaporation of water from a MWNT array, (c) The formation of ordered 'foam' from a patterned array consisting of circular holes (shown on the right).

work was an array of vertically aligned MWNTs grown on a silica substrate using CVD. The tubes were partially oxidized to aid wetting, then immersed in various liquids, which were allowed to evaporate at room temperature. The drying process resulted in the formation of random, foam-like patterns, as shown in Fig. 4.5. Ordered arrangements could also be formed, by using patterned substrates. Figure 4.5(c) shows an example, formed from a substrate containing an array of circular holes. It was shown that the nanotube foams could be elastically deformed, separated from the substrate to produce free-standing 'fabrics' or transferred to other substrates. It was suggested that the foams could have applications as shock-absorbent structural reinforcements or elastic membranes. At about the same time, and independently, Lei Jiang of the Chinese Academy of Sciences, Beijing, and colleagues produced similar patterns by applying drops of water to aligned MWNT films (4.40).

4.3.3 Pure MWNT fibres

The direct spinning of MWNT yarns using catalytic synthesis was discussed in the previous chapter (p. 51). Post-synthesis methods have also been developed for the production of continuous nanotube fibres. Perhaps the most successful technique involves pulling 'yarns' of nanotubes from arrays grown on flat substrates. This was first described by a team from Tsinghua University in 2002 (4.41). They were initially attempting to pull a bundle of nanotubes out of an array grown on a Si substrate. Instead, they managed to draw out a continuous yarn of nanotubes, a process they compared to drawing a thread from a silk cocoon. The yarns usually took the form of thin ribbons, and could be drawn to a length of 30 cm. Initially the threads were manually drawn with tweezers, but in later work a more controllable method, using an electric motor to perform the pulling with a constant speed was developed (4.42). A further advance on this process was described by Ray Baughman of University of Texas at Dallas and his colleagues in 2004 (4.43). In this work a twist was introduced as the yarns were drawn, thus applying a technique that has been used in textile production since ancient times. Scanning electron micrographs showing a nanotube yarn in the process of being simultaneously drawn and twisted are shown in Fig. 4.6 (the process was conducted outside the SEM and interrupted

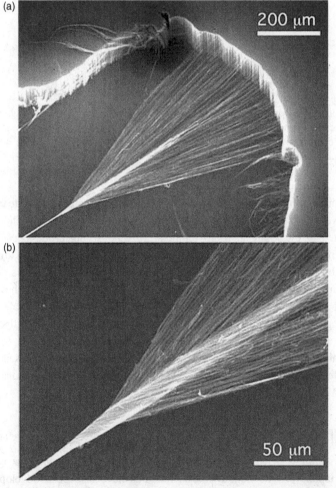

(a)

200 μm

(b)

50 μm

Fig. 4.6   SEM images illustrating the drawing of a nanotube yarn from a 'forest' grown on a flat substrate (4.43).

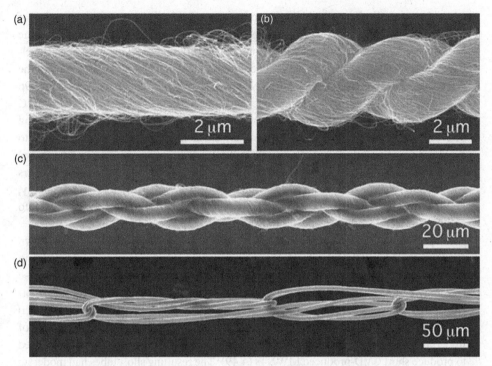

Fig. 4.7   SEM images of MWNT yarns (a) single, (b) two-ply, (c) four-ply and (d) knitted (4.43).

for SEM imaging). The yarns could be further twisted into two and four-ply MWNT threads, or formed into more complicated, knitted structures, as shown in Fig. 4.7. The single yarns had measured tensile strengths between 150 and 300 MPa, while higher strengths, between 250 and 460 MPa, were observed for two-ply yarns. It was shown that single yarns could be infiltrated with polyvinyl alcohol by soaking in a solution of the polymer and then drying. This increased the strength of the yarns to 850 MPa. Baughman and colleagues suggest that the MWNT yarns could have a host of applications in areas such as structural composites, protective clothing, artificial muscles and electronic textiles.

4.3.4   MWNT sheets

Baughman's team have also shown that MWNT sheets can be drawn from catalytically-grown nanotube forests (4.44). Again, the drawing process was initiated using an adhesive strip, and it was shown that sheets up to 1 m long and 5 cm wide could be made. The sheets were transparent, with highly anisotropic electronic properties and may have applications as electrodes or light-emitting diodes.

4.3.5   Breaking and cutting of MWNTs

There are several reasons why one might want to cut down MWNTs into short lengths. Broken or damaged MWNTs are more amenable to functionalization than pristine tubes, for example, due to the higher concentration of defects. Also the incorporation of catalytically-produced MWNTs into composites is often facilitated by breaking down the as-produced tubes into short lengths. A number of different methods have been developed for deliberately damaging MWNTs or cutting them into shorter lengths. One of the first successful methods of breaking MWNTs was described by the Oxford group in 1996 (4.45). In this work, samples of the tubes were suspended in CH_2Cl_2, cooled to $0\,°C$ and then subjected to high-energy ultrasound using an immersion horn (this method delivers much more energy to the tubes than the ultrasonic processing sometimes used to aid dispersion in solvents). Following this treatment a high proportion of the tubes contained defects such as bending and buckling. Stripping of the outer graphene layers was often observed, but completely fractured tubes were relatively rare. Subsequently, high-energy ultrasonic treatment has quite frequently been used to introduce defects into MWNTs to facilitate the attachment of catalytic metal particles to the tubes (4.46).

Janos Nagy and colleagues from Namur, Belgium, used ball-milling to break down catalytically-produced MWNTs (4.47). The untreated tubes were typically $50\ \mu m$ in length, while after $12\ h$ of milling the average length was around $0.8\ \mu m$. In 2006 Chinese researchers (4.48) showed that catalytically-produced MWNTs could be cut into short lengths by firstly depositing NiO particles onto them and then inducing a localized reaction between the carbon and the oxide particles. A short time later Milo Shaffer of Imperial College and colleagues showed that a simple oxidizing treatment could be used to produce short CVD-produced MWNTs (4.49). The resulting short tubes had moderate levels of functionalization, and showed enhanced dispersibility in organic solvents.

The most controlled method for cutting MWNTs was demonstrated by Alex Zettl and colleagues in 2005 (4.50). This group used the focused electron beam of an SEM to achieve precise cutting of arc-grown MWNTs. Partial cutting, creating hinge like geometries, as well as complete cutting was demonstrated.

4.4   Processing of single-walled tubes

4.4.1   Alignment and arrangement of SWNTs

Techniques for aligning nanotubes in polymer matrices are discussed in Chapter 9 (p. 229). Here we are concerned with methods for arranging and aligning pure single-walled nanotubes.

One way of organizing nanotubes into defined networks is to bond together nanotubes with functionalized tips. The first demonstration of this was given by the Smalley group in 1998 (4.51). In this work, SWNTs were firstly cut into short lengths ('fullerene pipes': see Section 4.4.4 below), then reacted firstly with thionyl chloride and secondly with $NH_2-(CH_2)_{11}-SH$ to produce an amide link between the alkanethiol and the nanotube ends. Gold particles were then attached to the thiol-derivitized tips of the tubes. The authors demonstrated that nanotubes could be connected together in this way, through bonding to a common gold particle. Subsequently, interesting work in this area has been carried out by Masahito Sano and colleagues (4.52, 4.53). In a 2001 *Science* paper (4.52) they reported the formation of rings from acid-oxidized SWNTs through esterification between the carboxylic acid and hydroxyl end-groups at the nanotube tips in the presence of a condensation reagent, 1,3-dicyclohexylcarbodiimide. The same group have grafted dendrimers or dendrons onto oxidized carbon nanotubes to form nanotube stars. Duesberg and colleagues have demonstrated the formation of junctions between SWNTs through amine linkages (4.54). The tubes could be joined in either an end-to-side or end-to-end configuration.

Similar methods have been used to produce arrays of SWNTs on surfaces. Zhongfan Liu and colleagues of Peking University have described a wet chemical technique for attaching SWNTs in a perpendicular manner to gold surfaces (4.55). The as-grown nanotubes were first chemically cut into 'pipes' and thiol-derivatized at the open ends. A gold crystal with a clean (111) surface was then dipped into an ethanol suspension of the thio-functionalized nanotubes, and assembly of the SWNTs occurred by spontaneous chemical adsorption to the gold surface through Au–S bonds. In an alternative method (4.56), a pre-treated gold surface was dipped into a suspension of carboxyl-terminated shortened SWNTs. This again resulted in an ordered assembly of perpendicularly oriented SWNTs. Justin Gooding from the University of New South Wales and colleagues used a similar method to produce vertically aligned arrays of SWNTs on gold, and then functionalized these with the enzyme microperoxidase MP-11 (4.57). The resulting arrays could be used as sensing devices (see also p. 282).

The use of dielectrophoresis to align MWNTs was mentioned above. The technique has also been successfully used to align and position SWNTs (e.g. 4.58, 4.59). Perhaps the most impressive demonstration of nanotube positioning using this method was given in 2007 by Ralph Krupke of the Institute for Nanotechnology in Karlsruhe and colleagues (4.59). In this work, nanotubes were deposited from an aqueous solution onto an array of electrodes, as illustrated in Fig. 4.8. An inhomogeneous electric field was generated by the two opposing needle-shaped electrodes, and the tubes were selectively deposited between these electrodes. An important aspect of the work was that just a single nanotube or nanotube bundle was deposited at the predefined locations. This was because the dielectrophoretic force field changed upon nanotube deposition, such that further tubes are repelled. The authors claimed that several million nanotube devices per cm^2 of substrate could be made in this way. Dielectrophoresis has also been used to separate metallic and semiconducting nanotubes as discussed in Section 4.5.2 below.

Another effective method for the assembly of large numbers of single-walled tubes has been described by Seunghun Hong and co-workers (4.60). Their approach was based on the formation of patterns of self-assembled monolayers (SAM) of molecules on a substrate, which were then used to guide the self-assembly of the tubes. Two types of surface region were created, one patterned with polar groups such as amino or carboxyl, and the other coated with non-polar groups such as methyl. When the substrate was placed in a suspension of SWNTs the tubes were found to be attracted to the polar regions and became aligned within them, as shown in Fig. 4.9. It was claimed that millions of individual tubes could be assembled in this way, covering an area of about 1 cm^2. Hongjie Dai's group has described a method for making devices based on short single-walled

Fig. 4.8   (a) A schematic illustration showing the deposition of SWNTs from an aqueous solution onto an array of electrodes, (b) a SEM image of the electrode array, with each electrode pair bridged by a nanotube (4.58).

carbon nanotubes which combines photolithography and shadow evaporation (4.61). In this way they produced a field-effect transistor based on a 50 nm semiconducting nanotube.

DNA can be used to position nanotubes (e.g. 4.62, 4.63), as discussed further in Chapter 8 (p. 217). Thus, Israeli researchers have demonstrated that the interactions of proteins and DNA can be used to assemble nanotubes into a field-effect transistor which operates at room temperature (4.62).

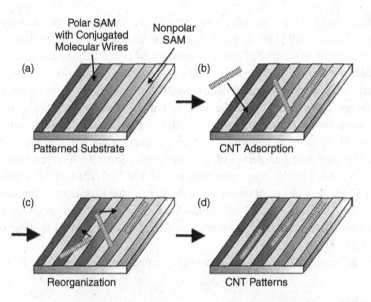

(a) Polar SAM with Conjugated Molecular Wires / Nonpolar SAM

Patterned Substrate

(b) CNT Adsorption

(c) Reorganization

(d) CNT Patterns

Fig. 4.9 A schematic diagram showing the directed assembly of carbon nanotubes using self-assembled monolayer (SAM) patterns (4.60).

Yet another approach to the arrangement of SWNTs involves the use of the Langmuir–Blodgett technique (e.g. 4.64, 4.65). This is a well established procedure for preparing monolayers of molecules on liquid surfaces. The Dai group has used the method to prepare densely packed monolayers of aligned SWNTs (4.64). The monolayers were transferred as a flat layer onto a SiO_2 substrate, and lithographic techniques and oxygen plasma etching were then used to form patterned arrays of the aligned SWNTs. Vertically aligned arrays of SWNTs have also been prepared on a substrate using the Langmuir–Blodgett method, and used to produce electrochemical sensors (4.65) (See p. ??).

The assembly of SWNTs is a highly active area of nanotube research at present, and only a snapshot is possible here. For a fuller discussion of the literature to 2007, the review by Yehai Yan *et al.* is recommended (4.66).

4.4.2 Pure SWNT strands

As already mentioned, nanotubes are far less amenable to processing than polymers, partly because of their tendency to stick together in bundles or ropes. In order to facilitate the application of processes such as extrusion to nanotubes it is necessary to break apart the bundles. In 2004, workers from Rice University showed that SWNT bundles could be effectively separated by treatment with 102% sulphuric acid (4.67, 4.68). This creates an aligned phase of positively charged nanotubes surrounded by acid anions, with around 8% nanotubes by weight. Following on from this work, the Rice group, with workers from the University of Pennsylvania, described a method for the large-scale production of pure SWNT fibres (4.69). For this work they used HiPco tubes, purified according to the Chiang protocol (4.16). The tubes were dispersed in the sulphuric acid, and then extruded through a capillary tube less than 125 μm in diameter into a coagulant bath containing either diethyl ether, 5 wt% aqueous sulphuric acid or water. The structure of the resulting carbon nanotube fibres depended on the coagulation conditions. Fibres spun into diethyl ether had a collapsed 'dogbone' structure, while fibres spun into dilute sulphuric acid or water retained their circular shape and were more dense. As a final step, the researchers heat-treated the fibres to remove water and residual acid.

Scanning electron microscopy showed a high degree of nanotube alignment within the fibres by SEM; this was confirmed by measurements using XRD and Raman spectroscopy. The mechanical properties of the fibres were good but not outstanding. Thus, the Young's modulus of the fibres was measured at 120 ± 10 GPa and the tensile strength 116 ± 10 MPa. The electrical resistivity of the fibres, after high-temperature annealing to remove residual acid, was ~ 2 milliohm cm. This is around two orders of magnitude higher than that of nanotube/polymer composite fibres, but similar to those of aligned mats of SWNTs.

For a fuller discussion of carbon nanotube suspensions and solutions, see Chapter 8.

4.4.3   SWNT sheets

In 1998, Andrew Rinzler of Rice University and co-workers described a simple method for purifying large quantities of single-walled nanotubes produced by laser vaporization (4.17). The method involved refluxing the raw soot in 2–3 M nitric acid for 45 h, centrifuging and washing with deionized water, filtering and vacuum baking. The end product consisted of a thin flexible film in which the tubes were held together like the fibres in a sheet of paper. Without really intending to do so, the authors had produced a new form of nanostructured carbon material: 'buckypaper' (an appropriate name, as the preparation process is similar to the ancient art of papermaking). Although the nanotubes making up the paper are not ultra-pure, buckypaper has proved to have some interesting properties (note that DWNT buckypaper has also been produced, as mentioned in the last chapter).

Some SEM images of SWNT thin-film samples produced by workers from Karlsruhe, Germany (4.70), using a similar technique to that of Rinzler *et al.*, are shown in Fig. 4.10. The pictures illustrate the structural integrity of the paper and the way it can be formed into curved shapes. The mechanical properties and failure mechanisms of buckypaper were studied by the same German group (4.71). Tensile tests were carried out on a 14 μm thick SWNT film, and strengths of the order of 10–20 MPa were measured. To examine failure mechanisms, the films were glued onto 3 mm copper TEM rings. Curing of the glue by drying at room temperature produced a contraction such that the films experienced tensile forces. A small hole was then punched into the centre of each specimen using a sharp needle, to induce tearing. Some images of the cracked films are shown in Fig. 4.11. The narrow regions of the cracks were often found to be bridged by taut SWNT strands, which terminated in branching structures, as can be seen in Fig. 4.11(b).

Several years after his initial work at Rice, Andrew Rinzler, now at the University of Florida, and his colleagues described a method for making ultra-thin buckypaper (4.72). The key to producing the thin film was to use a filter material (cellulose ester) that could

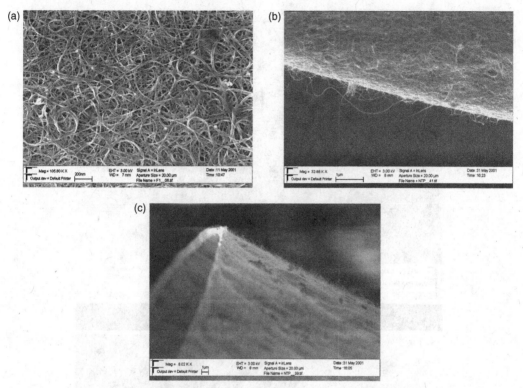

Fig. 4.10   SEM images of SWNT films produced by Hennrich *et al.* (4.70). (a), (b) Images of the surface that faced the suspension before filtering, (c) image of a curved sheet.

be dissolved away after depositing the nanotube layer. By using this approach they were able to produce films with thicknesses between 50 and 150 nm, which were transparent to visible light and electrically conductive. Other groups have explored applications of these films in a number of areas including light-emitting diodes and solar cells (4.73–4.75).

It has been shown that a degree of alignment can be induced in SWNT films by applying strong magnetic fields during the filtration process. This was demonstrated by researchers from Rice and the University of Pennsylvania (4.76–4.78). It was shown that this alignment led to anisotropic electrical and thermal transport properties: the parallel components of both the electrical and thermal conductivity increased with respect to unoriented material.

Filtration methods are not the only way to make SWNT films. It was mentioned above that the addition of sulphuric acid can aid the processing of SWNTs. Workers from the Georgia Institute of Technology and Rice have used this approach to facilitate film formation. The method involved dispersing SWNTs in oleum, pouring the dispersion into a Petri dish, then removing the acid and allowing the film to dry (4.79). The films seemed to have similar mechanical and electrical properties to those of buckypaper. Thin SWNT films can also be made by simply drying down aqueous solutions. Matteo Pasquali of Rice, and co-workers produced nanotube films on glass by evaporating droplets of water containing surfactant and single-walled tubes (4.80).

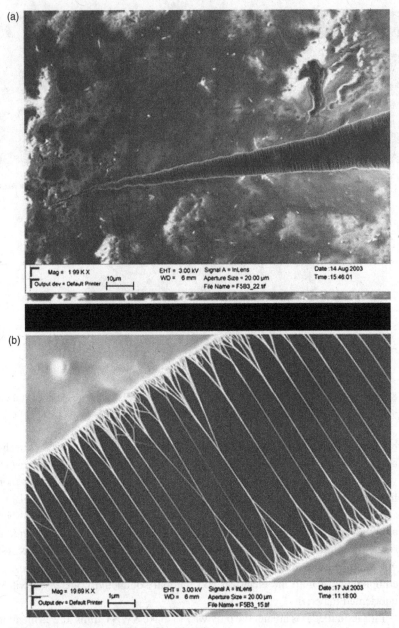

Fig. 4.11   SEM images of tears in SWNT films (4.71).

4.4.4   Length control of SWNTs

The first successful technique for cutting single-walled nanotubes into controlled lengths involved the use of a scanning tunnelling microscope tip to break individual tubes at defined points (4.81). While this approach may be useful for constructing devices using single tubes, methods for controlling the length of bulk samples of single-walled nanotubes into short lengths are also needed. As already mentioned, Smalley's group have developed such a method (4.51). They showed that samples of short tubes (which they named 'fullerene pipes') could be produced by prolonged sonication of the nanotube material in a mixture of concentrated sulphuric and nitric acids. During this treatment, it appears that localized sonochemistry produces holes in the tube sides, which are then further attacked by the acids to leave the open 'pipes'. Smalley and colleagues showed that the pipes could be sorted into different length fractions by a method known as field-flow fractionation. Other groups have also used acid treatments, sometime coupled with other procedures, to cut single-walled tubes into short lengths (e.g. 4.82).

Although effective in producing short tubes, the acid treatment method can be rather inefficient, due to high weight loss of SWNTs. As an alternative, fluorination has become a more popular method of cutting. Again the Rice group have taken the lead in this area. In 2002 (4.83) they reported that the pyrolysis of sidewall-fluorinated SWNTs at temperatures up to 1000 °C can cut them to lengths of ~50 nm. In later work, (4.84) they introduced a two-step process for cutting SWNTs which involved sidewall fluorination followed by treatment with strong oxidants such as H_2SO_5 (Caro's acid). This produced nanotubes with lengths of ~100 nm. As an alternative to fluorine, ozone (O_3) has also been used for cutting SWNTs (4.85). Here, the tubes were suspended in perfluoropolyether (a good solvent for ozone) and a mixture of 9 wt% of O_3 in O_2 was bubbled through the suspension. The lengths of the cut nanotubes depended on the length of the treatment. Thus, after 1 h the mean length was 92 nm, while after 8 h it was 59 nm.

4.5   Separating metallic and semiconducting single-walled nanotubes

4.5.1   Selective elimination

The first attempt to separate metallic (m) and semiconducting (s) nanotubes was reported by Phaedon Avouris's team at IBM (4.86). Their method relied on current-induced electrical breakdown to eliminate metallic tubes. Carbon nanotubes can withstand remarkable current densities, exceeding $10^9 \, A \, cm^{-2}$, but at high enough currents the nanotubes will burn up in air. In this way, metallic tubes can be removed selectively from SWNT ropes. The IBM team firstly used a gate electrode to deplete the electrical carriers (electrons or holes) from the semiconducting tubes within a SWNT rope. The metallic SWNTs within the rope could then be destroyed by current-induced oxidation, leaving the carrier-depleted semiconducting tubes, which carried no current, intact. Having eliminated the metallic tubes, the team were able to fabricate arrays of nanoscale field-effect transistors based solely on the remaining s-SWNTs, as discussed in Chapter 6 (p. 167). In the same study, they also showed that similar techniques could be used to remove individual shells one at a time from multiwalled tubes.

An alternative approach to the selective elimination of metallic SWNTs was introduced by Hongjie Dai's group in 2006 (4.87). This relied on the idea that m-SWNTs may have a higher chemical reactivity than s-SWNTs. The tubes were grown between patterned catalytic islands on a silicon wafer in the way described in Section 3.4.4. The tubes were treated with a methane plasma at 400 °C, to selectively hydrocarbonate (i.e. convert to hydrocarbons) the m-SWNTs, as well as small nanotubes of both types. A further annealing treatment at 600 °C removed unwanted functional groups from the remaining semiconducting tubes. In this way they were able to produce a large number of devices, each containing a small number of tubes.

4.5.2   Dielectrophoresis

Dielectrophoresis has been used to align and position both multiwalled and single-walled nanotubes, as discussed previously. In 2003, Ralph Krupke and colleagues showed that this technique could be used to separate metallic and semiconducting nanotubes (4.88). This is based on the fact that metallic and semiconducting tubes are expected to have different electrical polarizabilities. Krupke *et al.* calculated that that the dielectric constant of s-SWNTs is of the order of 5, while for m-SWNTs the value is at least 1000. For water the dielectric constant lies between these two figures, at approximately 80, so it should be possible to separate the two tube types in an aqueous suspension by exposing them to a strong and inhomogeneous electrical field. The method used by Krupke and colleagues involved suspending HiPco tubes in D_2O containing 1 wt% of SDS (D_2O was used rather than H_2O as it interferes less with the absorption spectroscopy of SWNTs). The suspension was then sonicated and centrifuged to further disperse and purify the tubes. Microelectrodes were prepared with electron-beam lithography and connected to a function generator which was operated at a frequency 10 MHz and a peak-to-peak voltage of 10 V. A drop of the suspension was applied to this chip, and it was found that the metallic tubes were attracted towards the microelectrode array, while the semiconducting tubes remained in the solution. Raman spectroscopy was used to confirm the effectiveness of the separation. A similar procedure, using repeated separation cycles, was demonstrated by Eleanor Campbell of Gothenburg University, Sweden and colleagues in 2005 (4.89).

In 2006 Howard Schmidt and colleagues from Rice used a dielectrophoresis-based technique to sort semiconducting SWNTs by size (4.90). This is possible because smaller diameter semiconducting tubes have a larger band gap, which affects their dielectric constant. The Rice group used a method called dielectrophoresis field flow fractionation, which involved injecting a suspension of SWNTs into a mobile phase which passed through a chamber, the bottom of which was fitted with an array of gold electrodes. A 1 MHz, 10 V peak-to-peak signal was applied to the electrodes, attracting the metallic tubes and leaving the semiconducting ones in solution. As the tubes eluted from the chamber, it was found that there were fewer smaller nanotubes as the retention time increased.

4.5.3   Selective functionalization

A group from the University of Connecticut used a method involving non-covalent functionalization with octadecylamine (ODA) to separate metallic and semiconducting nanotubes (4.91). This was based on the assumption that the physisorption of ODA on the sidewalls of semiconducting SWNTs will be stronger than on metallic ones. There seems little direct evidence for this, but the authors cite a demonstration by Kong and Dai that the electrical properties of s-SWNTs upon adsorption of linear alkylamines are significantly changed while those of m-SWNTs remain unaffected (4.92). Both HiPco and laser-ablated SWNTs were used in this study, and the tubes were carboxy-functionalized by acid treatment before treatment with ODA. It was hypothesized that the ODA treatment would selectively solubilize the s-SWNTs, leaving the m-SWNTs in the solid phase, and this seemed to be borne out in Raman studies.

In 2003 a collaborative group led by Ming Zheng of DuPont presented evidence that DNA can be used to separate metallic and semiconducting nanotubes. Initial work showed that that single-stranded DNA strongly interacts with SWNTs to form a stable DNA-carbon nanotube hybrid that effectively disperses carbon nanotubes in an aqueous solution (4.93), as also discussed in Chapter 8. In a subsequent study, the group found that a particular sequence of single stranded DNA self-assembles into a helical structure around individual carbon nanotubes (4.94). Importantly it was shown that that the electrostatics of the DNA-nanotube hybrid depended on tube diameter and electronic properties, enabling nanotube separation by anion exchange chromatography. Thus, early fractions were enriched in smaller diameter and metallic tubes, while late fractions were enriched in larger diameter and semiconducting tubes. This was confirmed using optical absorption and Raman spectroscopy. A review of the interactions between carbon nanotubes and nucleic acids has been given by Bibiana Onoa and colleagues (4.95).

In 2006, French researchers claimed that the covalent functionalization of SWNTs by azomethine ylides could be used to separate semiconducting from metallic tubes (4.96), as described in Chapter 8 (p. 209).

4.6   Discussion

This chapter began with a discussion of nanotube purification. A huge amount of work has been done in this area, and new purification protocols are continually being published. The methods that have been used include oxidative and acid treatments, high-temperature heating and physical techniques such as filtration, chromatography and centrifugation. Very often a combination of two or more of these methods is employed. Which one to choose depends on the types of nanotubes, and to some extent on the intended application, but a couple of protocols can be generally recommended. For multiwalled nanotubes produced catalytically, the high-temperature annealing process of Andrews and colleagues (4.7) is a highly effective way of both removing the catalyst impurities and improving the structural perfection of the tubes. For single-walled tubes, the methods described by Chiang and co-workers (4.15, 4.16) have proved very popular.

Methods of processing nanotubes were then considered and some impressive work has been carried out in this area. The production of 'buckypaper' from single-walled carbon nanotubes is a particularly interesting development. This is an extraordinary material which will surely find important applications. Also remarkable is Baughman's work on the spinning of MWNT yarns from arrays grown catalytically on substrates. It remains to be seen, however, whether this technique could be scaled up for commercial production. Great ingenuity has been used in developing methods for assembling nanotubes, particularly single-walled ones, into defined arrangements. The most promising methods seem to be those involving dielectrophoresis, although other approaches are being actively explored. Using techniques of this kind some groups have claimed to have arranged millions of tubes in a defined way. We are still some way from producing practical nanoelectronic devices based on arrays of tubes, however.

In the absence of a method of making nanotubes with a defined structure, methods for separating them according to their structure or properties are of great importance. Most work in this area has concentrated on separating metallic and semiconducting nanotubes, rather than isolating tubes with a specific structure. Some success has been achieved in the selective elimination of metallic SWNTs from mixtures of tubes, and promising results have also been reported using dielectrophoresis and selective functionalization. Of course, none of these methods would be needed if a method was available for synthesizing tubes with a defined structure, and this must remain the ultimate aim.

References

(4.1) T. W. Ebbesen, P. M. Ajayan, H. Hiura *et al.*, 'Purification of carbon nanotubes', *Nature*, **367**, 519 (1994).

(4.2) F. Ikazaki, S. Ohshima, K. Uchida *et al.*, 'Chemical purification of carbon nanotubes by the use of graphite intercalation compounds', *Carbon*, **32**, 1539 (1994).

(4.3) Y. J. Chen, M. L. H. Green, J. L. Griffin *et al.*, 'Purification and opening of carbon nanotubes via bromination', *Adv. Mater.*, **8**, 1012 (1996).

(4.4) Y. Sato, T. Ogawa, K. Motomiya *et al.*, 'Purification of MWNTs combining wet grinding, hydrothermal treatment, and oxidation', *J. Phys. Chem. B*, **105**, 3387 (2001).

(4.5) J.-M. Bonard, T. Stora, J.-P. Salvetat *et al.*, 'Purification and size-selection of carbon nanotubes', *Adv. Mater.*, **9**, 827 (1997).

(4.6) G. S. Duesberg, M. Burghard, J. Muster *et al.*, 'Separation of carbon nanotubes by size exclusion chromatography', *Chem. Commun.*, 435 (1998).

(4.7) R. Andrews, D. Jacques, D. Qian *et al.*, 'Purification and structural annealing of multiwalled carbon nanotubes at graphitization temperatures', *Carbon*, **39**, 1681 (2001).

(4.8) X. H. Chen, C. S. Chen, Q. Chen *et al.*, 'Non-destructive purification of multi-walled carbon nanotubes produced by catalyzed CVD', *Mater. Lett.*, **57**, 734 (2002).

(4.9) W. Huang, Y. Wang, G. H. Luo *et al.*, '99.9% purity multi-walled carbon nanotubes by vacuum high-temperature annealing', *Carbon*, **41**, 2585 (2003).

(4.10) Y. Wang, J. Wu and F. Wei, 'A treatment method to give separated multi-walled carbon nanotubes with high purity, high crystallization and a large aspect ratio', *Carbon*, **41**, 2939 (2003).

(4.11) H. Zhang, C. H. Sun, F. Li *et al.*, 'Purification of multiwalled carbon nanotubes by annealing and extraction based on the difference in van der Waals potential', *J. Phys. Chem. B*, **110**, 9477 (2006).

(4.12) R. C. Haddon, J. Sippel, A. G. Rinzler *et al.*, 'Purification and separation of carbon nanotubes', *MRS Bulletin*, **29**, 252 (2004).

(4.13) T. J. Park, S. Banerjee, T. Hemraj-Benny *et al.*, 'Purification strategies and purity visualization techniques for single-walled carbon nanotubes', *J. Mater. Chem.*, **16**, 141 (2006).

(4.14) P. X. Hou, C. Liu and H. M. Cheng, 'Purification of carbon nanotubes', *Carbon*, **46**, 2003 (2008).

(4.15) I. W. Chiang, B. E. Brinson, R. E. Smalley *et al.*, 'Purification and characterization of single-wall carbon nanotubes', *J. Phys. Chem. B*, **105**, 1157 (2001).

(4.16) I. W. Chiang, B. E. Brinson, A. Y. Huang *et al.*, 'Purification and characterization of single-wall carbon nanotubes (SWNTs) obtained from the gas-phase decomposition of CO (HiPco process)', *J. Phys. Chem. B*, **105**, 8297 (2001).

(4.17) A. G. Rinzler, J. Liu, H. J. Dai *et al.*, 'Large-scale purification of single-wall carbon nanotubes: process, product, and characterization', *Appl. Phys. A*, **67**, 29 (1998).

(4.18) V. Georgakilas, D. Voulgaris, E. Vazquez *et al.*, 'Purification of HiPco carbon nanotubes via organic functionalization', *J. Amer. Chem. Soc.*, **124**, 14 318 (2002).

(4.19) S. Banerjee and S. S. Wong, 'Rational sidewall functionalization and purification of single-walled carbon nanotubes by solution-phase ozonolysis', *J. Phys. Chem. B*, **106**, 12 144 (2002).

(4.20) S. Bandow, A. M. Rao, K. A. Williams *et al.*, 'Purification of single-wall carbon nanotubes by microfiltration', *J. Phys. Chem. B*, **101**, 8839 (1997).

(4.21) K. B Shelimov, R. O. Esenaliev, A. G. Rinzler *et al.*, 'Purification of single-wall nanotubes by ultrasonically assisted filtration', *Chem. Phys. Lett.*, **282**, 429 (1998).

(4.22) G. S. Duesberg, J. Muster, V. Krstic *et al.*, 'Chromatographic size separation of single-wall carbon nanotubes', *Appl. Phys. A*, **67**, 117 (1998).

(4.23) S. Niyogi, H. Hu, M. A. Hamon *et al.*, 'Chromatographic purification of soluble single-walled carbon nanotubes (s-SWNTs)', *J. Amer. Chem. Soc.*, **123**, 733 (2001).

(4.24) A. P. Yu, E. Bekyarova, M. E. Itkis *et al.*, 'Application of centrifugation to the large-scale purification of electric arc-produced single-walled carbon nanotubes', *J. Amer. Chem. Soc.*, **128**, 9902 (2006).

(4.25) K. E. Hurst, A. C. Dillon, D. A. Keenan *et al.*, 'Cleaning of carbon nanotubes near the π-plasmon resonance', *Chem. Phys. Lett.*, **433**, 301 (2007).

(4.26) R. Sen, S. M. Rickard, M. E. Itkis *et al.*, 'Controlled purification of single-walled carbon nanotube films by use of selective oxidation and near-IR spectroscopy', *Chem. Mater.*, **15**, 4273 (2003).

(4.27) M. E. Itkis, D. E. Perea, S. Niyogi *et al.*, 'Purity evaluation of as-prepared single-walled carbon nanotube soot by use of solution-phase near-IR spectroscopy', *Nano Lett.*, **3**, 309 (2003).

(4.28) H. Hu, B. Zhao, M. E. Itkis *et al.*, 'Nitric acid purification of single-walled carbon nanotubes', *J. Phys. Chem. B*, **107**, 13 838 (2003).

(4.29) A. C. Dillon, T. Gennett, K. M. Jones *et al.*, 'A simple and complete purification of single-walled carbon nanotube materials', *Adv. Mater.*, **11**, 1354 (1999).

(4.30) M. E. Itkis, D. E. Perea, R. Jung *et al.*, 'Comparison of analytical techniques for purity evaluation of single-walled carbon nanotubes', *J. Amer. Chem. Soc.*, **127**, 3439 (2005).

(4.31) M. S. P. Shaffer, X. Fan and A. H. Windle, 'Dispersion and packing of carbon nanotubes', *Carbon*, **36**, 1603 (1998).

(4.32) M. S. P. Shaffer and A. H. Windle, 'Analogies between polymer solutions and carbon nanotube dispersions', *Macromolecules*, **32**, 6864 (1999).

(4.33) W. H. Song, I. A. Kinloch and A. H. Windle, 'Nematic liquid crystallinity of multiwall carbon nanotubes', *Science*, **302**, 1363 (2003).

(4.34) W. H. Song and A. H. Windle, 'Isotropic–nematic phase transition of dispersions of multi-wall carbon nanotubes', *Macromolecules*, **38**, 6181 (2005).

(4.35) K. Yamamoto, S. Akita and Y. Nakayama, 'Orientation of carbon nanotubes using electro-phoresis', *Jpn. J. Appl. Phys.*, **35**, L917 (1996).

(4.36) K. Yamamoto, S. Akita and Y. Nakayama, 'Orientation and purification of carbon nano-tubes using ac electrophoresis', *J. Phys. D*, **31**, L34 (1998).

(4.37) J. Y. Chung, K. H. Lee, J. H. Lee *et al.*, 'Toward large-scale integration of carbon nano-tubes', *Langmuir*, **20**, 3011 (2004).

(4.38) Y. Lan, E. B. Wang, Y. H. Song *et al.*, 'Covalent assembly of shortened multiwall carbon nanotubes on polyelectrolyte films and relevant electrochemistry study'. *J. Colloid Interface Sci.*, **284**, 216 (2005).

(4.39) N. Chakrapani, B. Q. Wei, A. Carrillo *et al.*, 'Capillarity-driven assembly of two-dimensional cellular carbon nanotube foams', *Proc. Natl. Acad. Sci. USA*, **101**, 4009 (2004).

(4.40) H. Liu, S. H. Li, J. Zhai *et al.*, 'Self-assembly of large-scale micropatterns on aligned carbon nanotube films', *Angew. Chem. Int. Ed.* **43**, 1146 (2004).

(4.41) K. L. Jiang, Q. Q. Li and S. S. Fan, 'Spinning continuous carbon nanotube yarns', *Nature*, **419**, 801 (2002).

(4.42) B. Zhang, K. L. Jiang, C. Feng *et al.*, 'Spinning and processing continuous yarns from 4-inch wafer scale super-aligned carbon nanotube arrays', *Adv. Mater.*, **18**, 1505 (2006).

(4.43) M. Zhang, K. R. Atkinson and R. H. Baughman, 'Multifunctional carbon nanotube yarns by downsizing an ancient technology', *Science*, **306**, 1358 (2004).

(4.44) M. Zhang, S. Fang, A. A. Zakhidov *et al.*, 'Strong, transparent, multifunctional, carbon nanotube sheets', *Science*, **309**, 1215 (2005).

(4.45) K. L. Lu, R. M. Lago, Y. K. Chen *et al.*, 'Mechanical damage of carbon nanotubes by ultrasound', *Carbon*, **34**, 814 (1996).

(4.46) C. W. Yang, X. G. Hu, D. L. Wang *et al.*, 'Ultrasonically treated multi-walled carbon nanotubes (MWCNTs) as PtRu catalyst supports for methanol electrooxidation', *J. Power Sources*, **160**, 187 (2006).

(4.47) N. Pierard, A. Fonseca, Z. Kónya *et al.*, 'Production of short carbon nanotubes with open tips by ball milling', *Chem. Phys. Lett.*, **335**, 1 (2001).

(4.48) X. X. Wang, J. N. Wang, L. F. Su *et al.*, 'Cutting of multi-walled carbon nanotubes by solid-state reaction', *J. Mater. Chem.*, **16**, 4231 (2006).

(4.49) M. Q. Tran, C. Tridech, A. Alfrey *et al.*, 'Thermal oxidative cutting of multi-walled carbon nanotubes', *Carbon*, **45**, 2341 (2007).

(4.50) T. D. Yuzvinsky, A. M. Fennimore, W. Mickelson *et al.*, 'Precision cutting of nanotubes with a low-energy electron beam', *Appl. Phys. Lett.*, **86**, 053109 (2005).

(4.51) J. Liu, A. G. Rinzler, H. J. Dai *et al.*, 'Fullerene pipes', *Science*, **280**, 1253 (1998).

(4.52) M. Sano, A. Kamino, J. Okamura *et al.*, 'Ring closure of carbon nanotubes', *Science*, **293**, 1299 (2001).

(4.53) M. Sano, A. Kamino and S. Shinkai, 'Construction of carbon nanotube "stars" with dendrimers', *Angew. Chem. Int. Ed.*, **40**, 4661 (2001).

(4.54) P. W. Chiu, G. S. Duesberg, U. Dettlaff-Weglikowska *et al.*, 'Interconnection of carbon nanotubes by chemical functionalization', *Appl. Phys. Lett.*, **80**, 3811 (2002).

(4.55) Z. F. Liu, Z. Y. Shen, T. Zhu *et al.*, 'Organizing single-walled carbon nanotubes on gold using a wet chemical self-assembling technique', *Langmuir*, **16**, 3569 (2000).

(4.56) X. L. Nan, Z. N. Gu and Z. F. Liu, 'Immobilizing shortened single-walled carbon nanotubes (SWNTs) on gold using a surface condensation method', *J. Colloid Interface Sci.*, **245**, 311 (2002).

(4.57) J. J. Gooding, R. Wibowo, J. Q. Liu *et al.*, 'Protein electrochemistry using aligned carbon nanotube arrays', *J. Amer. Chem. Soc.*, **125** 9006 (2003).

(4.58) S. Banerjee, B. E. White, L. M. Huang *et al.*, 'Precise positioning of single-walled carbon nanotubes by ac dielectrophoresis', *J. Vac. Sci. Tech. B*, **24**, 3173 (2006).

(4.59) A. Vijayaraghavan, S. Blatt, D. Weissenberger *et al.*, 'Ultra-large-scale directed assembly of single-walled carbon nanotube devices', *Nano Lett.*, **7**, 1556 (2007).

(4.60) S. G. Rao, L. Huang, W. Setyawan *et al.*, 'Large-scale assembly of carbon nanotubes', *Nature*, **425**, 36 (2003).

(4.61) A. Javey, P. F. Qi, Q. Wang *et al.*, '10- to 50-nm-long quasi-ballistic carbon nanotube devices obtained without complex lithography', *Proc. Natl. Acad. Sci. USA*, **101**, 13 408 (2004).

(4.62) K. Keren, R. S. Berman, E. Buchstab *et al.*, 'DNA-templated carbon nanotube field-effect transistor', *Science*, **302**, 1380 (2003).

(4.63) H. J. Xin and A. T. Woolley, 'DNA-templated nanotube localization', *J. Amer. Chem. Soc.*, **125**, 8710 (2003).

(4.64) X. L. Li, L. Zhang, X. R. Wang *et al.*, 'Langmuir–Blodgett assembly of densely aligned single-walled carbon nanotubes from bulk materials', *J. Amer. Chem. Soc.*, **129**, 4890 (2007).

(4.65) X. Yu, B. Munge, V. Patel *et al.*, 'Carbon nanotube amplification strategies for highly sensitive immunodetection of cancer biomarkers', *J. Amer. Chem. Soc.*, **128**, 11 199 (2006).

(4.66) Y. Yan, M. B. Chan-Park and Q. Zhang, 'Advances in carbon-nanotube assembly', *Small*, **3**, 24 (2007).

(4.67) S. Ramesh, L. M. Ericson, V. A. Davis *et al.*, 'Dissolution of pristine single walled carbon nanotubes in superacids by direct protonation', *J. Phys. Chem. B*, **108**, 8794 (2004).

(4.68) V. A. Davis, L. M. Ericson, A. N. G. Parra-Vasquez *et al.*, 'Phase behavior and rheology of SWNTs in superacids', *Macromolecules*, **37**, 154 (2004).

(4.69) L. M. Ericson, H. Fan, H. Q. Peng *et al.*, 'Macroscopic, neat, single-walled carbon nanotube fibers', *Science*, **305**, 1447 (2004).

(4.70) F. Hennrich, S. Lebedkin, S. Malik *et al.*, 'Preparation, characterization and applications of free-standing single walled carbon nanotube thin films', *Phys. Chem. Chem. Phys.*, **4**, 2273 (2002).

(4.71) S. Malik, H. Rosner, F. Hennrich *et al.*, 'Failure mechanism of free standing single-walled carbon nanotube thin films under tensile load', *Phys. Chem. Chem. Phys.*, **6**, 3540 (2004).

(4.72) Z. C. Wu, Z. H. Chen, X. Du *et al.*, 'Transparent, conductive nanotube films', *Science*, **305**, 1273 (2004).

(4.73) D. H. Zhang, K. Ryu, X. L. Liu *et al.*, 'Transparent, conductive, and flexible carbon nanotube films and their application in organic light-emitting diodes', *Nano Lett.*, **6**, 1880 (2006).

(4.74) J. van de Lagemaat, T. M. Barnes, G. Rumbles *et al.*, 'Organic solar cells with carbon nanotubes replacing In_2O_3:Sn as the transparent electrode', *Appl. Phys. Lett.*, **88**, 233503 (2006).

(4.75) T. M. Barnes, J. van de Lagemaat, D. Levi *et al.*, 'Optical characterization of highly conductive single-wall carbon-nanotube transparent electrodes', *Phys. Rev. B*, **75**, 235410 (2007).

(4.76) J. Hone, M. C. Llaguno, N. M. Nemes *et al.*, 'Electrical and thermal transport properties of magnetically aligned single wall carbon nanotube films', *Appl. Phys. Lett.*, **77**, 666 (2000).

(4.77) D. A. Walters, M. J. Casavant, X. C. Qin *et al.*, 'In-plane-aligned membranes of carbon nanotubes', *Chem. Phys. Lett.*, **338**, 14 (2001).

(4.78) J. E. Fischer, W. Zhou, J. Vavro *et al.*, 'Magnetically aligned single wall carbon nanotube films: preferred orientation and anisotropic transport properties', *J. Appl. Phys.*, **93**, 2157 (2003).

(4.79) T. V. Sreekumar, T. Liu, S. Kumar *et al.*, 'Single-wall carbon nanotube films', *Chem. Mater.* **15**, 175 (2003).

(4.80) R. Duggal, F. Hussain and M. Pasquali, 'Self-assembly of single-walled carbon nanotubes into a sheet by drop drying', *Adv. Mater.*, **18**, 29 (2006).

(4.81) L. C. Venema, J. W. G. Wildoer, H. L. J. T. Tuinstra *et al.*, 'Length control of individual carbon nanotubes by nanostructuring with a scanning tunnelling microscope', *Appl. Phys. Lett.*, **71**, 2629 (1997).

(4.82) T. J. Aitchison, M. Ginic-Markovic, J. G. Matisons *et al.*, 'Purification, cutting, and sidewall functionalization of multiwalled carbon nanotubes using potassium permanganate solutions', *J. Phys. Chem. C*, **111**, 2440 (2007).

(4.83) Z. Gu, H. Peng, R. H. Hauge, *et al.*, 'Cutting single-wall carbon nanotubes through fluorination', *Nano Lett.*, **2**, 1009 (2002).

(4.84) K. J. Ziegler, Z. Gu, J. Shaver *et al.*, 'Cutting single-walled carbon nanotubes', *Nanotechnology*, **16**, S539 (2005).

(4.85) Z. Y. Chen, K. J. Ziegler, J. Shaver *et al.*, 'Cutting of single-walled carbon nanotubes by ozonolysis', *J. Phys. Chem. B*, **110**, 11 624 (2006).

(4.86) P. G. Collins, M. S. Arnold and P. Avouris, 'Engineering carbon nanotubes and nanotube circuits using electrical breakdown', *Science*, **292**, 706 (2001).

(4.87) G. Y. Zhang, P. F. Qi, X. R. Wang *et al.*, 'Selective etching of metallic carbon nanotubes by gas-phase reaction', *Science*, **314**, 974 (2006).

(4.88) R. Krupke, F. Hennrich, H. von Lohneysen *et al.*, 'Separation of metallic from semiconducting single-walled carbon nanotubes', *Science*, **301**, 344 (2003).

(4.89) D. S. Lee, D. W. Kim, H. S. Kim *et al.*, 'Extraction of semiconducting CNTs by repeated dielectrophoretic filtering', *Appl. Phys. A*, **80**, 5 (2005).

(4.90) H. Q. Peng, N. T. Alvarez, C. Kittrell *et al.*, 'Dielectrophoresis field flow fractionation of single-walled carbon nanotubes', *J. Amer. Chem. Soc.*, **128**, 8396 (2006).

(4.91) D. Chattopadhyay, I. Galeska and F. Papadimitrakopoulos, 'A route for bulk separation of semiconducting from metallic single-wall carbon nanotubes', *J. Amer. Chem. Soc.*, **125**, 3370 (2001).

(4.92) J. Kong and H. J. Dai, 'Full and modulated chemical gating of individual carbon nanotubes by organic amine compounds', *J. Phys. Chem. B*, **105**, 2890 (2001).

(4.93) M. Zheng, A. Jagota, E. D. Semke *et al.*, 'DNA-assisted dispersion and separation of carbon nanotubes', *Nat. Mater.*, **2**, 338 (2003).

(4.94) M. Zheng, A. Jagota, M. S. Strano *et al.*, 'Structure-based carbon nanotube sorting by sequence-dependent DNA assembly', *Science*, **302**, 1545 (2003).

(4.95) B. Onoa, M. Zheng, M. S. Dresselhaus *et al.*, 'Carbon nanotubes and nucleic acids: tools and targets', *Physica Status Solidi A*, **203**, 1124 (2006).

(4.96) C. Menard-Moyon, N. Izard, E. Doris *et al.*, 'Separation of semiconducting from metallic carbon nanotubes by selective functionalization with azomethine ylides', *J. Amer. Chem. Soc.*, **128**, 6552 (2006).

5   Structure

Soon after the discovery of fullerene-related carbon nanotubes it became clear that a completely new framework would be needed to analyse the structures and symmetries of these new materials. Although theoretical methods have been developed for analysing cylindrical arrays in biology (5.1), these are insufficient for a full analysis of nanotube structure. The challenge of formulating the new approaches necessary for classifying nanotube structure was taken up by Mildred Dresselhaus and co-workers, and by Carter White and colleagues, among others. The techniques developed by these groups have been essential in determining the electronic and vibrational properties of nanotubes, as discussed in later chapters. Theoretical discussions have also been given of the layer structure of multiwalled tubes, of tube capping and of other aspects of nanotube structure such as elbow connections.

Experimental studies of nanotube structure have mainly been carried out using microscopy. X-ray and neutron diffraction have generally been of less value, since samples of nanotubes always contain tubes with a wide range of different structures. High-resolution transmission electron microscopy (HRTEM) has been by far the most widely used and most valuable technique. Recent improvements in the resolution of HRTEM have meant that the atomic network which makes up individual tubes can be imaged directly, as well as the layer structure, opening the way for a deeper understanding of their structure. The use of spectroscopic techniques to probe nanotube structure is discussed in Chapter 7.

The present chapter begins with a brief discussion of bonding in graphite and fullerenes. Theoretical models of carbon nanotube structure are then summarized. Experimental studies of the structure of nanotubes, mainly using HRTEM, are then covered, beginning with multiwalled tubes and then moving on to single-walled tubes. Finally, the application of neutron diffraction to the study of carbon nanotube structure is briefly discussed.

5.1   Bonding in carbon materials

A free carbon atom has the electronic structure $1s^2 2s^2 2p^2$. In order to form covalent bonds, one of the 2s electrons is promoted to 2p, and the orbitals are then hybridized in one of three possible ways. In graphite, one of the 2s electrons hybridizes with two of the 2p electrons to give three sp^2 orbitals at 120° to each other in a plane, with the remaining

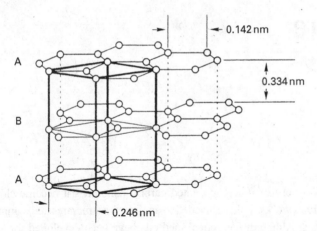

A

B

A

← 0.142 nm

0.334 nm

← 0.246 nm

Fig. 5.1   The structure of hexagonal (Bernal) graphite, showing the unit cell.

orbital having a p_z configuration, at 90° to this plane. The sp^2 orbitals form the strong σ bonds between carbon atoms in the graphite planes, while the p_z, or π, orbitals provide the weak van der Waals bonds between the planes. The overlap of π orbitals on adjacent atoms in a given plane provides the electron bond network which gives graphite its relatively high electrical conductivity. In naturally occurring or high-quality synthetic graphite, the stacking sequence of the layers is generally ABAB, with an interlayer {002} spacing of approximately 0.334 nm, as shown in Fig. 5.1. This structure is often known as Bernal graphite after John D. Bernal who first proposed it in 1924 (5.2). The unit cell contains four atoms, and the space group is $P6_3/mmc$ (D_{6h}). In less perfect graphites, the interplanar spacing is found to be significantly larger than the value for single crystal graphite (typically ~0.344 nm), and the layer planes are randomly rotated with respect to each other about the c axis. Such graphites are termed turbostratic.

In diamond, each carbon atom is joined to four neighbours in a tetrahedral structure. The bonding here is sp^3 and results from the mixing of one 2s and three 2p orbitals. Diamond is less stable than graphite, and is converted to graphite at a temperature of 1700 °C at normal pressures. Disordered carbons containing sp^3-bonded atoms are also rapidly transformed into graphite at high temperatures.

The C_{60} molecule, shown in Fig. 1.1, consists of carbon atoms bonded in an icosahedral structure made up of twenty hexagons and twelve pentagons. Each of the carbon atoms in C_{60} is joined to three neighbours, so the bonding is essentially sp^2, although there is a small amount of sp^3 character due to the curvature: theory suggests that the hybridization in C_{60} is $sp^{2.28}$ (5.3). Note that all 60 carbon atoms are identical, so that the strain is evenly distributed over the molecule. The bonding in carbon nanoparticles and nanotubes is also primarily sp^2, although once again there may be some sp^3 character in regions of high curvature.

A new chapter in the science of carbon opened in 2004 with the isolation of individual sheets of graphene by Andre Geim of the University of Manchester, with colleagues from Russia (5.4). The technique they used was amazingly simple, and just involved rubbing a freshly cleaved graphite surface against another surface. This left a variety of flakes attached to the surface, including single graphene sheets. The sheets were characterized by AFM, TEM and optical microscopy. This discovery has attracted huge interest, comparable with that which surrounded nanotubes following Iijima's 1991 paper, owing to the extraordinary electronic properties of graphene and its potential in nanoscale devices (5.5).

5.2   The structure of carbon nanotubes: theoretical discussion

5.2.1   Vector notation for carbon nanotubes

As mentioned in Chapter 1, there are two possible high-symmetry structures for nanotubes, known as 'zigzag' and 'armchair'. These are illustrated in Fig. 1.3. In practice, it is believed that most nanotubes do not have these highly symmetric forms but have structures in which the hexagons are arranged helically around the tube axis, as in Fig. 5.2. These structures are generally known as chiral, since they can exist in two mirror-related forms.

The simplest way of specifying the structure of an individual tube is in terms of a vector, which we label $\mathbf{C}$, joining two equivalent points on the original graphene lattice. The cylinder is produced by rolling up the sheet such that the two end-points of the vector are superimposed. Because of the symmetry of the honeycomb lattice, many of the cylinders produced in this way will be equivalent, but there is an 'irreducible wedge' comprising one-twelfth of the graphene lattice, within which unique tube structures are defined. Figure 5.3 shows a small part of this irreducible wedge, with points on the lattice labelled according to the notation of Dresselhaus et al. (5.6, 5.7). Each pair of integers (n, m) represents a possible tube structure. Thus the vector $\mathbf{C}$ can be expressed as

$$\mathbf{C} = n\mathbf{a}_1 + m\mathbf{a}_2$$

where $\mathbf{a}_1$ and $\mathbf{a}_2$ are the unit cell base vectors of the graphene sheet, and $n \geq m$. It can be seen from Fig. 5.3 that $m = 0$ for all zigzag tubes, while $n = m$ for all armchair tubes. All other tubes are chiral. In the case of the two 'archetypal' nanotubes which can be capped by one-half of a C_{60} molecule, the zigzag tube is represented by the integers $(9, 0)$ while

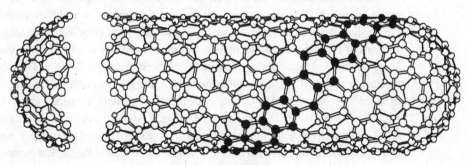

Fig. 5.2   A drawing of a chiral nanotube (adapted from ref. 5.6).

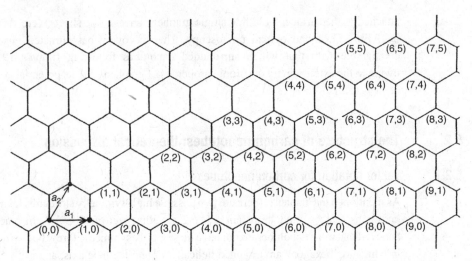

Fig. 5.3   Graphene layer with atoms labelled using (n, m) notation. Unit vectors of the 2D lattice are also
shown.

the armchair tube is denoted by $(5, 5)$. Since $|\mathbf{a}_1| = |\mathbf{a}_2| = 0.246$ nm, the magnitude of $\mathbf{C}$ in
nanometres is $0.246\sqrt{(n^2 + nm + m^2)}$, and the diameter d_t is given by

$$d_t = 0.246\sqrt{(n^2 + nm + m^2)}/\pi$$

The chiral angle, Θ, is given by

$$\Theta = \sin^{-1}\frac{\sqrt{3}m}{2\sqrt{(n^2 + nm + m^2)}}$$

5.2.2   Unit cells of nanotubes

If we think of a nanotube as a 'one-dimensional crystal', we can define a translational unit
cell along the tube axis. For all nanotubes, the translational unit cell has the form of a
cylinder. Considering again the two archetypal tubes that can be capped by one-half of a
C_{60} molecule, the 'unrolled' cylindrical unit cells for both of these are shown in Fig. 5.4.
For the armchair tube, the width of the cell is equal to the magnitude of $\mathbf{a}$, the unit vector
of the original 2D graphite lattice, while for the zigzag tube the width of the cell is $\sqrt{3}\mathbf{a}$.
Larger diameter armchair and zigzag nanotubes have unit cells which are simply longer
versions of these. For chiral nanotubes, the lower symmetry results in larger unit cells. A
simple method of constructing these cells has been described by Jishi, Dresselhaus and
colleagues (5.6–5.10). This involves drawing a straight line through the origin O of the
irreducible wedge normal to $\mathbf{C}$, and extending this line until it passes exactly through an
equivalent lattice point. This is illustrated in Fig. 5.5 for the case of a $(6, 3)$ nanotube. The
length of the unit cell in the tube axis direction is the magnitude of the vector $\mathbf{T}$. Expressions
can be derived for this in terms of C, the magnitude of $\mathbf{C}$, and the highest common divisor
of n and m, which we denote d_H (5.6, 5.7). If $n - m \neq 3rd_H$, where r is some integer, then

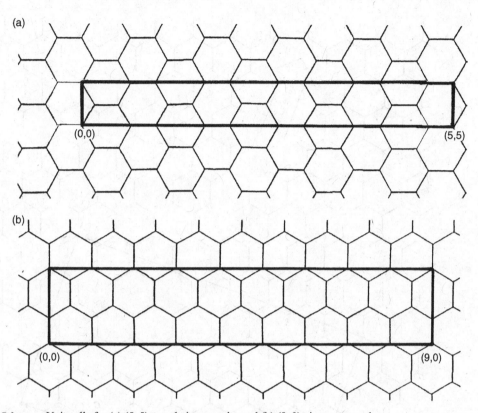

Fig. 5.4   Unit cells for (a) $(5, 5)$ armchair nanotube and (b) $(9, 0)$ zigzag nanotube.

$$T = \sqrt{3}\, C/d_{\mathrm{H}}$$

while if $n - m = 3rd_{\mathrm{H}}$, then

$$T = \sqrt{3}\, C/3d_{\mathrm{H}}$$

It can also be shown that the number of carbon atoms per unit cell of a tube specified by (n, m) is $2N$ such that

$$N = 2(n^2 + m^2 + nm)/d_{\mathrm{H}} \quad \text{if } n - m \neq 3rd_{\mathrm{H}}$$

and

$$N = 2(n^2 + m^2 + nm)/3d_{\mathrm{H}} \quad \text{if } n - m = 3rd_{\mathrm{H}}$$

These simple expressions enable the diameters and unit cell parameters of nanotubes to be readily calculated. For nanotubes in the diameter range that is typically observed experimentally, i.e. ~ 2–$30\,$nm, the unit cells can be very large. For example, the tube denoted $(80, 67)$, which has a diameter of approximately $10\,$nm, has a unit cell $54.3\,$nm in length containing $64\,996$ atoms. These large unit cells can present problems in calculating the

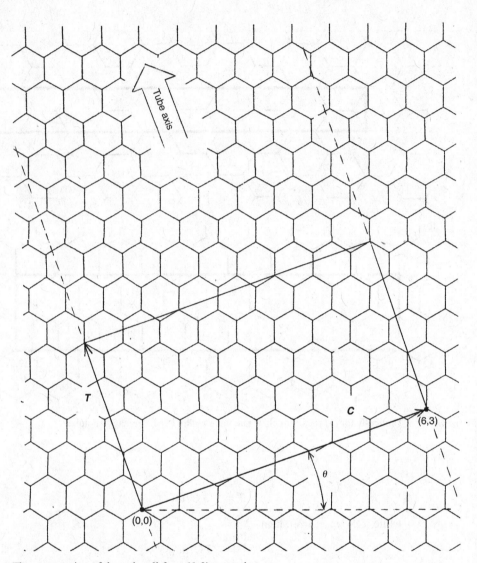

Fig. 5.5   The construction of the unit cell for a (6, 3) nanotube.

electronic and vibrational properties of nanotubes. For this reason, White, Mintmire and colleagues have proposed an alternative method of generating tube structures, which makes use of helical operators rather than a translational unit cell. This will not be detailed here, but has been described in a number of papers and reviews (e.g. 5.11, 5.12).

5.2.3   Symmetry classification of nanotubes

We now consider the symmetry classification of carbon nanotubes, once again following the work of Mildred Dresselhaus and co-workers (5.6–5.10). The symmetry of armchair and zigzag nanotubes is considered first. Such tubes can be represented by *symmorphic* groups, that is groups in which the rotations can be treated by simple point group representations. This differentiates them from chiral tubes, in which the symmetry operations involve both translations and rotations.

In determining the symmetry classification of nanotubes, we assume that the tube length is much greater than the diameter, so that the caps can be neglected. Since all armchair and zigzag nanotubes have a rotational symmetry axis and may additionally have either a mirror plane at right angles to this axis or an inversion centre, they fall into either the D_{nh} or the D_{nd} group. In deciding between these two groups, we follow Dresselhaus and colleagues in making the assumption that all tubes have an inversion centre. Now inversion is an element of D_{nh} only for even n, and is an element of D_{nd} only for odd n. It follows that the symmetry group for armchair or zigzag tubes with n even is D_{nh} while the group for armchair or zigzag tubes with n odd is D_{nd}.

For chiral tubes the symmetry groups are non-symmorphic, i.e. the symmetry operations involve both translations and rotations. Thus, the basic symmetry operation $R = (\psi, \tau)$, involves a rotation by an angle ψ followed by a translation τ. This operation corresponds to the vector $\mathbf{R} = p\mathbf{a}_1 + q\mathbf{a}_2$. Thus (p, q) denotes the coordinates reached when the symmetry operation (ψ, τ) acts on an atom at $(0, 0)$. Dresselhaus *et al.* show that the values of p and q are given by

$$mp - nq = d_{\mathrm{H}}$$

with the conditions $q < m/d_{\mathrm{H}}$ and $p < n/d_{\mathrm{H}}$. It can also be shown that the parameters ψ and τ are given by

$$\psi = 2\pi \frac{\Omega}{N d_{\mathrm{H}}}$$

and

$$\tau = \frac{T d_{\mathrm{H}}}{N}$$

where the quantity Ω is defined as

$$\Omega = \{p(m + 2n) + q(n + 2m)\}/(d_{\mathrm{H}}/d_{\mathrm{R}})$$

with

$$d_R = \begin{cases} d_{\mathrm{H}} & \text{if} \quad n - m \text{ is not a multiple of } 3d_{\mathrm{H}} \\ 3d_{\mathrm{H}} & \text{if} \quad n - m \text{ is a multiple of } 3d_{\mathrm{H}} \end{cases}$$

We are now in a position to consider the symmetry group for chiral nanotubes. Unlike armchair and zigzag nanotubes, chiral tubes contain no mirror planes, and therefore belong to C symmetry groups. Considering first tubes with $d_{\mathrm{H}} = 1$, the order of the rotational axis is equal to 2π divided by the number of rotation operations required to reach a lattice vector, i.e. $2\pi/\psi = N/\Omega$, so that the symmetry group becomes $C_{N/\Omega}$. For tubes with $d_{\mathrm{H}} \neq 1$, the symmetry group is expressed as a direct product, $C_{d\mathrm{H}} \otimes C'_{N/\Omega}$.

We can illustrate the meaning of some of these parameters with reference to Fig. 5.6, which relates to the chiral $(4, 2)$ nanotube. In this case, the parameters p and q are 1 and 0

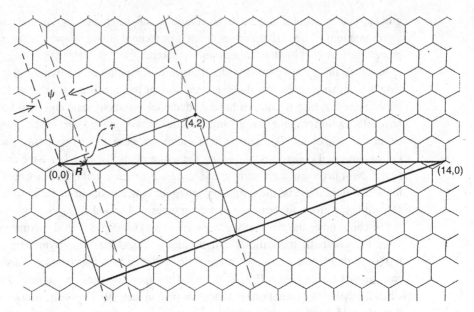

Fig. 5.6   A diagram illustrating symmetry operations for chiral nanotubes.

respectively, so that **R** is the vector joining $(0, 0)$ to $(1, 0)$. If we now imagine the 2D sheet being rolled up to form the tube, the line joining $(0, 0)$ to $(1, 0)$, $(2, 0)$, etc. becomes a helix running around the cylinder. Eventually this line intersects a lattice point at a distance T along the tube. The length of the solid line at the point of intersection is then $N/d_H \times R$, where N is one-half of the number of atoms in the unit cell and d_H is the highest common divisor of n and m. For the $(4, 2)$ nanotube, $N = 28$, $d_H = 2$, so the length of the line at the point of intersection is $14R$. This is represented in Fig. 5.6 by the intersection of the line joining $(0, 0)$ to $(1, 0)$, $(2, 0)$ ... to a line drawn through the lower edge of the unit cell. The intersection occurs at the point denoted by $(14, 0)$. Now, for the $(4, 2)$ tube the quantity Ω is equal to 10, so that in this case the symmetry group is $C_2 \otimes C'_{28/10}$.

For a fuller treatment of the symmetry classification of nanotubes, the reader should consult references 5.6–5.10 and other papers by Dresselhaus, Jishi and co-workers.

5.2.4   Defects in the hexagonal lattice

The discussion so far has assumed that the hexagonal carbon network making up the nanotubes' sidewalls is perfect. In practice, this is never the case, and it is important to consider the types of defects that can exist (5.13).

Nanotubes containing abrupt elbow-like bends are quite often observed in samples produced by arc-evaporation (see Section 5.3.5). Connections of this type have been analysed by a number of workers (e.g. 5.14–5.18). It has been established that armchair tubes can be joined to zigzag tubes by elbow connections involving a pentagonal ring on the outer side of the elbow and a heptagon on the inner side. As an example, Fig. 5.7 shows a connection between a $(5, 5)$ armchair tube and a $(9, 0)$ zigzag tube (5.17).

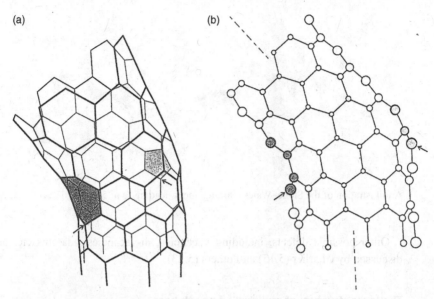

Fig. 5.7   An illustration of an 'elbow connection' between a (5, 5) armchair and a (9, 0) zigzag nanotube (5.17). (a) A perspective drawing with pentagonal and heptagonal rings shaded, (b) The structure projected on symmetry plane of elbow.

According to Dunlap (5.15), the optimal angle between tubes joined by a pentagon–heptagon connection should be 150°, but model-building exercises by Fonseca and colleagues (5.16) produced an angle of 144°. It was pointed out by the Lucas and Cohen groups in 1996 (5.17, 5.18) that elbow connections of this kind could constitute metal/semiconductor or semiconductor/semiconductor junctions. This is discussed further in Chapter 6 (p. 164).

As well as analysing single elbow connections, Fonseca *et al.* discussed the formation of nanotube tori and helices by the inclusion of a number of pentagon–heptagon pairs. Nanotube tori, or 'hoops', have now been observed experimentally (5.19). Helical tubes are often observed in catalytically-produced nanotube samples, as discussed in Chapter 3.

Another class of defect that can occur in the hexagonal carbon network involves adjacent pentagon–heptagon pairs. Individual pentagons and heptagons introduce positive or negative curvature into the hexagonal network as discussed in Section 5.2.6 below. In adjacent pairs the effects are cancelled out, so the geometry of the network is retained. The defects can occur singly, as first discussed by Jean-Christophe Charlier and colleagues (5.20), or in a '5-7-7-5' rearrangement. The latter forms as a result of the Stone–Wales transformation. This mechanism, first put forward by Anthony Stone and David Wales in 1986 to explain interconversion between fullerene isomers (5.21), involves a 90° bond rotation, as illustrated in Fig. 5.8. The Stone–Wales transformation may be important in the growth of nanotubes in the solid state and in the ductile fracture of nanotubes. In 2007, Kazu Suenaga, Sumio Iijima and their colleagues directly imaged pentagon–heptagon pair defects in an SWNT (5.22), as discussed in Section 5.5.3 below.

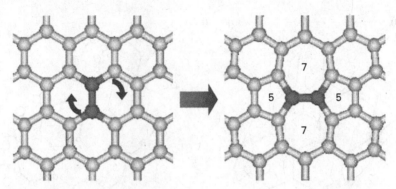

Fig. 5.8   An illustration of the Stone–Wales rearrangement, leading to a '5-7-7-5' defect (5.21, 5.22).

Other possible defects, including vacancies, di-vacancies, adatoms etc. have been discussed by Charlier (5.13) and others (5.23).

5.2.5   The layer structure of multiwalled nanotubes

The most basic question concerning the layer structure of multiwalled tubes is whether they have a scroll-like, 'Swiss-roll', structure, or whether they instead consist of a 'Russian doll', or nested, arrangement of discrete tubes. These two possible arrangements are illustrated in Fig. 5.9. Alternatively, the structure might consist of a mixture of these two arrangements, as discussed by several authors (e.g. 5.24, 5.25). Experimental studies generally point to a Russian doll structure, at least for MWNTs produced by arc-evaporation, as outlined in Section 5.3.1. Assuming this assumption to be correct, we now address the question of the structural relationship between successive cylinders, following discussions that have been given by Zhang and colleagues (5.26) and by Reznik *et al.* (5.27).

If the concentric graphene tubes are separated by a distance of approximately 0.334 nm, then successive tubes should differ in circumference by $(2\pi \times 0.334)$ nm $\approx$ 2.1 nm. It can readily be seen that this is not possible for zigzag tubes, since 2.1 nm is not a precise multiple of 0.246 nm, the width of one hexagon. The closest approximation to the 'correct' separation is obtained if two successive cylinders differ by nine rows of hexagons, which produces an inter-tube distance of 0.352 nm. A schematic section through a three layer zigzag tube, reproduced from Zhang *et al.* (5.26), is shown in Fig. 5.10. Here, the bold lines indicate the 9 and 18 extra rows of atoms that have added to the centre and outer tubes respectively. The inclusion of these extra rows is similar to the introduction of a Shockley partial dislocation. It is clear from Fig. 5.10 that, for the most part, the ABAB stacking of perfect graphite is not present in concentric zigzag tubes. However, short regions exist half-way between each 'dislocation' in which there is a good approximation to ABAB stacking.

In the case of armchair tubes, multiwalled structures can be assembled in which the ABAB arrangement is maintained and the interlayer distance is 0.34 nm. This is because 2.1 nm is close to 5 × 0.426 nm, the length of the repeat unit from which armchair tubes

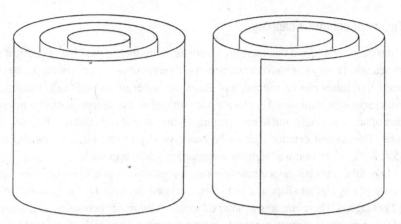

Fig. 5.9   A schematic illustration of 'Swiss roll' and 'Russian doll' models for multiwalled nanotubes.

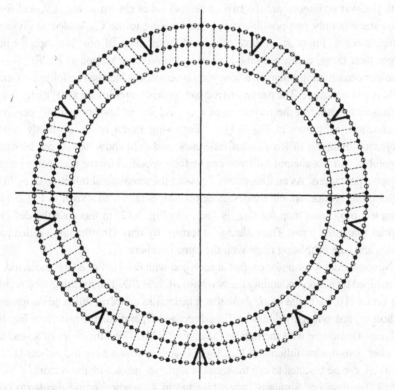

Fig. 5.10   A schematic illustration of a 3-layer nanotube showing how 'interfacial dislocations' (bold lines) can be introduced to accommodate strains. Full circles represent atoms in the plane of paper, open circles represent atoms out of plane of the paper (5.26).

are constructed. For chiral nanotubes, the situation is complicated, but in general it is not possible to have two tubes with exactly the same chiral angle separated by the graphite interplanar distance. Overall it seems unlikely that the ABAB stacking of single-crystal graphite will be present in cylindrical carbon nanotubes, except possibly in small areas.

5.2.6   Theory of nanotube capping

It has been established that there are a very large number of possible cylindrical graphene structures. Theory has also determined that all nanotubes larger than the archetypal (5, 5) and (9, 0) tubes can be capped, and that the number of possible caps increases rapidly with increasing diameter. There have been several different approaches to the representation of nanotube caps, and to enumerating the number of caps that will fit onto a particular tube. The earliest detailed discussion was given by Fujita, Dresselhaus and colleagues (5.8, 5.28, 5.29), and we begin by summarizing their approach.

Like fullerenes, all capped nanotubes must obey Euler's law. This states that a hexagonal lattice of any size or shape can only form a closed structure by the inclusion of precisely 12 pentagons. Therefore, any nanotube cap must contain six pentagons, and considerations of strain dictate that these pentagons must be isolated from each other (neglecting, for the moment, caps containing heptagons). As noted above, the smallest tubes that can be capped with *isolated* pentagons are the two archetypal tubes shown in Fig. 1.2, and for each of these there is only one possible cap, corresponding to the C_{60} molecule divided in two different ways. Fujita *et al.* have calculated the number of possible caps for nanotubes larger than these, using a method based on 'projection mapping' (5.28). This method involves constructing a map on a honeycomb network, which can be folded to form a given fullerene or nanotube. The pentagons are constructed by removing a 60° triangular segment of lattice, resulting in the formation of a conical defect known as a 60° positive wedge disclination, as shown in Fig. 5.11(a). Following Fujita *et al.* we firstly consider the projection mapping of icosahedral fullerenes, and then show how it can be extended to nanotubes. An icosahedral fullerene can be fully specified by the vector that connects two adjacent pentagons. As an illustration, consider the icosahedral fullerene C_{140}. In this case the defining vector, which we designate (n_f, m_f), is (2, 1), as shown in Fig. 5.11(b). The complete projection map for C_{140} is shown in Fig. 5.12; in this case the defects form a regular triangular array. The fullerene is formed by removing the non-shaded part of the lattice and superimposing rings with the same numbers.

Now consider a nanotube capped at each end with one-half of a C_{140} molecule. This can be mapped by simply extending the two lines *AC* and *BD*. The resulting tube is chiral, with the vector (10, 5). Fujita *et al.* show that a general icosahedral fullerene designated by the indices (n_f, m_f), when divided in half in a direction perpendicular to one of the five-fold axes, will cap a nanotube with the indices $(5n_f, 5m_f)$ (5.28). Thus, the series of so-called 'magic number' icosahedral fullerenes, C_{60}, C_{240}, C_{540}, ..., which have the indices (1, 1), (2, 2), (3, 3) ..., can be bisected to cap the series of armchair tubes with the vectors (5, 5), (10, 10), (15, 15) and so on. Similarly, when bisected in a direction perpendicular to one of the three-fold axes, these fullerenes will cap the tubes (9, 0), (18, 0), (27, 0), etc.

As already indicated, all nanotubes larger than the (5, 5) and (9, 0) tubes (with one exception) can be capped in more than one way. This is illustrated in Fig. 5.13 which shows two different ways of capping the chiral nanotube defined by the vector (7, 5). In fact, Dresselhaus and colleagues calculated that there are 13 possible caps for this tube.

The theory of nanotube capping has also been discussed by Patrick Fowler, David Manolopoulos and their colleagues (5.30). These researchers used a method based on

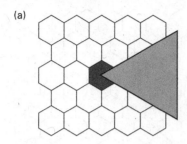

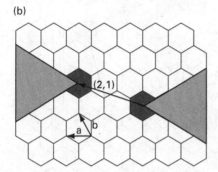

Fig. 5.11   (a) The creation of a pentagonal defect in a hexagonal lattice by removal of the shaded area.
(b) The vector (n_f, m_f) connecting two pentagonal defects which specify the icosahedral
fullerene C_{140} (5.28).

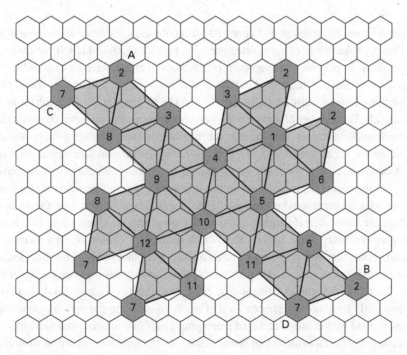

Fig. 5.12   A projection map for C_{140} (5.28).

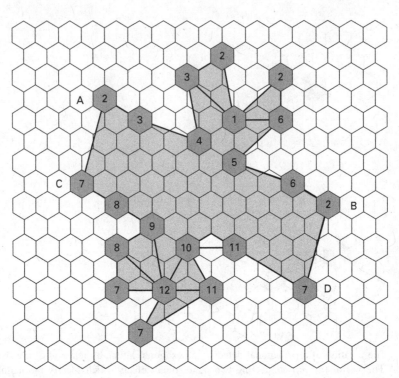

Fig. 5.13   A projection map illustrating two different ways of capping the chiral nanotube defined by the vector (7, 5) (5.28).

graph theory to represent and enumerate nanotube caps for tubes up to 3 nm in diameter. Their representations of the isolated-pentagon caps for (9, 0) and (10, 0) tubes are shown in Fig. 5.14. In general, their approach produced even larger numbers of possible caps for a given tube than did the method of Dresselhaus and colleagues. Thus, they found 39 possible isolated-pentagon caps for (9, 0), (10, 0) and (11, 0) tubes, while Dresselhaus *et al.* found only 21. The number of caps rapidly becomes huge as the diameter increases. Stephanie Reich and colleagues showed that the number of caps, including those with adjacent pentagons, varied with $d^{7.8}$ (5.31). For caps fulfilling the isolated pentagon rule the number of caps was smaller for small diameters, but the power-law behaviour was recovered for larger diameters. The explanation for this is that for large tube diameters the fraction of caps with adjacent pentagons becomes negligible. It should be noted that for a given tube diameter there are fewer caps for armchair and zigzag tubes than for chiral tubes. This is due to the higher symmetry of the achiral tubes, which reduces the choices of caps. Reich *et al.* have also shown that although a given nanotube can have thousands of distinct caps, quite the opposite is true for the inverse problem: a given cap only fits onto one particular nanotube.

Experimental studies of nanotube caps, described in Section 5.3.4, show that they frequently have conical shapes, so it is worth considering the possible cone angles that are formed by the introduction of pentagonal rings into a hexagonal network. A cone is formed by the introduction of fewer pentagons than the six needed to form a cylinder. It can be shown quite easily that the cone angle, α, is given by:

Table 5.1 Cone angles for graphitic cones wtih various numbers of pentagons

Number of pentagons	Cone angles in degrees
1	112.9
2	83.6
3	60.0
4	38.9
5	19.2
6	0.0

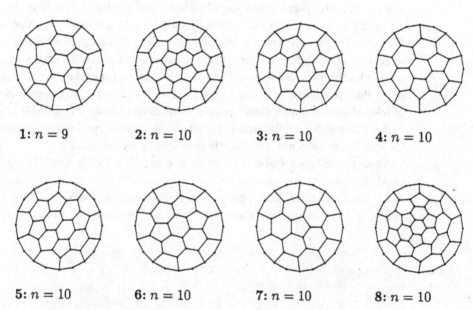

1: $n = 9$ 2: $n = 10$ 3: $n = 10$ 4: $n = 10$

5: $n = 10$ 6: $n = 10$ 7: $n = 10$ 8: $n = 10$

Fig. 5.14 Diagrams representing isolated-pentagon caps for $(9, 0)$ and $(10, 0)$ tubes, from the work of Fowler and Manolopoulos and colleagues (5.30).

$$\sin (\alpha/2) = 1 - (n_p/6)$$

where n_p is the number of pentagons in the cone. Values for the opening angles of cones containing 1–5 pentagonal rings are given in Table 5.1.

5.3 Experimental studies: multiwalled nanotubes produced by arc-evaporation

5.3.1 The layer structure: experimental observations

As discussed in Section 5.2.5, there are two possible arrangements for the layer structure of multiwalled nanotubes: the 'Swiss-roll', structure or the 'Russian doll' configuration. The most direct way to determine the multilayer structure would be to prepare cross-sections of MWNTs, by setting the nanotubes in a resin and then sectioning and imaging using HRTEM, but this has proved extremely difficult to achieve. Therefore, we have to infer the structure from more indirect measurements. A consideration of nanotube reactivity would seem to point to a 'closed', Russian doll structure. As discussed in Chapter 10, a great deal of work has been carried out on the reaction of nanotubes with gas- or liquid-phase oxidants, with the aim of opening the tubes. These studies invariably show that the tubes are preferentially attacked in the cap region, with the main body of the tubes being left intact. This seems to be inconsistent with a scroll model, which would have a reactive surface all along the length of the tube, due to the terminating graphene layers.

Turning now to high-resolution TEM images of multiwalled nanotubes, these generally show evenly spaced lattice fringes with an equal number of fringes on either side of the central core. Examples taken from Iijima's 1991 *Nature* paper (5.32) are shown in Fig. 5.15. The fact that micrographs of multiwalled nanotubes almost always show the same number of fringes on either side of the central cavity could be taken as providing support for the Russian doll model, although it certainly cannot be taken as unequivocal proof. Perhaps stronger evidence for a Russian doll structure is the presence in the tubes of internal caps or closed compartments, as discussed below. It is difficult to reconcile such features with a scroll structure. For all of these reasons, most researchers now favour the Russian doll structure for MWNTs produced by arc-evaporation.

Although the lattice spacings on either side of the cavities in MWNTs are generally evenly spaced, this is not always the case. It is quite often found that the fringes on one or both sides of the cavity contain 'gaps', or anomalously large spacings, as shown in Fig. 5.16. The possible reasons for this are discussed in subsequent sections.

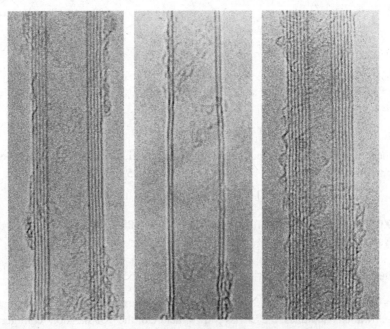

Fig. 5.15   Some of Iijima's first images of arc-grown multiwalled nanotubes (5.32).

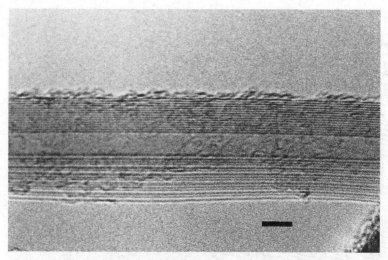

Fig. 5.16 A micrograph of a multiwalled nanotube showing uneven spacings in the layer structure on either side of the central core. Scale bar 5 nm.

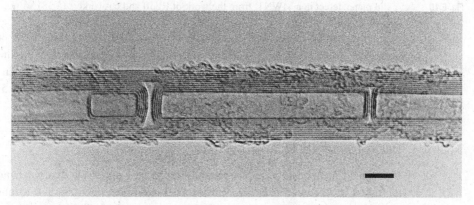

Fig. 5.17 A micrograph showing internal compartments in a multiwalled nanotube. Scale bar 5 nm.

Most X-ray diffraction measurements on nanotube samples give interlayer {002} spacings of approximately 0.344 nm (e.g. 5.24, 5.33, 5.34), which is very similar to the value found in turbostratic graphites, although figures ranging from 0.342 (5.27) to 0.375 nm (5.35) have been obtained. For double-walled nanotubes, interlayer distances can range from 0.34 to 0.41 nm (5.36, 5.37).

A frequent observation in multilayered tubes is the presence of one or more layers traversing the central core. More complicated internal structures, sometimes involving the formation of closed compartments, are also quite commonly seen. An example is shown in Fig. 5.17. Such features can represent a barrier to the filling of nanotubes with foreign materials, as discussed in Chapter 10.

Edge dislocation-type defects are quite commonly seen. An example, taken from the work of Rodney Ruoff's group (5.38), is shown in Fig. 5.18. A pair of terminating lattice fringes can be seen on the inner and outer sides of the tube. Ruoff *et al.* suggested that

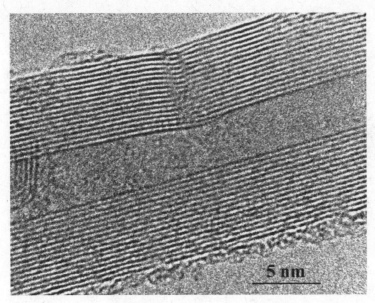

Fig. 5.18   A slip plane defect in a MWNT, from the work of Ruoff and co-workers (5.38).

these edge defects represent a changeover from scroll-like to a Russian doll configuration. Malcolm Heggie and colleagues have analysed in detail the way in which screw dislocation type defects can separate scroll from Russian doll structures within the same nanotube (5.25). However, it is possible that defects such as that shown in Fig. 5.18 could simply be discontinuities in a nested structure.

5.3.2   Electron diffraction of MWNTs

Electron diffraction has been quite usefully applied to single-walled carbon nanotubes, as discussed in Section 5.5.2, but has been of less value when applied to multiwalled nanotubes. This is because electron diffraction patterns of MWNTs contain spots from tubes with different chiralities, and are therefore difficult to interpret. This could be taken as further evidence for a Russian doll structure, since a 'scroll' nanotube would have the same helicity throughout.

Iijima included electron diffraction patterns of individual MWNTs in his original *Nature* paper (5.32), but there have been relatively few such studies since that time. Perhaps the most detailed work in this area was carried out by Severin Amelinckx and co-workers in 1993–94 (5.26, 5.39, 5.40), who found evidence for zigzag, armchair and chiral tubes within the same multiwalled nanotube. References on the theory of electron diffraction by nanotubes are given in Section 5.5.2.

5.3.3   The cross-sectional shape of multiwalled nanotubes

Direct observations of the cross-sectional shape of nanotubes in the electron microscope have proved very difficult to achieve. Therefore, inferences about the cross-sectional shape of multiwalled tubes have to be made from images recorded perpendicular to the tube axis. It should be noted that cap structure can influence the likely cross-sectional shapes. For example, Ebbesen has argued that a cap with five-fold symmetry will impose a faceted shape on the tube (5.41). However, this effect is likely to be slight, and will diminish as one moves further away from the cap region. A stronger effect on nanotube shape is probable with 'asymmetric cone' caps, of the kind discussed in the next section. Caps of this type will result in an 'egg shaped' cross-section, although once again this effect may diminish in regions well removed from the caps.

It was noted above that high-resolution electron micrographs of multiwall nanotubes often show unevenly spaced lattice fringes on one or both sides of the core, as in Fig. 5.16. Amelinckx and colleagues suggested that this may be due to scroll-like elements in the multilayer structure, but an alternative explanation has been suggested by Mingqi Liu and John Cowley of Arizona State University (5.42). These researchers carried out a detailed analysis of high-resolution images of multiwall nanotubes, and concluded that in many cases the tubes had polygonal cross-sections made up of flat regions joined by regions of high curvature. In images of the regions where the two planar sheets join, Liu and Cowley argue that fringes with spacings greater than 0.34 nm would be observed. This situation is illustrated schematically in Fig. 5.19. Note that their model assumes a relatively 'perfect' multilayer structure, and the observed gaps are a consequence of the joining together of idealized flat regions with the 0.34 nm spacing. The maximum interlayer spacing predicted by this model is 0.41 nm. In reality spacings considerably larger than this are observed (in Fig. 5.16 spacings greater than 0.6 nm are present). In such cases the multilayer structure is probably rather more imperfect than envisaged by Liu and Cowley.

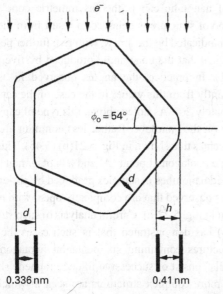

e^-

$\phi_0 = 54°$

d

d h

0.336 nm   0.41 nm

Fig. 5.19   A schematic drawing of a polygonalized nanotube, as envisaged by Liu and Cowley (5.42).

Another factor that must be taken into account when considering the cross-sectional shapes of nanotubes is the presence of distorting forces resulting from contacts between adjacent tubes. This effect was discussed by Rodney Ruoff and colleagues (5.43). These workers described high-resolution TEM observations of nanotubes in contact along one edge. In micrographs of adjacent nanotubes, {002} fringes were found to be more intense along the inner region, where the two tubes make contact, than at the outer edges, indicating a flattening of the tubes along the contact region. Ruoff *et al.* carried out calculations for the interaction of a pair of double-layer tubes using a Lennard-Jones model for the van der Waals interaction. These showed considerable flattening in the contact area, in agreement with the experimental observations. Ruoff and colleagues also found that the {002} interlayer spacings on the sides of the tubes adjacent to the contact region were reduced by about 0.008 nm compared with those on the outer sides.

5.3.4   MWNT cap structure

Theoretical work on the capping of nanotubes was discussed in Section 5.2.6, where it was noted that tubes with diameters larger than about 1 nm can be capped in a large number of different ways. Experimental studies show that multilayer nanotube caps do indeed have a wide range of different structures. In the great majority of cases the cap structures are unsymmetrical, but caps with higher symmetry are sometimes seen. Two beautiful micrographs, by Iijima (5.44), of symmetrical tube caps are shown in Fig. 5.20, together with diagrams indicating the approximate positions of the pentagonal rings in each case. It is notable that the degree of faceting increases as one moves from the inner graphene layers to the outer layers. This is in general agreement with predictions about the shapes of higher fullerenes, namely that they become less spherical as the size increases. Iijima estimates that the largest cap of the left-hand tube corresponds to a one-half of the icosahedral fullerene C_{6000}.

A commonly observed type of nanotube cap is the 'asymmetric cone' structure, illustrated in Fig. 5.21(a). This type of structure is believed to result from the presence of a single pentagon at the position indicated by the arrow, with five further pentagons at the apex of the cone. Theory predicts that the cone angle produced by five pentagons should be 19.19° (see Table 5.1). In practice, the angles observed in asymmetric cone caps can differ quite significantly from this value. In the case of the cap shown in Fig. 5.21(a), the angle is approximately 26°. As noted above, this type of cap imposes a non-circular cross-sectional shape on the nanotube. Rather less common are caps displaying a 'bill-like' morphology such as that shown in Fig. 5.21(b) (5.45). This structure results from the presence of a single pentagon at point 'A' and a heptagon at point 'B'.

Although virtually all multiwalled nanotubes in samples produced by arc-evaporation are closed, examples are sometimes observed that are completely open, with no obvious cap structure. An example is shown in Fig. 5.22(a). Careful analysis of such structures by Iijima and colleagues (5.44, 5.46) has demonstrated that in such cases the tubes are terminated with semitoroidal structures containing six pentagon–heptagon pairs, as illustrated in Fig. 5.22(b). The walls consist of successive folded graphene sheets, with no dangling edges. Occasionally, more complex structures are seen in which an inner tube extends beyond a semitoroidal tube termination (5.46).

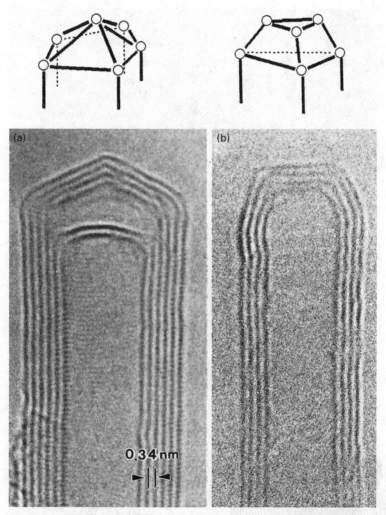

Fig. 5.20   Micrographs showing symmetrical nanotube caps, with drawings indicating the location of the pentagons (5.44).

5.3.5   Elbow connections and branching structures

As noted above, nanotubes containing abrupt elbow-like bends are sometimes observed in samples prepared using the conventional arc-evaporation method. An example is shown in Fig. 5.23. In most cases the tubes on either side of the elbow are of different diameters, and the joints seem to be invariably associated with 'internal caps'. It has been suggested that this type of structure results from the presence of a pentagonal ring on the inner side of the elbow and a heptagon on the inner side, as discussed in Section 5.2.4. However, in experimental images, there are often discontinuities in the layer structure on the inner side of the elbow joint, as can be seen in Fig. 5.23. Therefore, the structure of the connections may be less perfect than envisaged by theoreticians. Measurements of the

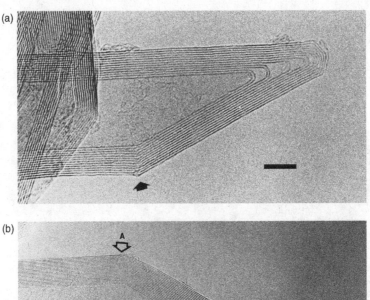

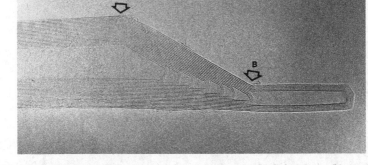

Fig. 5.21   (a) A nanotube cap with an asymmetric cone structure. Scale bar 5 nm. (b) A nanotube cap with a bill-like structure (5.45).

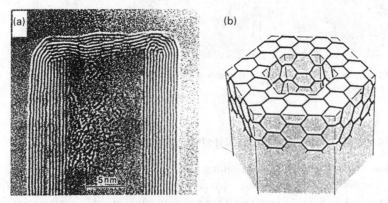

Fig. 5.22   (a) A micrograph of a semitoroidal tube termination, (b) a schematic drawing of the structure (5.44).

exact angle on electron micrographs is difficult because it is almost impossible to know whether the tube axis is precisely perpendicular to the electron-beam direction. However, angles differing quite considerably from the 150° predicted theoretically have been measured, even where the joint is believed to be approximately perpendicular to the beam.

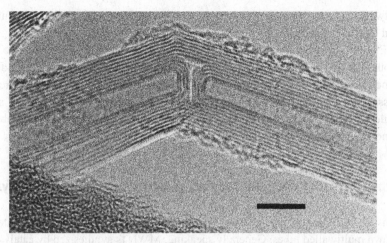

Fig. 5.23   An elbow connection joining two multiwalled nanotubes. Scale bar 5 nm.

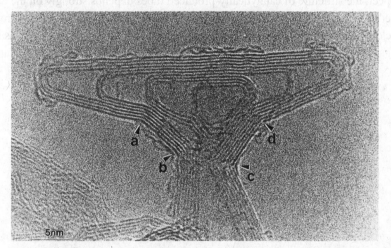

Fig. 5.24   An example of a branching nanotube structure, produced by the arc-evaporation method with modified electrodes (5.47).

More complex, branching nanotube structures, which also apparently contained negative curvature, were observed in 1995 by Dan Zhou and Supapan Seraphin of the University of Arizona (5.47). These structures were produced under the usual arc-evaporation conditions, but with a graphite anode which had been drilled out to leave a hollow core approximately 0.32 cm in diameter. Three types of branched structure were described, with 'L', 'Y' and 'T' configurations. An example of the T type is shown in Fig. 5.24 with the negatively curved points labelled a, b, c and d. As with the elbow connections, the angles made by these junctions varied quite considerably. The reason for the formation of such structures is not clear at present, but Zhou and Chow have argued that they provide evidence for a 'crystallization' model of nanotube growth (5.48) (see Section 2.2.5). Jason Qiu's group has produced more extensive branched MWNT networks by carrying out arc-evaporation using an anode containing a mixture of coal and CuO powder (5.49).

Elbow connections and branching structures of the kind described here could be thought of as constituting the first steps towards building 'molecular scaffolding' from nanotubes. Elbow junctions might also have interesting electronic properties, as discussed in Chapter 6 (Section 6.4.1), and 3-point nanotube junctions are of interest as building blocks of nanoelectronics (5.50). Branching structures are also observed in nanotubes grown by catalysis as discussed in the next section.

5.4   Experimental studies: multiwalled nanotubes produced by catalysis

This section gives a brief overview of some of the structural features of catalytically-grown multiwalled tubes. Generally speaking, MWNTs synthesized by catalysis tend to be less perfect in structure than nanotubes produced by arc-evaporation. As one would expect, the structure of catalytically-produced tubes depends strongly on the conditions used, and in particular on the temperature. When prepared at relatively low temperatures (below about 600 °C) the tubes tend to be irregular in form and imperfect, as shown in Fig. 5.25. In such tubes the degree of graphitization is low. At higher temperatures,

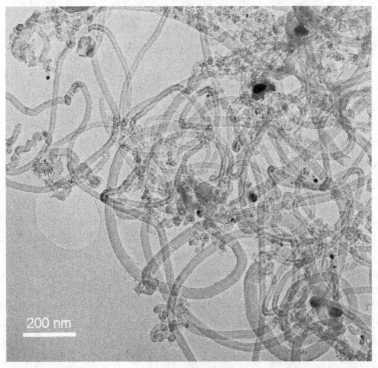

Fig. 5.25   Multiwalled nanotubes produced from ethyne using a Ni(II)-exchanged zeolite catalyst at 288 °C (5.51).

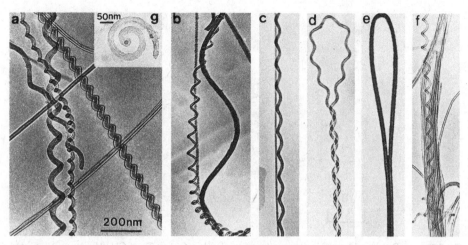

Fig. 5.26   Examples of helical nanotubes grown by the cobalt-catalysed decomposition of ethyne (5.52).

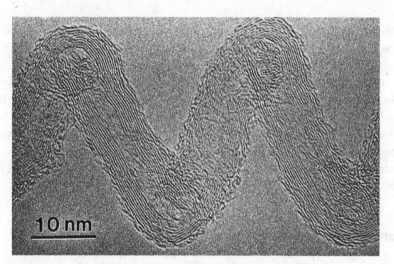

Fig. 5.27   A high-resolution micrograph of a helical nanotube (5.52).

relatively straight and perfect nanotubes can be produced as in Fig. 3.3. Helically coiled tubes are often observed in samples produced catalytically, as mentioned in Chapter 3. Examples taken from the work of Severin Amelinckx and his colleagues at the University of Antwerp are shown in Figs. 5.26 and 5.27 (5.52). Amelinckx *et al.* have discussed the growth mechanism of these structures in detail (see p. 54).

Another commonly observed feature of catalytically-grown nanotubes is a bamboo-like structure, with many 'internal caps', as shown in Fig. 5.28. An early description of such structures was given in 1981 by Marc Audier and colleagues who observed them in tubes produced by the disproportionation of carbon monoxide at 400 °C on an Fe–Co alloy catalyst (5.54). Experimental results show that the bamboo tubes nucleate at higher carbon feedstock pressures than those required for SWNT and MWNT nucleation (5.55),

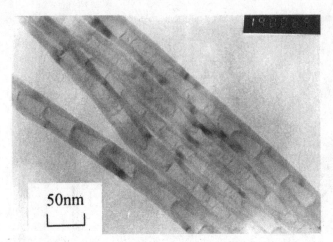

50nm

Fig. 5.28   Bamboo-like nanotubes produced by the decomposition of CH_4/H_2 over a Ni–Cu/Al$_2$O$_3$ catalyst at 770 °C (5.53).

although the reasons for this are not yet fully understood. Bamboo-like structures can also be produced when arc-discharge is carried out with doped electrodes (5.56). The growth mechanism of these structures was discussed in Section 3.3.

Branching structures are sometimes observed in nanotubes grown by catalysis. Tree-like formations were described by Baker in early work (5.57). In 2005 Patrick Bernier of the University of Montpellier and his colleagues produced branched MWNTs in higher yield using a method in which an aqueous catalyst-precursor solution was sprayed into a furnace (5.58). They proposed a mechanism in which the branching resulted from a restructuring of the catalytic particles into lobed forms. Branched nanotubes have also been produced by using Y-shaped nanochannel alumina as a template (5.59).

5.5   Experimental studies: single-walled nanotubes

5.5.1   General features

Samples of single-wall nanotubes tend to be far more homogenous than samples of multiwall nanotubes, with a much narrower range of diameters, and contain fewer obvious defects. High-resolution electron micrographs of single-wall nanotubes generally show 'featureless' narrow tubes, and are therefore relatively uninformative compared with images of multiwalled nanotubes. Some of the general features of single-wall nanotubes have been discussed in previous chapters. The tubes often form bundles, and images of these bundles viewed end-on show close-packed arrays of tubes, as shown in Fig. 3.14. Both the tube bundles and individual tubes are frequently curled and looped. However, regular helical structures such as those observed in catalytically-produced multilayer tubes are not observed in single-walled nanotubes. The caps of SWNTs, like those of multilayer tubes, can have various shapes although most of them appear to be simple domes. Asymmetric cone caps are quite common; an example can be seen at the

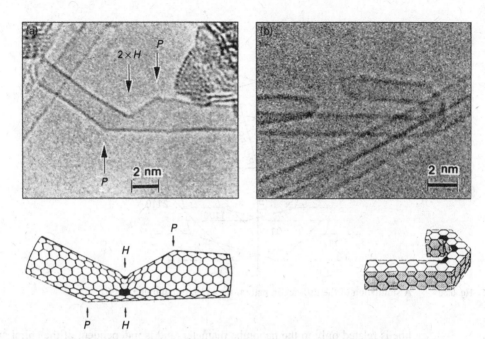

Fig. 5.29 Single-walled nanotubes displaying changes in growth direction, with drawings showing the positions of heptagons and pentagons, from the work of Iijima (5.60).

bottom left-hand side of Fig. 2.10. Elbow connections appear to be rare in single-walled tubes, but they are sometimes observed, and SWNTs containing such junctions have been used as nanoscale diodes (see Chapter 6, p. 164). Iijima has also observed some striking structures involving changes in the growth direction of SWNTs, as shown in Fig. 5.29 (5.60). The first of these, Fig. 5.29(a), can be understood in terms of pentagonal and heptagonal defects in the positions indicated, while the second 'candy-cane' structure may result from pentagon–heptagon pairs such as those which are present in the semitoroidal caps discussed above.

5.5.2 Electron diffraction of SWNTs

Thorough theoretical discussions of electron diffraction from single-walled nanotubes have been given by a number of authors. This theory will not be considered in detail here, but interested readers should consult the papers by Lu-Chang Qin (5.61–5.63) and Philippe Lambin and Amand Lucas (5.64, 5.65). In general, the diffraction pattern of a single-walled nanotube will have the form shown in Fig. 5.30. This diagram, adapted from the work of Jannik Meyer of the Max Planck Institute for Solid State Research, Stuttgart, and colleagues (5.66), shows the simulated diffraction pattern for a (15, 6) nanotube. Two separate sets of peaks are present, corresponding to the upper and lower graphene layers of the nanotube. The peaks appear as streaks due to the curvature of the graphene sheet. For armchair and zigzag tubes the two sets of peaks coincide. Along the centre of the pattern runs the equatorial line. The periodicity of the intensities on this

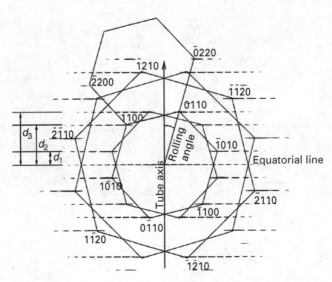

Fig. 5.30   A simulation of the diffraction pattern for a (15, 6) nanotube (5.66).

line is related only to the nanotube diameter, and is independent of the chiral angle Θ. It might be thought that the angle between the two sets of peaks, or 'rolling angle', would be equal to the chiral angle, but this is not necessarily the case. The difference between the two angles is a result of the cylindrical curvature of the diffracting nanotube. A further discrepancy also arises if the electron beam is not exactly normal to the tube. Qin, Ichihashi and Iijima have calculated the correction required to determine the true helicity from the measured rolling angle (5.67). Meyer *et al.* employed an alternative approach, in which the chiral angle is determined from the relative distances of the peaks to the equatorial line, indicated by d_1, d_2 and d_3 in Fig. 5.30. The advantage of this method is that these distances are independent of the angle of incidence of the electron beam.

Experimentally, obtaining electron diffraction patterns of single-walled nanotubes presents a considerable challenge. Quite apart from the small number of electrons that are diffracted by an individual tube, there are problems with beam damage and specimen drift. The fact that tubes may not be straight, or exactly perpendicular to the electron beam, can also complicate the interpretation of diffraction data. Considering these problems, it is remarkable that Iijima and Ichihashi were able to include a diffraction pattern of an individual single-walled tube in the 1993 *Nature* paper announcing their synthesis of SWNTs (5.68). This was interpreted in terms of a chiral tube structure. Since that early work, there have been a number of studies in which electron diffraction has been used to determine SWNT structure. In 1997, Iijima, Ichihashi and Qin (5.67) reported electron diffraction patterns of two individual tubes, and determined the indices to be (12, 1) and (31, 13). Lambin and Lucas and colleagues compared electron diffraction and scanning tunnelling microscopy as methods of determining SWNT structure in 2000 (5.69). More recent experimental studies of SWNT structure using electron diffraction have been described by Qin's group (e.g. 5.70), and by Meyer and

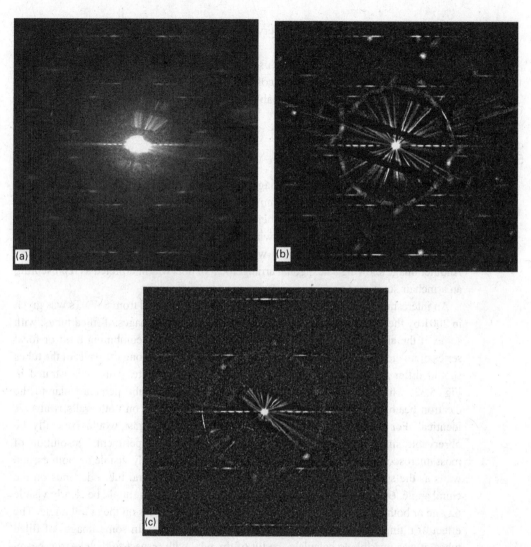

Fig. 5.31   Electron diffraction patterns of individual SWNTs recorded by Meyer and colleagues. (a) (24, 11) nanotube, (b) (16, 09) nanotube, (c) (13, 13) 'armchair' nanotube (5.66).

colleagues (5.66). The Meyer group used a novel sample preparation procedure, which involved the *in situ* growth of SWNTs on an Si grid. Nickel nanoparticles (diameter 4–5 nm) were used as the catalyst, and methane as the carbon feedstock; the synthesis temperature was 900 °C. In this way, well-separated, long and straight individual tubes ideally suited for diffraction could be grown between the grid bars. Examples of the excellent diffraction patterns they were able to record using this method are shown in Fig. 5.31. In all, diffraction patterns were recorded from 28 nanotubes grown by the CVD process. The chiral angles of these tubes were not randomly distributed. Six of the tubes had the armchair structure, while no zigzag tubes were found. Of the remaining tubes, most had chiral angles close to the armchair value of 30°. A bias towards the armchair structure has also been observed in other studies using fluorescence spectroscopy (see Chapter 7, p. 190).

Electron diffraction has also been used to study bundles of single-walled tubes. As noted in Section 3.4.5, Colomer and colleagues have recorded diffraction patterns from bundles of SWNTs produced by laser ablation and CVD, and found evidence of a narrow range of chiralities (5.71).

5.5.3   HRTEM of SWNTs

Achieving atomic resolution TEM images of single-walled nanotubes is just as difficult as recording electron diffraction patterns, if not more so. It appears that the first such images were obtained by Dmitri Golberg and colleagues from Tsukuba in 1999 (5.72). These workers published images showing fringes with a spacing of 0.21 nm running at right angles to the tube axis. These were interpreted in terms of a zigzag structure. Another image showing a hexagonal arrangement of dots was interpreted as representing an armchair structure.

An interesting analysis of the kind of contrast to be expected from SWNTs was given in 2001 by the Oxford group (5.73). In this work, enhanced images of filled tubes, with some of the aberrations removed, were obtained by digitally combining a tilt or focal series of images. They showed that the periodicities observed along the walls of the tubes should differ quite markedly depending on the tube structure. This is illustrated in Fig. 5.32, which shows the contrast to be expected for a tube perpendicular to the electron beam. In case of achiral tubes the observed contrast on both walls is always identical. For the armchair tube (Fig. 5.32a), no wall contrast would generally be observable since the spacing of 0.125 nm is beyond the experimental resolution of most microscopes. For the zigzag tube, fringes should be clearly visible on both carbon walls as the spacing is 0.216 nm. The contrast visible for chiral tubes depends on the chiral angle, Θ. If this is small (less than about 10°) then fringes should be clearly visible on one or both of the walls. The spacing of the fringes depends on the chiral angle. The effect of tilting on the observed contrast was also simulated. In some images of filled tubes, it was possible to calculate the tilt of the tube with respect to the electron beam, and then determine the structure of the tube from the contrast observed in the walls (5.73–5.75). The Oxford group's work on HRTEM imaging of filled SWNTs is covered in Section 10.5.

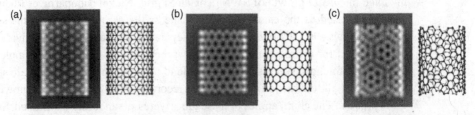

(a)   (b)   (c)

Fig. 5.32   Simulated HRTEM images of SWNTs. (a) armchair structure, (b) zigzag, (c) chiral. From the work of Friedrichs *et al.* (5.73).

Some beautiful high-resolution TEM studies of single-walled tubes have been carried out by Suenaga, Iijima and colleagues. Images from a paper they published in *Nature* in 2004 (5.76) are shown in Figs. 5.33(a) and (b). Atomic resolution contrast is clearly visible between the two dark lines corresponding to the vertical tube walls. This represents a superposition of the contrast from the upper and lower walls. In the Fourier transform (or optical diffraction pattern) of this image (Fig. 5.33(b) inset) two regular hexagons can be identified, representing the upper and lower tube walls. By measuring the angles that these hexagons make with the equatorial line, and accurately measuring the tube diameter, the chiral indices could be determined. In this case, the indices were found to be (13, 8). Defects were also deliberately introduced into tubes by electron bombardment, and the resulting structures imaged. Figure 5.33(d) shows a junction between two tubes with slightly different diameters that was produced in this way. By analysing the Fourier transform of this image, Iijima and colleagues were able to assign the indices (17, 0) to the upper tube and (18, 0) to the lower tube. Since the (17, 0) tube is semiconducting and the (18, 0) tube metallic, the junction constitutes a nanodiode.

Perhaps the most impressive HRTEM images of nanotube structure were published by the Suenaga–Iijima group in 2007 (5.22). In this work, topological defects were deliberately introduced into single-walled tubes by heating at 2000 °C and cooling rapidly. These were then imaged in an aberration-corrected TEM at 120 kV. Among the images they obtained was one of a '5-7-7-5' defect; this is reproduced in Fig. 5.34. Note that the bright spots here do not represent individual carbon atoms but rather the centre of carbon rings. An accumulation of topological defects was observed near the kink of a deformed nanotube, suggesting that dislocation motions or active topological defects are responsible for the plastic deformation of SWNTs.

5.5.4 Scanning tunnelling microscopy of SWNTs

If high-resolution TEM imaging of nanotube structure is hugely challenging, achieving atomic resolution scanning tunnelling microscope (STM) images is no easier. First of all there is the practical problem of anchoring the tubes sufficiently firmly to a flat support material. Highly oriented pyrolytic graphite (HOPG) has often been employed as a substrate, but here there is a risk of misinterpretation, since the atomic structure of HOPG is identical to that of nanotubes. For these reasons, there are advantages in alternative substrates such as gold. Even when the atomic resolution image of a nanotube has been recorded, extracting structural information is not always straightforward. For example, the measured pitch angle can be strongly affected by distortions due to a torsional twist (5.77). Distortions can also result from the way the STM image is formed. As Meunier and Lambin and their co-workers have pointed out (5.78–5.80), the tunnelling current flowing from the tip to the nanotube follows the shortest path, i.e. perpendicular to the nanotube surface, rather than vertically downwards. The effect of this is that the image exhibits a stretching of the lattice in the direction normal to the tube axis. In an undistorted image, the angle between the 'zigzag' and 'armchair' directions i.e. the angle between the line connecting the (1, 0), (2, 0), ... atoms and that connecting the

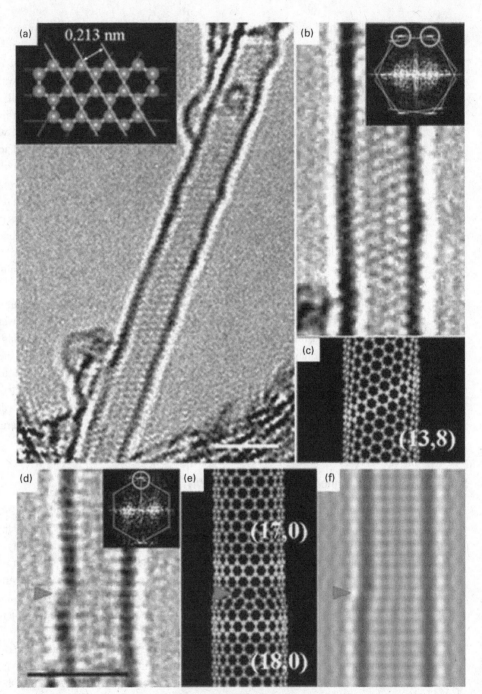

Fig. 5.33   HRTEM images of single-walled tubes, from the work of Suenaga, Iijima *et al.* (5.76). (a) Typical HRTEM image of a SWNT with enhanced contrast of the zigzag chain (inset). (b) Moiré pattern and its optical diffraction, (c) best-fit model of SWNT structure, (d) cross-sectional view of a topological defect in SWNT (e) pentagon–heptagon pair forming junction of (17, 0) and (18, 0) tubes, (f) simulated image for the SWNT with defect rotated by 90°.

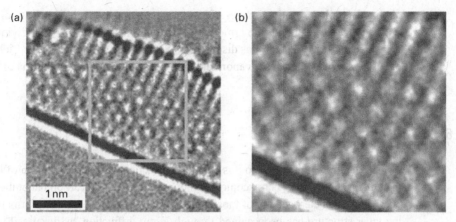

Fig. 5.34 HRTEM image of pentagon–heptagon pair defect in SWNT, following heating at 2000 °C (5.22).

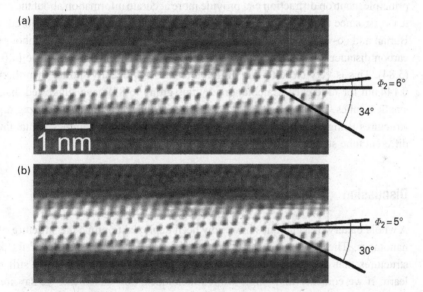

Fig. 5.35 Scanning tunnelling microscope image of SWNT, from the work of Venema *et al.* (5.80). (a) Uncorrected image with an angle between the armchair and zigzag directions of 34° instead of 30°, (b) the same image, corrected for the distortion.

(1, 1), (2, 2), ... atoms (see Fig. 5.3) should be 30°. In raw STM images it is often found that the angle differs from this value. This is illustrated in Fig. 5.35, taken from the work of Liesbeth Venema *et al.* (5.80). Figure 5.35(a) is an uncorrected image of a nanotube, showing a 'zigzag–armchair' angle of 34°, and a chiral angle of 6°. In the corrected image, with the zigzag–armchair angle adjusted to 30°, the chiral angle becomes 5°. In early STM studies of SWNTs, this correction was not always applied, possibly resulting in slightly erroneous (n, m) assignments.

Most experimental STM images of nanotubes have produced evidence for a random distribution of structures. In a classic study, Cees Dekker and co-workers (5.81) obtained atomic resolution images of 20 tubes, prepared by laser vaporization, of which only one was zigzag and one armchair. Electronic measurements on individual tubes were also made in this work, as discussed in Section 6.3.2. Studies by the Lieber group of SWNTs produced by laser vaporization also found a random distribution of helicities (5.82, 5.83).

5.6   Neutron diffraction

As noted in the introduction to this chapter, diffraction methods have been of less value than microscopic and spectroscopic techniques in the study of carbon nanotubes, mainly because of the heterogeneity of most nanotube samples. Some useful information about the layer structure has been gained through X-ray diffraction, as discussed in Section 5.3.1, but the (hkl) peaks in X-ray powder patterns provide little useful information. In principle, neutron diffraction can provide more accurate information about the hexagonal network, since it allows a wider range of scattering vectors, Q, to be explored. Andrzej Burian and co-workers used neutron scattering to determine nearest-neighbour carbon–carbon distances in single- and multi walled nanotubes, and found a value of 0.141 nm (5.84). This is very close to that of graphite and significantly shorter than the value of 0.144 nm for fullerenes. Calculations by this group (5.85) have suggested that, for the smallest tubes, it may be possible to distinguish between armchair, zigzag and chiral structures using neutron diffraction, although the data from samples containing many different tube structures would be difficult to interpret.

5.7   Discussion

A whole battery of techniques have now been deployed to probe the structure of carbon nanotubes. These techniques are slowly giving us a clearer picture of the kinds of structures that are present in real nanotube samples, although there is still much to learn. If we consider first single-walled nanotubes, how much can we say for certain about their structure? We know that chiral structures are far more common than achiral, but this is unsurprising, since the number of possible chiral forms greatly exceeds the number of zigzag and armchair structures. More interesting is the evidence for a bias towards near-armchair structures, at least for catalytically-produced nanotubes. Both electron diffraction (5.66) and spectrofluorimetric (Chapter 7 p. 190) studies have pointed towards this trend. The possible reasons for this are not well understood. It should be borne in mind, however, that some studies using Raman spectroscopy have found a random distribution of SWNT structures (Section 7.3), so the preference for near-armchair structures cannot be said to have been definitely confirmed.

How much can we say about the structural perfection of single-walled carbon nanotubes? In general, transmission electron microscope images suggest a high degree of perfection; defects such as those shown in Fig. 5.34 are rare. STM images that directly show the atomic structure of the SWNTs also indicate defect-free structures, although inevitably microscopic images only show short lengths of tube. However, as pointed out in Section 3.9, there are other reasons to believe that SWNTs typically exhibit a high degree of perfection. The evidence for this comes partly from mechanical measurements (see Section 7.1.3) and partly from the work of Collins *et al.* who demonstrated that SWNTs have a very low defect density, as already discussed (5.86).

Turning now to multiwalled nanotubes, it seems fair to say that these are generally more defective than their single-walled counterparts. This is true even of those produced by arc-evaporation, which remains the best method for making high-quality MWNTs. Transmission electron microscope images of these tubes quite often show features such as terminating lattice fringes and irregularly spaced gaps. Elbow connections are commonly observed, and the tubes often contain internal compartments. Catalytically-produced MWNTs tend to be still less perfect in structure, often being helically coiled or exhibiting bamboo-like structures, although the quality of catalytically-made MWNTs can be improved by increasing the synthesis temperature.

The differing degrees of structural perfection of single- and multiwalled nanotubes affect the kind of applications in which they can be used. Thus, the low defect density of SWNTs means that it is feasible to use them in nanoelectronic devices, where the presence of defects could have a serious effect on performance. Multiwalled tubes, with their more complex and relatively defective structures, would be of little use in such devices. On the other hand, the presence of defects in multiwalled tubes may be less damaging to their mechanical properties than to those of single-walled tubes. For a SWNT, a single defect would be a point of weakness, whereas for a MWNT, a defect in one layer could be compensated by a more perfect structure in other layers.

References

(5.1) R. O. Erickson, 'Tubular packing of spheres in biological fine structure', *Science*, **181**, 705 (1973).

(5.2) J. D. Bernal, 'The structure of graphite', *Proc. Roy. Soc. A*, **106**, 749 (1924).

(5.3) R. C. Haddon, 'The fullerenes: powerful carbon-based electron-acceptors', *Phil. Trans. Roy. Soc. A*, **343**, 53 (1993).

(5.4) K. S. Novoselov, A. K. Geim, S. V. Morozov *et al.*, 'Electric field effect in atomically thin carbon films', *Science*, **306**, 666 (2004).

(5.5) A. K. Geim and K. S. Novoselov, 'The rise of graphene', *Nat. Mater.*, **6**, 183 (2007).

(5.6) R. Saito, M. Fujita, G. Dresselhaus *et al.*, 'Electronic structure of chiral graphene tubules', *Appl. Phys. Lett.*, **60**, 2204 (1992).

(5.7) M. S. Dresselhaus, G. Dresselhaus and R. Saito, 'Physics of carbon nanotubes', *Carbon*, **33**, 883 (1995).

(5.8) M. S. Dresselhaus, G. Dresselhaus and P. C. Eklund, *Science of Fullerenes and Carbon Nanotubes*, Academic Press, San Diego, 1996.

(5.9) R. A. Jishi, M. S. Dresselhaus and G. Dresselhaus, 'Symmetry properties of chiral carbon nanotubes', *Phys. Rev. B*, **47**, 16 671 (1993).

(5.10) R. A. Jishi, D. Inomata, K. Nakao *et al.*, 'Electronic and lattice properties of carbon nanotubes', *J. Phys. Soc. Jpn.*, **63**, 2252 (1994).

(5.11) C. T. White, D. H. Robertson and J. W. Mintmire, 'Helical and rotational symmetries of nanoscale graphitic tubules', *Phys. Rev. B*, **47**, 5485 (1993).

(5.12) J. W. Mintmire and C. T. White, *Carbon Nanotubes: Preparation and Properties*, ed. T. W. Ebbesen, CRC Press, Boca Raton, 1997, p. 191.

(5.13) J. C. Charlier, 'Defects in carbon nanotubes', *Acc. Chem. Res.*, **35**, 1063 (2002).

(5.14) L. A. Chernozatonskii, 'Carbon nanotube elbow connections and tori', *Phys. Lett. A*, **170**, 37 (1992).

(5.15) B. I. Dunlap, 'Relating carbon tubules', *Phys. Rev. B*, **49**, 5643 (1994).

(5.16) A. Fonseca, K. Hernadi, J. B. Nagy *et al.*, 'Model structure of perfectly graphitizable coiled carbon nanotubes', *Carbon*, **33**, 1759 (1995).

(5.17) P. Lambin, A. Fonseca, J. P. Vigneron *et al.*, 'Structural and electronic properties of bent carbon nanotubes', *Chem. Phys. Lett.* **245**, 85 (1995).

(5.18) L. Chico, V. H. Crespi, L. X. Benedict *et al.*, 'Pure carbon nanoscale devices: nanotube heterojunctions', *Phys. Rev. Lett.*, **76**, 971 (1996).

(5.19) J. Liu, H. J. Dai, J. H. Hafner *et al.*, 'Fullerene "crop circles" ', *Nature*, **385**, 780 (1997).

(5.20) J. C. Charlier, T. W. Ebbesen and P. Lambin 'Structural and electronic properties of pentagon–heptagon pair defects in carbon nanotubes', *Phys. Rev. B*, **53**, 11 108 (1996).

(5.21) A. J. Stone and D. J. Wales, 'Theoretical studies of icosahedral C_{60} and some related species', *Chem. Phys. Lett.*, **128**, 501 (1986).

(5.22) K. Suenaga, H. Wakabayashi, M. Koshino *et al.*, 'Imaging active topological defects in carbon nanotubes', *Nat. Nanotech.*, **2**, 358 (2007).

(5.23) J. Kotakoski, A. V. Krasheninnikov and K. Nordlund, 'Energetics, structure, and long-range interaction of vacancy-type defects in carbon nanotubes: atomistic simulations', *Phys. Rev. B*, **74**, 245420 (2006).

(5.24) O. Zhou, R. M. Fleming, D. W. Murphy *et al.*, 'Defects in carbon nanostructures', *Science*, **263**, 1744 (1994).

(5.25) I. Suarez-Martinez, G. Savini, A. Zobelli *et al.*, 'Dislocations in carbon nanotube walls', *J. Nanosci. Nanotech.*, **7**, 3417 (2007).

(5.26) X. F. Zhang, X. B. Zhang, G. Van Tendeloo *et al.*, 'Carbon nano-tubes; their formation process and observation by electron microscopy', *J. Cryst. Growth*, **130**, 368 (1993).

(5.27) D. Reznik, C. H. Olk, D. A. Neumann *et al.*, 'X-ray powder diffraction from carbon nanotubes and nanoparticles', *Phys. Rev. B*, **52**, 116 (1995).

(5.28) M. Fujita, R. Saito, G. Dresselhaus *et al.*, 'Formation of general fullerenes by their projection on a honeycomb lattice', *Phys. Rev. B*, **45**, 13 834 (1992).

(5.29) M. S. Dresselhaus, G. Dresselhaus and P. C. Eklund, 'Fullerenes', *J. Mater. Res.*, **8**, 2054 (1993).

(5.30) G. Brinkman, P. W. Fowler, D. E. Manolopoulos *et al.*, 'A census of nanotube caps', *Chem. Phys. Lett.*, **315**, 335 (1999).

(5.31) S. Reich, L. Li and J. Robertson, 'Structure and formation energy of carbon nanotube caps', *Phys. Rev. B*, **72**, 165423 (2005).

(5.32) S. Iijima, 'Helical microtubules of graphitic carbon', *Nature*, **354**, 56 (1991).

(5.33) Y. Maniwa, R. Fujiwara, H. Kira *et al.*, 'Multiwalled carbon nanotubes grown in hydrogen atmosphere: An X-ray diffraction study', *Phys. Rev. B*, **64**, 073105 (2001).

(5.34) Y. Saito, T. Yoshikawa, S. Bandow *et al.*, 'Interlayer spacings in carbon nanotubes', *Phys. Rev. B*, **48**, 1907 (1993).

(5.35) M. Bretz, B. G. Demczyk and L. Zhang, 'Structural imaging of a thick-walled carbon microtubule', *J. Cryst. Growth*, **141**, 304 (1994).

(5.36) J. L. Hutchison, N. A. Kiselev, E. P. Krinichnaya *et al.*, 'Double-walled carbon nanotubes fabricated by a hydrogen arc discharge method', *Carbon*, **39**, 761 (2001).

(5.37) W. C. Ren, F. Li, J. Chen *et al.*, 'Morphology, diameter distribution and Raman scattering measurements of double-walled carbon nanotubes synthesized by catalytic decomposition of methane', *Chem. Phys. Lett.*, **359**, 196 (2002).

(5.38) J. G. Lavin, S. Subramoney, R. S. Ruoff *et al.*, 'Scrolls and nested tubes in multiwall carbon nanotubes', *Carbon*, **40**, 1123 (2002).

(5.39) X. F. Xhang, X. B. Xhang, D. Bernaerts *et al.*, 'The texture of catalytically grown coil-shaped carbon nanotubules', *Europhys. Lett.*, **27**, 141 (1994).

(5.40) X. B. Zhang, X. F. Zhang, S. Amelinckx *et al.* 'The reciprocal space of carbon tubes: a detailed interpretation of the electron diffraction effects', *Ultramicroscopy*, **54**, 237 (1994).

(5.41) T. W. Ebbesen, 'Carbon nanotubes', *Ann. Rev. Mater. Sci.*, **24**, 235 (1994).

(5.42) M. Liu and J. M. Cowley, 'Structures of carbon nanotubes studied by HRTEM and nanodiffraction', *Ultramicroscopy*, **53**, 333 (1994).

(5.43) R. S. Ruoff, J. Tersoff, D. C. Lorents *et al.*, 'Radial deformation of carbon nanotubes by van der Waals forces', *Nature*, **364**, 514 (1993).

(5.44) S. Iijima, 'Growth of carbon nanotubes', *Mater. Sci. Eng. B*, **19**, 172 (1993).

(5.45) S. Iijima, T. Ichihashi and Y. Ando, 'Pentagons, heptagons and negative curvature in graphite microtubule growth', *Nature*, **356**, 776 (1992).

(5.46) S. Iijima, P. M. Ajayan and T. Ichihashi, 'Growth model for carbon nanotubes', *Phys. Rev. Lett.*, **69**, 3100 (1992).

(5.47) D. Zhou and S. Seraphin, 'Complex branching phenomena in the growth of carbon nanotubes', *Chem. Phys. Lett.*, **238**, 286 (1995).

(5.48) D. Zhou and L. Chow, 'Complex structure of carbon nanotubes and their implications for formation mechanism', *J. Appl. Phys.*, **93**, 9972 (2003).

(5.49) Z. Y. Wang, Z. B. Zhao and J. S. Qiu, 'Synthesis of branched carbon nanotubes from coal', *Carbon*, **44**, 1321 (2006).

(5.50) M. Menon and D. Srivastava, 'Carbon nanotube "T junctions": nanoscale metal–semiconductor–metal contact devices', *Phys. Rev. Lett.*, **79**, 4453 (1997).

(5.51) D. J. Cardin, A. K. Lay and A. Gilbert, 'Low-temperature synthesis of ordered graphite nanofibres using nickel(II)-exchanged zeolites', *J. Mater. Chem.*, **15**, 403 (2005).

(5.52) D. Bernaerts, X. B. Zhang, X. F. Zhang *et al.*, 'Electron microscopy study of coiled carbon tubules', *Phil. Mag. A*, **71**, 605 (1995).

(5.53) J. L. Chen, Y. D. Li, Y. M. Ma *et al.*, 'Formation of bamboo-shaped carbon filaments and dependence of their morphology on catalyst composition and reaction conditions', *Carbon*, **39**, 1467 (2001).

(5.54) M. Audier, A. Oberlin, M. Oberlin *et al.*, 'Morphology and crystalline order in catalytic carbons', *Carbon*, **19**, 217 (1981).

(5.55) W. Z. Li, J. G. Wen, Y. Tu *et al.*, 'Effect of gas pressure on the growth and structure of carbon nanotubes by chemical vapor deposition', *Appl. Phys. A*, **73**, 259 (2001).

(5.56) Y. Saito and T. Yoshikawa, 'Bamboo-shaped carbon tube filled partially with nickel', *J. Cryst. Growth.* **134**, 154 (1993).

(5.57) R. T. K. Baker, 'Catalytic growth of carbon filaments', *Carbon*, **27**, 315 (1989).

(5.58) O. T. Heyning, P. Bernier and M. Glerup, 'A low cost method for the direct synthesis of highly Y-branched nanotubes', *Chem. Phys. Lett.*, **409**, 43 (2005).

(5.59) J. Li, C. Papadopoulos and J. Xu, 'Growing Y-junction carbon nanotubes', *Nature*, **402**, 253 (1999).

(5.60) S. Iijima, 'TEM characterisation of graphitic structures', *Proc. 13th Int. Congr. Electron Microscopy*, Paris, 1994, p. 295.

(5.61) L. C. Qin, 'Electron diffraction from cylindrical nanotubes', *J. Mater. Res.*, **9**, 2450 (1994).

(5.62) L. C. Qin, 'Electron diffraction from carbon nanotubes', *Rep. Prog. Phys.*, **69**, 2761 (2006).

(5.63) L. C. Qin, 'Determination of the chiral indices (n, m) of carbon nanotubes by electron diffraction', *Phys. Chem. Chem. Phys.*, **9**, 31 (2007).

(5.64) P. Lambin and A. A. Lucas, 'Quantitative theory of diffraction by carbon nanotubes', *Phys. Rev. B*, **56**, 3571 (1997).

(5.65) A. A. Lucas and P. Lambin, 'Diffraction by DNA, carbon nanotubes and other helical nanostructures', *Rep. Prog. Phys.*, **68**, 1181 (2005).

(5.66) J. C. Meyer, M. Paillet, G. S. Duesberg *et al.*, 'Electron diffraction analysis of individual single-walled carbon nanotubes', *Ultramicroscopy*, **106**, 176 (2006).

(5.67) L. C. Qin, T. Ichihashi and S. Iijima, 'On the measurement of helicity of carbon nanotubes', *Ultramicroscopy*, **67**, 181 (1997).

(5.68) S. Iijima and T. Ichihashi, 'Single-shell carbon nanotubes of 1-nm diameter', *Nature*, **363**, 603 (1993).

(5.69) P. Lambin, V. Meunier, L. Henrard *et al.*, 'Measuring the helicity of carbon nanotubes', *Carbon*, **38**, 1713 (2000).

(5.70) Z. J. Liu, Q. Zhang and L. C. Qin, 'Determination and mapping of diameter and helicity for single-walled carbon nanotubes using nanobeam electron diffraction', *Phys. Rev. B*, **71**, 245413 (2005).

(5.71) J. F. Colomer, L. Henrard, P. Lambin *et al.*, 'Electron diffraction and microscopy of single-wall carbon nanotube bundles produced by different methods', *Eur. Phys. J. B*, **27**, 111 (2002).

(5.72) D. Golberg, Y. Bando, L. Bourgeois *et al.*, 'Atomic resolution of single-walled carbon nanotubes using a field emission high-resolution transmission electron microscope', *Carbon*, **37**, 1858 (1999).

(5.73) S. Friedrichs, J. Sloan, M. L. H. Green *et al.*, 'Simultaneous determination of inclusion crystallography and nanotube conformation for a Sb_2O_3/SWNT single-walled nanotube composite', *Phys. Rev. B*, **64**, 045406 (2001).

(5.74) R. R. Meyer, S. Friedrichs, A. I. Kirkland *et al.*, 'A composite method for the determination of the chirality of single walled carbon nanotubes', *J. Microscopy*, **212**, 152 (2003).

(5.75) A. I. Kirkland and J. Sloan, 'Direct and indirect electron microscopy of encapsulated nanocrystals', *Topics Catalysis*, **21**, 139 (2002).

(5.76) A. Hashimoto, K. Suenaga, A. Gloter *et al.*, 'Direct evidence for atomic defects in graphene layers', *Nature*, **430**, 870 (2004).

(5.77) W. Clauss, D. J. Bergeron and A. T. Johnson, 'Atomic resolution STM imaging of a twisted single-wall carbon nanotube', *Phys. Rev. B*, **58**, R4266 (1998).

(5.78) V. Meunier and P. Lambin, 'Scanning tunnelling microscopy of carbon nanotubes', *Phil. Trans. Roy. Soc. A*, **362**, 2187 (2004).

(5.79) V. Meunier and P. Lambin, 'Tight-binding computation of the STM image of carbon nanotubes', *Phys. Rev. Lett.*, **81**, 5588 (1998).

(5.80) L. C. Venema, V. Meunier, P. Lambin *et al.*, 'Atomic structure of carbon nanotubes from scanning tunneling microscopy', *Phys. Rev. B*, **61**, 2991 (2000).

(5.81) J. W. G. Wildöer, L. C. Venema, A. G. Rinzler *et al.*, 'Electronic structure of atomically resolved carbon nanotubes', *Nature*, **391**, 59, (1998).

(5.82) T. W. Odom, J.-L. Huang, P. Kim *et al.*, 'Atomic structure and electronic properties of single-walled carbon nanotubes', *Nature*, **391**, 62, (1998).

(5.83) T. W. Odom, J.-L. Huang, P. Kim *et al.*, 'Structure and electronic properties of carbon nanotubes', *J. Phys. Chem. B*, **104**, 2794 (2000).

(5.84) A. Burian, J. Koloczek, J. C. Dore *et al.*, 'Radial distribution function analysis of spatial atomic correlations in carbon nanotubes', *Diam. Relat. Mater.*, **13**, 1261 (2004).

(5.85) J. Koloczek, Y. K. Kwon and A. Burian, 'Characterization of spatial correlations in carbon nanotubes-modelling studies', *J. Alloys Compounds*, **328**, 222 (2001).

(5.86) Y. W. Fan, B. R. Goldsmith and P. G. Collins, 'Identifying and counting point defects in carbon nanotubes', *Nat. Mater.*, **4**, 906 (2005).

6 Physical properties I: electronic

One of the most amazing characteristics of carbon nanotubes is that they can be metallic, like copper, or semiconducting, like silicon, depending on their structure. These properties were first predicted theoretically and then confirmed experimentally by the extraordinarily skilful application of techniques such as scanning tunnelling microscopy. By the late 1990s, nanotube-based devices such as diodes and field effect transistors were being constructed, and in 2001 the first logic gate based on a single nanotube bundle was reported. At the same time, many groups around the world have explored the field emission properties of nanotubes, for potential applications in display devices. Research into the electronic properties of carbon nanotubes represents a remarkably dynamic and fast-moving field, keeping abreast of which is no easy task. Nevertheless, some of the early work in this area, both theoretical and experimental, has stood the test of time, and many of the pioneering papers have become established classics.

This chapter begins with a brief summary of the electronic structure of graphite and then shows how this has been used as a basis for a theory of the electronic properties of carbon nanotubes. Experimental measurements on the electronic properties of nanotubes are then reviewed and the use of nanotubes in electronic devices described. The magnetic properties of nanotubes are summarized. Finally, the field emission properties of carbon nanotubes are discussed.

6.1 Electronic properties of graphite

As one would expect from its structure, the electronic properties of graphite are highly anisotropic. Electron mobility within the planes is high, as a result of overlap between the π orbitals on adjacent atoms, and the room temperature in-plane resistivity of high-quality single crystal graphite is approximately $0.4\,\mu\Omega\,\text{m}$. However, mobility perpendicular to the planes is relatively low. The first detailed band structure calculations for graphite, by P. R. Wallace in 1947 (6.1), were carried out for conduction solely in the planes, and ignored any interactions between planes. The following expression for the energy, E_{2D}, of an electron at a point defined by the wavevectors k_x, k_y was obtained:

$$E_{2D}(k_x k_y) = \pm\gamma_0 \left\{ 1 + 4\cos\left(\frac{\sqrt{3}\,k_x a}{2}\right) \cos\left(\frac{k_y a}{2}\right) + 4\cos^2\left(\frac{k_y a}{2}\right) \right\}^{1/2} \qquad [6.1]$$

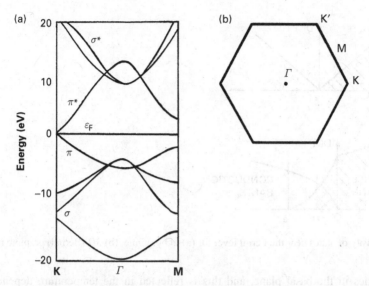

Fig. 6.1 (a) The dispersion relation for 2D graphite along the directions Γ–K and Γ–M in the Brillouin zone, (b) a sketch of the Brillouin zone for 2D graphite.

where γ_0 is the nearest-neighbour transfer integral and $a = 0.246$ nm is the in-plane lattice constant.

The unit cell for 2D graphite contains two atoms, so we have four valence bands, three σ and one π. The above expression produces bonding and antibonding π bands which just touch at the corners of the hexagonal 2D Brillouin zone, so that at zero Kelvin the bonding π band would be completely full and the antibonding π band completely empty. Figure 6.1(a) shows the E vs. k curves for 2D graphite along the direction Γ–K in the Brillouin zone; a sketch of the Brillouin zone for 2D graphite is given in Fig. 6.1(b). The density of states near the Fermi level for 2D graphite is shown in Fig. 6.2(a).

The band structure of three-dimensional graphite was calculated by Slonczewski, Weiss and McClure in the mid 1950s (6.2–6.4). The model shows that the π bands overlap by ~40 MeV, making graphite a semi-metal with free electrons and holes at all temperatures. This results in about $10^{-4}N$ electrons lying in the conduction band at 0 K, where N is the number of atoms, leaving the same number of holes in the valence band. A sketch of the density of states near the Fermi level for 3D graphite is shown in Fig. 6.2(b). A detailed discussion of the band structure of graphite is not necessary here, but excellent reviews have been given by a number of authors (6.2, 6.5–6.7).

The Slonczewski–Weiss–McClure (SWMcC) model allows us to calculate the electronic transport properties of graphite. However, such calculations are difficult, and a close agreement with experiment is not always obtained (6.6). The calculations show that graphite has a carrier density of the order of 10^{18} cm^{-3}, i.e. about one carrier per 10^4 atoms. Thus the conductivity will be very low compared with, say, copper which has one free carrier per atom. The low carrier density is partly offset by relatively high carrier

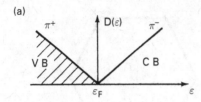

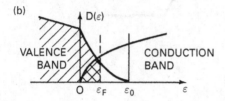

Fig. 6.2   The density of states near the Fermi level for (a) 2D graphite, (b) 3D (Bernal) graphite (6.2).

mobilities in the basal plane, and this is reflected in the temperature dependence of resistivity.

6.2   Electronic properties of nanotubes: theory

6.2.1   Band structure of single-walled tubes

In determining the band structure of graphite, it was assumed that the graphene planes are infinite in two directions, and artificial boundary conditions are introduced on a macroscopic scale in order to determine the band structure. For carbon nanotubes, we have a structure that is macroscopic along the fibre axis, but with a circumference of atomic dimensions. Therefore, while the number of allowed electron states in the axial direction will be large, the number of states in the circumferential direction will be very limited. The allowed states can be thought of as lying on a number of parallel lines within the 2D graphene Brillouin zone. The nanotube BZ is then constructed by 'compressing' these lines into a single line. This will now be discussed in more detail, drawing on the work of Mildred Dresselhaus and co-workers from MIT, and Noriaki Hamada and colleagues from Iijima's laboratory in Tsukuba (6.8–6.11) (although, as noted in Chapter 1, it is recognized that the first electronic structure calculations for carbon nanotubes were actually carried out by a group from the Naval Research Laboratory in Washington (6.12)).

We consider armchair tubes first. Using the well-known expression for a periodic boundary condition (6.13), allowed values for the wave-vector in the circumferential direction can then be written as,

$$k_x^\upsilon = \frac{\upsilon}{N_x}\frac{2\pi}{\sqrt{3}a}$$
[6.2]

for $\upsilon = 1, \ldots, N_x$

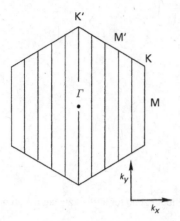

Fig. 6.3   An illustration of the allowed k-values in the Brillouin zone for a $(5, 5)$ armchair nanotube.

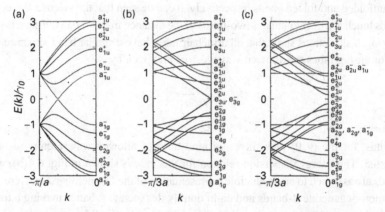

Fig. 6.4   The dispersion relations for (a) armchair $(5, 5)$ nanotube, (b) zigzag $(9, 0)$ nanotube, (c) zigzag $(10, 0)$ nanotube (6.10).

Taking the example of the 'archetypal' $(5, 5)$ armchair tube, υ has the values $1, \ldots, 5$. Thus, there are five allowed modes in the y direction in this case, so that the one-dimensional energy dispersion relations lie along five lines on either side of the centre of the Brillouin zone with a further line passing through the centre, as shown in Fig. 6.3. Now, it was noted above that the valence and conduction bands for graphite are degenerate at the K point. Therefore, nanotubes with a set of wave vectors which include the K point should be metallic. For armchair tubes, it is clear from Fig. 6.3 that the orientation of the Brillouin zone means that there will always be one set of allowed vectors passing through the K point, which leads to the conclusion that all armchair tubes are metallic.

The energy dispersion relation for a $(5, 5)$ armchair tube is obtained by substituting the allowed values of k_x^{υ} into equation [6.1], and is shown in Fig. 6.4(a). Each band can be assigned to an irreducible representation of the D_{5d} point group, and is labelled accordingly in the figure. The A bands are non-degenerate and the E bands are doubly

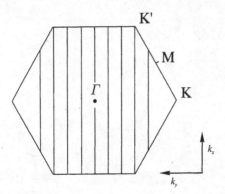

Fig. 6.5   An illustration of the allowed k-values in the Brillouin zone for a $(9, 0)$ zigzag nanotube.

degenerate, so the total number of valence bands in this case is 10; the $+$ and $-$ labels denote the unfolded and folded bands respectively. It can be seen that the valence and conduction bands touch at a position that is two-thirds of the distance from $k = 0$ to the zone boundary at $k = \pi/a_0$. Calculations show that all armchair tubes have a similar band structure.

For zigzag tubes the allowed wavevectors are given by

$$k_y^v = \frac{v}{N_y} \frac{2\pi}{a} \tag{6.3}$$

for $v = 1, \ldots, N_y$.

Thus, for the $(9, 0)$ tube there are nine lines of allowed wavevectors, as shown in Fig. 6.5. The energy dispersion relation for this case is shown in Fig. 6.4(b), where the bands are assigned to an irreducible representation of the D_{9d} point group. Here, there are two non-degenerate A-bands and eight doubly degenerate E bands making a total of 18. The valence and conduction bands touch at $k = 0$, so that in this case the tube is a metal. The reason for this is clear from Fig. 6.5, where it can be seen that one of the lines of allowed wavevectors for this tube passes through a K-point. This is not the case for all zigzag tubes, and only occurs when n is divisible by 3. Thus, for a $(10, 0)$ tube there is an energy gap between the valence and conduction bands at $k = 0$, as shown in Fig. 6.4(c), and the tube would be expected to be a semiconducting. The electronic density of states for the $(9, 0)$ and $(10, 0)$ zigzag tubes have been calculated by the Dresselhaus group (6.14), and are shown in Fig. 6.6. It can be seen that there is a finite density of states at the Fermi level for the metallic $(9, 0)$ tube, and a vanishing density of states for the semiconducting $(10, 0)$ tube.

Chiral nanotubes may also be either metallic or semiconducting, depending on chiral angle and tube diameter. Dresselhaus *et al.* (6.9, 6.10, 6.14) show that metallic conduction occurs when

$$n - m = 3q \tag{6.4}$$

where n and m are the integers that specify the tube's structure and q is an integer. Thus roughly one-third of chiral tubes are metallic and two-thirds are semiconducting.

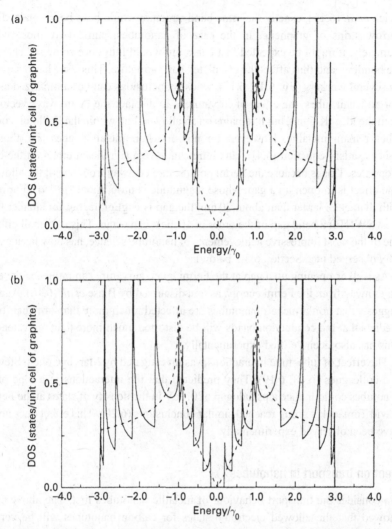

(a)

(b)

Fig. 6.6 The electronic 1D density of states per unit cell for two zigzag tubes (6.14). (a) The (9, 0) metallic tube, (b) the (10, 0) metallic tube. The dotted line shows the density of states for a 2D graphene sheet.

In summarizing the above discussion we can say that all armchair single-walled tubes are expected to be metallic, while approximately one-third of zigzag and chiral tubes should be metallic, with the remainder being semiconducting.

6.2.2 Effect of curvature and of tube–tube interactions

In the previous section the band structure of nanotubes was derived from that of graphene by limiting the number of allowed electron states in the circumferential direction. No allowance was made for the fact that nanotubes are curved. As pointed out by

Reich *et al.* (6.15), exactly the same band structure would have been obtained for long, narrow strips of graphene. In the case of nanotubes, particularly those with small diameters, it might be expected that curvature would introduce some sp^3 character into the bonding, and thus affect the calculated band structure. This issue has been addressed by several workers (e.g. 6.16, 6.17), whose conclusions can be summarized as follows. For armchair tubes, the effect of curvature is to displace the Fermi wave vector slightly from the ideal K point, but it remains on an allowed line within the Brillouin zone, so the tubes remain metallic. However, for non-armchair metallic tubes (i.e. those whose indices satisfy equation [6.4]), curvature can have a significant effect on the electronic properties. This is because the Fermi point is moved away from one of the allowed lines. The effect is to open up a gap, whose magnitude is proportional to $1/d^2$. For nanotubes with diameters larger than about 20 nm, the gap is negligible, but for smaller tubes (i.e. most SWNTs) a band gap of the order of 10 meV can occur. This is a small effect but, in one of the most impressive achievements of nanotube science, has now been experimentally observed (see Section 6.3.2 below).

As well as opening up a gap at the Fermi level, curvature can influence the electronic states away from the Fermi energy, as first discussed by Blase *et al.* (6.16). Again theory suggests that non-armchair nanotubes are affected much more than armchair tubes, and predicts that the conduction bands will be distorted much more than the valence bands. This has also been verified experimentally.

The effect of tube–tube interactions was investigated by Marvin Cohen, Steven Louie and colleagues (6.18, 6.19). They predicted that the interactions in a rope of $(10, 10)$ nanotubes could induce a pseudogap of 0.1 eV in the density of states at the Fermi level. David Tománek's group reached similar conclusions (6.20). This effect does not seem to have been observed experimentally.

6.2.3   Electron transport in nanotubes

We consider the transport behaviour of metallic nanotubes first. The above discussion showed that the allowed electronic states for carbon nanotubes will be very limited compared with those for bulk graphite. The consequence of this is that the transport behaviour of metallic nanotubes will be essentially that of a quantum wire, so that conduction occurs through well separated, discrete electron states. Thus, the resistance does not increase smoothly as the length of the wire is increased, but is the same independent of length, assuming no scattering. An important aspect of quantum wire behaviour is that transport along the tubes is ballistic in nature. Ballistic transport occurs when electrons pass along a conductor without experiencing any scattering from impurities or phonons; effectively, the electrons encounter no resistance, and dissipate no energy in the conductor (6.21). In other words, the material can conduct a large current without getting hot – a highly desirable characteristic for the construction of nanoscale circuits. It should be noted, however, that the conductance is not infinite as in a super-conductor; the magnitude of the conductance quantum, G_0, is given by $2e^2/h$. The theory of quantized conduction in nanotubes has been discussed by a number of groups (e.g. 6.22–6.25), who have shown that that conducting single-shell nanotubes have two conductance channels, so that the conductance of a SWNT should be $2G_0$. This equates to a resistance of ~ 6.5 kΩ, again assuming no scattering and perfect contacts. Quantum wire behaviour has now been observed in both multiwalled tubes and single-walled tubes, as discussed in Section 6.3.3.

Transport in semiconducting SWNTs is more complicated, and appears to be diffusive rather than ballistic (6.26). However, experiments have found evidence for extremely high mobilities in semiconducting SWNTs (6.27). The precise nature of the scattering processes in semiconducting tubes has not yet been fully established, and a detailed discussion would be beyond the scope of this book. The interested reader should see references 6.28–6.30.

6.2.4   Effect of a magnetic field

The effect of a magnetic field on the electronic properties of carbon nanotubes in a magnetic field was discussed in the mid 1990s by Hiroshi Ajiki and Tsuneya Ando of the University of Tokyo (6.31, 6.32), and this work will now be summarized.

Consider first a magnetic field applied parallel to the tube axis. Calculations, using **k.p** perturbation theory (6.31, 6.32), indicate that in this case the band gap would oscillate with increasing magnetic field, so that a metallic tube would become firstly semiconducting and then metallic again, with a period dependent on the magnetic field strength. This behaviour is a consequence of the Aharonov–Bohm effect, which is a phenomenon characteristic of quantum wires. In the case of nanotubes, the Aharonov–Bohm effect means that a magnetic field changes the boundary conditions that determine how the 2D graphene energy bands are cut. The variation of energy gap with magnetic flux for an initially metallic nanotube is shown in Fig. 6.7 (6.31). Here, the magnetic flux is given in units of φ_0, the flux quantum defined by $\varphi_0 = hc/e$, where h is Planck's constant, c is the velocity of light and e is the electronic charge. The magnitude of the magnetic field required to deliver a flux quantum deceases rapidly with increasing nanotube diameter. Thus, for a tube with diameter 0.7 nm the magnetic field required is 10 700 T, while for a 30 nm tube the field would be 5.85 T. Since fields larger than about 30 T cannot be easily achieved, full Aharonov–Bohm oscillations would probably only be observable in practice for relatively large-dia-meter tubes.

Oscillations of electron energy bandgap with increasing magnetic field are also predicted when the field is applied *perpendicular* to the tube axis. This case was also considered by Ajiki and Ando using the **k.p** method (6.31), and the tight-binding approximation (6.32). Figure 6.8 shows the band gap as a function of magnetic field for three zigzag nanotubes with different circumferences. Here the energy gap is expressed in units of $\gamma_0(a/L)^2$ where γ_0 is the nearest-neighbour transfer integral, a is the unit cell base vector and L is the nanotube circumference, and the magnetic field in units of $L/2\pi l$ where $l = \sqrt{c\hbar/eH}$ and H is the magnitude of the magnetic field. The three zigzag tubes chosen would all be metallic in the absence of a magnetic field, and it can be seen that the variation of band gap with magnetic field is identical in each case. The field required to produce the maximum bandgap for a (60, 0) tube (diameter = 4.7 nm) would

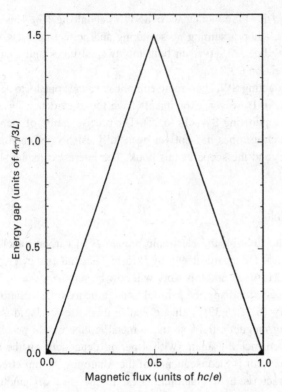

Fig. 6.7   The variation of the energy gap with magnetic flux parallel to the tube axis for an initially metallic nanotube (6.31).

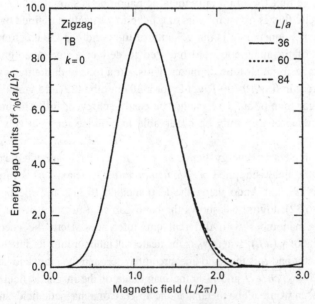

Fig. 6.8   The band gap as a function of a magnetic field perpendicular to the tube axis for three zigzag nanotubes with different diameters (6.32).

be 220 T. Experimental studies of the electronic properties of nanotubes in a magnetic field are discussed in Section 6.3.4.

It has only been possible here to give a fairly brief summary of the theory of carbon nanotubes' electronic properties. For more detailed discussions, a number of excellent reviews are available (e.g. 6.33–6.37).

6.3   Electronic properties of nanotubes: experimental measurements

6.3.1   Early studies of multiwalled nanotubes

The first attempts to make electrical measurements on individual nanotubes were carried out on MWNTs. In 1996, researchers from the Catholic University of Louvain, Belgium, described an experiment on a single multiwalled tube (prepared by arc-evaporation) supported on a an oxidized silicon wafer (6.38). Electrical resistance was determined as a function of temperature, down to $T = 30$ mK. Resistance was found to rise with falling temperature, indicating that the tube was semiconducting.

A short time later Thomas Ebbesen of NEC and colleagues described a series of resistance measurements on eight different nanotubes (6.39). Before carrying out electrical measurements, these workers annealed the nanotubes at 2850 °C, with the aim of removing defects. The tubes were then deposited onto an oxidized silicon surface between gold pads. A focused ion beam microscope was used to image the supported nanotubes, and when a suitable tube had been located, four 80 nm wide tungsten wires were deposited to produce an arrangement such as the one illustrated in Fig. 6.9. The

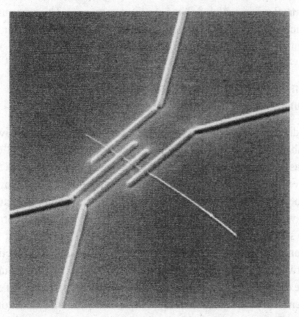

Fig. 6.9   A focused ion beam image of four tungsten wires connected to an individual nanotube, from the work of Ebbesen *et al.* (6.39). Each tungsten wire is 80 nm wide.

tungsten leads could then be connected to gold pads to enable four-probe resistance measurements to be carried out. The distance between contacts on the tubes was in the range 0.3–1.0 μm. In order to measure the temperature effects on resistance, the samples were mounted on a cryostat.

The electronic properties of eight different nanotubes differed widely. The highest measured resistance, for a tube with a diameter of 10 nm, was greater than 10^8 Ω, while the lowest resistance, for a 18.2 nm tube, was 2×10^2 Ω. In both cases the distance between leads onto the tubes was 1.0 μm. Ebbesen and colleagues estimate that these figures translate to resistivity values of 8 mΩ m and 0.051 μΩ m respectively. Although these values are only very approximate, they show that the room temperature resistivity of nanotubes can in some cases be comparable, or lower, than the in-plane resistivity of graphite, which is approximately 0.4 μΩ m.

The temperature dependence of resistivity of the tubes also differed widely. In addition to the variation from tube to tube, different segments of a single tube could sometimes have different temperature profiles. The commonest type of behaviour was a consistent slight increase in resistivity with decreasing temperature. Ebbesen and colleagues did not consider that these tubes should be thought of as semiconducting, particularly in view of their low resistivities. Instead they suggested that the tubes are essentially metallic, and the variations in their resistivities and temperature dependence are due to the interplay of changes in carrier concentration and mobilities. In other cases, however, tubes did display clear semiconducting behaviour. Experiments apparently demonstrating quantum transport in multiwalled tubes are discussed in Section 6.3.3.

Phaedon Avouris' group has investigated the limits of electronic transport in MWNTs (6.40). Electrical breakdown was found to occur in a series of sharp current steps, in contrast to metal wires, which fail in a continuous manner. This was attributed to the sequential destruction of individual nanotube shells. The failure was found to occur much more readily in air than in a vacuum.

6.3.2 Correlation between electronic properties and structure of single-walled nanotubes

The first study to directly correlate the electronic properties of SWNTs with their structure was described in *Nature* by Cees Dekker and co-workers in 1998 (6.41). In this work, which was also discussed in the previous chapter (Section 5.5.4), the tubes were probed by a combination of STM and scanning tunnelling spectroscopy (STS). Imaging of the nanotubes by STM showed that a variety of nanotube structures were present: most tubes were found to be chiral, with only a minority having armchair or zigzag structures. In scanning tunnelling spectroscopy, the tip is positioned at an appropriate point on the sample and the tunnelling current (I) is measured as a function of the bias voltage (V) between the tip and the sample. According to a simple but realistic model, the derivative of the I–V curve gives an approximation to the density of electronic states (DOS). Current–voltage curves obtained from eight individual nanotubes are shown in Fig. 6.10(a). Tubes 1–6 were chiral, tube 7 zigzag and tube 8 armchair. Most of the curves show a low conductance at low bias, followed by several kinks at larger bias

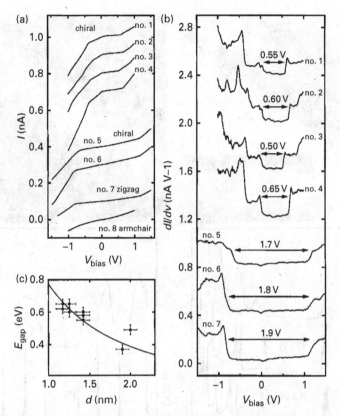

Fig. 6.10   Electronic measurements on single-walled nanotubes by Dekker and colleagues (6.41).
(a) Current–voltage curves for individual nanotubes. (b) Derivatives dI/dV. (c) Energy gap vs.
diameter for semiconducting chiral tubes.

voltages. The derivatives dI/dV for tubes 1–7, which give an indication of the densities of
states, are shown in Fig. 6.10(b). Two categories of tube can be distinguished here:
one with gap values around 0.5–0.6 eV, the other with larger gap values, in the range
1.7–1.9 eV. Dekker *et al.* identified the first category as semiconducting, the second as
metallic. Thus, the chiral tubes were found to be either semiconducting (tubes 1–4) or
metallic (tubes 5 and 6), while the zigzag tube (tube 7) was metallic. These observations
are consistent with theoretical predictions, although in the *Nature* paper the (n, m) indices
for the eight tubes were not precisely determined, so an exact comparison between
experimental measurements with theory was not possible. Such a comparison was
made in a subsequent paper (6.42), in which Dekker and colleagues determined the
(n, m) indices for a number of nanotubes (using the correction methods outlined in
Section 5.5.4), and confirmed that the electronic characteristics of the tubes were exactly
as predicted by theory.

Similar studies were made by Charles Lieber's group at about the same time as
the Dekker work (6.43). Some of their results are reproduced in Fig. 6.11. The image
in Fig. 6.11(a) shows a tube that was indexed as $(13, 7)$, while the upper trace in Fig. 6.11(c)

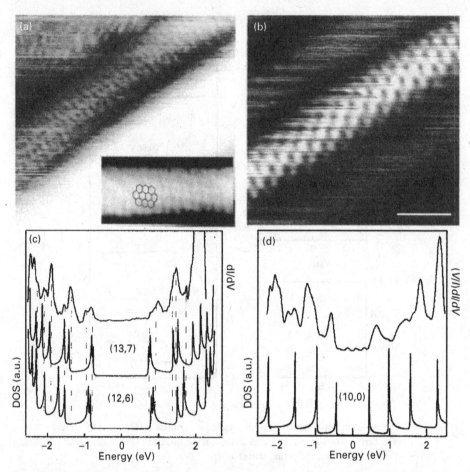

Fig. 6.11   STM images and STS spectra of SWNTs, from the work of Lieber *et al.* (6.44). (a) An atomic
resolution image of a (13, 7) tube, (b) an image of a (10, 0) tube, (c) a comparison of the DOS
obtained from experiment (upper curve) and tight-binding calculation for the (13, 7) SWNT (second
curve from top); the calculated DOS for a (12, 6) tube is included for comparison, (d) a comparison
of the DOS obtained from experiment (upper curve) and calculation for the (10, 0) SWNT.

shows the experimental DOS from this tube (i.e. the *dI/dV* plot). It can be seen that there
is a good agreement with the calculated DOS for the (13, 7) tube, especially below the
Fermi energy. The DOS for a (12, 6) tube, which is the next closest metallic SWNT, is
also shown, and the agreement here is much less good. Figure 6.11(b) shows a (10, 0)
tube (6.44), with its experimental DOS given in Fig. 6.11(d). Here there is reasonable
agreement with the calculated DOS below the Fermi energy, but poor agreement above.
The discrepancy between theory and experiment above E_F, is a consequence of tube
curvature, as discussed in Section 6.2.2. Taken together, the work of the Dekker and
Lieber groups provided strong support for the theories of nanotube electronic properties developed by Dresselhaus, Hamada and others, and represented a landmark in
nanotube science.

In later work, Lieber and colleagues confirmed a further theoretical prediction, namely that for some metallic tubes, curvature can have the effect of opening up a gap at the Fermi level, effectively destroying the metallic character (6.45). Figure 6.12 illustrates some of the gaps observed in metallic zigzag tubes. For isolated armchair nanotubes, no gap was seen, again confirming theory. However, for armchair tubes in bundles, a small suppression in the density of states, termed a pseudogap, was found. Lieber's group have also demonstrated that the DOS of single-walled nanotubes is drastically modified at localized structures, such as bends and caps (6.43, 6.44).

While STM and STS have been much the most powerful techniques for correlating the structure of SWNTs with their electronic characteristics, spectroscopic methods can also provide information about electronic properties. Raman spectroscopy has been the most useful technique in this area (6.46–6.50). The main features of Raman spectra of SWNTs are described in the next chapter. Both the G-band and the radial breathing feature can be used to distinguish between metallic and semiconducting nanotubes. In SWNTs the G-band is composed of two components, one peaked at $1590\,\mathrm{cm}^{-1}$ (G^+) and the other peaked at about $1570\,\mathrm{cm}^{-1}$ (G^-). The Dresselhaus group and others have shown that the lineshape of the G^- feature is highly sensitive to whether the SWNT is metallic or semiconducting (6.48, 6.49). The same group showed that the radial breathing mode can be used to obtain values of E_{ii}, the electronic transition energies, for individual tubes. This was achieved through a careful analysis of the laser energy dependence of the Stokes and anti-Stokes Raman spectra of nanotubes (6.50).

6.3.3 Quantum conductance

A direct demonstration of quantum transport in nanotubes was given in 1998 by Walt de Heer and colleagues (6.51), in experiments using bundles of multiwalled tubes. The set-up they employed is shown in Fig. 6.13(a). A nanotube bundle (at the extremity of which was a single tube) is dipped into a heatable reservoir containing mercury. When a circuit is established, the current is measured as a function of the position of the nanotube in the mercury. It was found that the conductance did not change smoothly with position, as would be expected for a classical conductor, but instead jumped to a constant value as soon as the nanotube entered the mercury and remained at this value as it was dipped further in. The conductance remained constant until the nanotube was withdrawn from the mercury, as shown in Fig. 6.13(b). The value of the conductance in these experiments was found to be approximately equal to the conductance quantum, G_0. Subsequent work by this group produced similar results (6.52, 6.53). It was noted in Section 6.2.3 that the conductance of a nanotube should be $2G_0$ rather than G_0. The reason for this discrepancy between theory and experiment has been explained in terms of interwall interactions (6.25).

A significant feature of the studies by de Heer *et al*. was that the nanotubes were found to be undamaged, even at relatively high voltages (6 V) for extended times. It was calculated that the power dissipation at these voltages would produce enormously high temperatures (up to $20\,000\,\mathrm{K}$) in the tubes if they were acting as classical resistors. The survival of the nanotubes therefore provided strong evidence for ballistic transport.

(a)

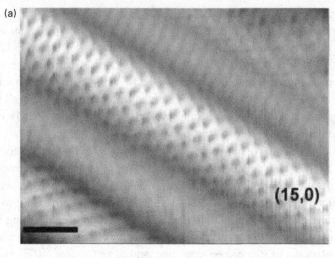

(15,0)

(b)

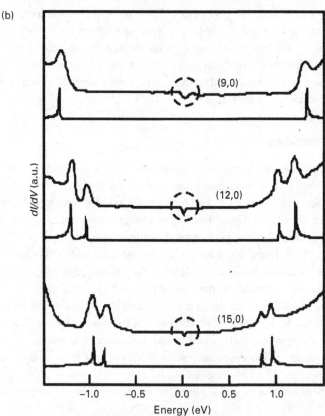

Fig. 6.12   Observations of 'gaps' in densities of states for metallic tubes, from work by Lieber *et al.* (6.45).
(a) STM image of (15, 0) zigzag nanotube, (b) tunnelling conductance data, *dI/dV*, for different
zigzag SWNTs, with corresponding theoretical DOS.

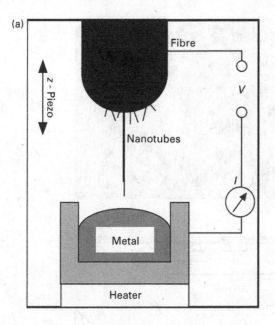

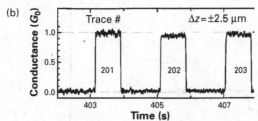

Fig. 6.13   (a) The experimental arrangement used by de Heer *et al.* to measure resistance of individual MWNTs, (b) The conductance of a nanotube moved at constant speed into and out of mercury as a function of time (6.51).

   The first experiments apparently demonstrating single-electron transport in SWNTs were carried out using 'ropes' rather than individual tubes (6.54), and will not be discussed here. Quantum wire behaviour in individual SWNTs was first observed by a group including Cees Dekker and Richard Smalley in 1997 (6.55). This group demonstrated the phenomenon of Coulomb blockade in individual tubes supported on Pt electrodes on a Si/SiO$_2$ substrate. Coulomb blockade is observed in a nanostructure connected to two electrodes when two conditions are met (6.56). The first is that the connections are relatively poor, so that the contact resistance is larger than R_Q, the resistance quantum, given by $h/2e^2$. The second is that the nanostructure is sufficiently small, so that the capacitance is small and the energy needed for adding an electron to the system e^2/C is larger than the thermal energy $K_B T$. The arrangement used by the Dekker–Smalley group is shown in Fig. 6.14. The distance between the contacts was approximately 140 nm. A bias voltage was applied between these contacts, and a gate voltage was applied to the third electrode in the upper-left corner of the image, to vary the

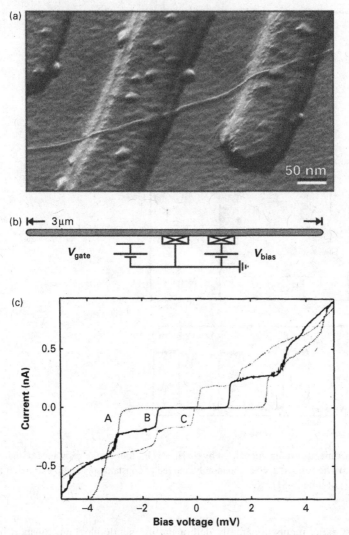

Fig. 6.14   Measurements of the electronic properties of individual SWNTs, by Tans *et al.* (6.55). (a) An
AFM image of a SWNT on a Si/SiO$_2$ substrate, with two 15 nm Pt electrodes, (b) circuit diagram,
(c) current–voltage curves at different gate voltages.

electrostatic potential of the tube. Measurements were made at a temperature of 5 mK,
and at a range of gate voltages. As can be seen in Fig. 6.14(c), it was found that the *I–V*
curves displayed a flat region near *V* = 0. This suppressed conductance, or 'gap', is
indicative of Coulomb blockade.

   A significant advance was made in 2001 when Postma and colleagues described
SWNT-based single-electron transistors which operated at room temperature (6.57).
This group deposited SWNTs onto a Si/SiO$_2$ substrate and used an AFM tip to create
sharp bends, or buckles, in the tubes, as described earlier. These buckles worked as the

barriers, only allowing single electrons through under the right voltages. The whole device was only 1 nm wide and 20 nm long.

Dai's group have also observed quantum conductance in metallic SWNTs (6.58). In this work they fabricated SWNT devices by the patterned CVD growth of SWNTs on Si/SiO_2 wafers, followed by deposition of metal source/drain contacts. By using Pd electrodes they were able to achieve ohmic contacts with the nanotubes. The tubes exhibited room-temperature conductance near the ballistic transport limit of $2G_0$ and high current-carrying capability ($\sim 25\,\mu A$ per tube). The mean free paths for acoustic phonon scattering were ~ 500 nm at room temperature and $\gg 4\mu m$ at low temperatures. Similar techniques were used to make field-effect transistors using semiconducting SWNTs, as discussed in Section 4.4.1.

In addition to the early work by de Heer *et al.*, a number of later studies have also demonstrated quantum wire behaviour in MWNTs (6.59–6.61). An example is the work by Ahlskog and colleagues (6.59). This involved fabricating a three-terminal nanotube device from two multiwalled nanotubes by positioning one on top of the other using an atomic-force microscope. The lower nanotube, with gold contacts at both ends, acted as the central island of a single-electron transistor while the upper one functioned as a gate electrode. Coulomb blockade oscillations were observed on the nanotube at sub-Kelvin temperatures. A different approach was used by Yoneya *et al.* (6.61). Here, very small MWNT islands were produced by etching using an oxygen plasma. This was achieved as follows. Firstly, MWNTs (prepared by arc-evaporation) were deposited onto a Si/SiO_2 substrate. Three Pt/Au electrodes were deposited by thermal evaporation, two to form contacts with the MWNT and one to act as a gate. The gate electrode was approximately 1.5 μm away from the MWNT. A resist material was then spin coated over the whole device. Two narrow trenches were then etched into the resist using an oxygen plasma, eventually cutting two gaps into the nanotube. Before cutting, the tubes generally displayed normal metallic behaviour, but when they had been completely etched through, Coulomb blockade behaviour was observed. Lizzie Brown, of the UK's National Physical Laboratory and colleagues have recorded conductance steps in MWNTs using a modified scanning probe microscope (6.62). Thermal measurements were made at the same time, and evidence for ballistic transport of phonons was found (see also p. 197).

6.3.4 Electronic properties of nanotubes in a magnetic field

As noted in Section 6.2.4, Ajiki and Ando predicted that a magnetic field can have a profound effect on the electronic properties of carbon nanotubes. Some of these predictions have now been confirmed. Among the first experimental work in this area was that carried out by Christian Schönenberger of the University of Basel, Switzerland, and colleagues, who reported the observation of Aharonov–Bohm oscillations in carbon nanotubes in 1999 (6.63). Multiwalled nanotubes were used, since the magnetic fields required to induce the effect in relatively large-diameter tubes are experimentally achievable. The electrical resistances of MWNTs with diameters of around 16 nm were measured as a function of magnetic flux up to a maximum of about 14 T. Resistance peaks were observed at $B = 0$, and smaller peaks at $B = \pm 8.5$ T, in good agreement with the predictions of Ajiki and Ando. In 2004, Alexey Bezryadin's group at the University of Illinois showed that short multiwalled nanotubes with a diameter of about 30 nm could be converted from semiconducting to metallic by the application of a magnetic field parallel to the tube axis (6.64). The magnetic field required for one Aharonov–Bohm 'period' was about 6 T. More recently, researchers from France and Switzerland have observed the Aharonov–Bohm effect in MWNTs threaded by magnetic fields as large as 55 T (6.65).

For a while it seemed that it might not be possible to observe these effects in single-walled nanotubes, owing to the enormous fields that would be needed. However, in 2004 Junichiro Kono at Rice University and his team reported measurements that provided evidence for Aharonov–Bohm behaviour in SWNTs (6.66). They used special facilities at the high-field magnet laboratory in Florida State University to apply magnetic fields as high as 45 T, corresponding to 1% of a full Aharonov–Bohm period. Optical absorption and photoluminescence spectroscopy were used to determine the electronic properties of the tubes, and showed clear evidence that the band gap of semiconducting nanotubes shrank in the presence of a magnetic field. At about the same time, Hongjie Dai's group found that relatively low magnetic fields applied parallel to the axis of a single-walled tube caused large modulations to the valence band conductance of the nanotube (6.67).

6.3.5   Superconductivity

There are a few reports claiming that pure carbon nanotubes can display superconductivity. As mentioned on p. 16, in 2000 a group from the Hong Kong University of Science and Technology reported the pyrolytic synthesis of SWNTs with a diameter of just 0.4 nm (6.68). The tubes were produced by pyrolysing tripropylamine in the channels of the nanoporous aluminophosphate $AlPO_4$. In a subsequent paper (6.69), the same group reported superconducting characteristics at the relatively high temperature of 15 K. This was attributed to an enhancement of electron–phonon coupling in these ultra-small nanotubes. Superconductivity in ropes of single-walled tubes has also been reported by Hélène Bouchiat of the Université Paris-Sud and colleagues This group exploited the proximity effect, in which an ordinary metal can carry a supercurrent when placed between two superconductors (6.70). Superconductivity in multiwalled tubes at temperatures up to 12 K was also reported by Japanese workers in 2006 (6.71). In general, however, the investigation of superconducting behaviour in nanotubes has only generated moderate interest.

6.4   Nanoelectronic devices

6.4.1   Diodes

In 1996 theorists pointed out (6.72, 6.73) that 'elbow connections' joining tubes of different structures could constitute nanoscale heterojunctions (the structure of elbow connections was discussed in Chapter 4, p. 115). Thus, a connection between a metallic

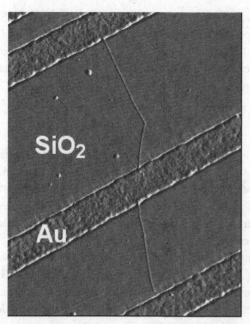

Fig. 6.15   An atomic force microscope image of a diode constructed using a kinked nanotube (6.75).

tube and a semiconducting one could form a rectifying diode: higher energy electrons from the semiconducting side of the junction could flow 'downhill' to the metallic side, but they could not travel the other way. Work by Collins *et al.* in 1998 (6.74) provided the first evidence for this kind of behaviour in single-walled tubes. In this work, an STM tip was moved along the length of individual tubes, and positions were found where the current transport behaviour changed abruptly. Effectively, the tubes passed current in only one direction. This was attributed to pentagon–heptagon defects, although no direct proof was given. A year later, clear experimental evidence for the rectifying behaviour of elbow connections was given by Cees Dekker and colleagues (6.75). The setup used was similar to that employed to demonstrate quantum wire behaviour in individual SWNTs (6.55). Thus, the nanotubes were dispersed on Pt electrodes on a Si/SiO$_2$ substrate. A few tubes were found that naturally contained elbow connections or 'kinks'; an example is shown in Fig. 6.15. Electrical transport measurements across the kink showed that the junction did indeed constitute a nanoscale diode, which passed current in one direction but not the other.

Another approach to producing nanotube diodes is to use doping to create a *p–n* junction, as in a conventional silicon diode. Hongjie Dai's team were the first to demonstrate this approach, in 2000 (6.76). This was achieved by doping one half of a semiconducting SWNT with K, while leaving the other half undoped. The undoped segment acted as a *p*-type semiconductor, for reasons which are discussed in the next section, while the doped part became *n*-type due to electron donation from the adsorbed K atoms. Under certain conditions the junctions displayed behaviour consistent with that of a tunnel diode. For example, negative differential conductance (i.e. a decrease in current as a function of voltage) was observed over certain voltage ranges. A SWNT *p–n* junction diode with rather superior characteristics was described in 2004 by Ji Ung Lee and colleagues of the General Electric Company (6.77). In this work the 'doping' was carried out by applying different charges to different parts of the tube rather than actually depositing a dopant onto the tube.

6.4.2   Field effect transistors

Several experiments that involved the construction of single-electron transistors using metallic SWNTs were discussed in Section 6.3.3. This section is concerned with field effect transistors (FETs), constructed using both semiconducting and metallic tubes. The field effect transistor (FET), which forms the basis for modern integrated circuits, consists of two metal electrodes designated 'source' and 'drain' connected by a semiconducting channel. In conventional devices, the channel is made of Si. A third electrode, the 'gate', is separated from the channel by a thin insulator film. Normally, if no charge is placed on the gate, no charge flows between the source and drain. In a carbon nanotube field effect transistor (CNTFET), the channel between the source and the drain is an individual semiconducting single-walled nanotube. The first such device was produced by the Dekker group in 1998 (6.78). The structure of the device, which operated at room temperature, is shown in Fig. 6.16(a). The distance between the Pt electrodes was approximately 280 nm and the thermally grown SiO_2 layer was 300 nm thick. The Si substrate served as a back-gate. Figure 6.16(b) shows the conductance through the nanotube at zero bias as a function of the gate voltage. When a negative gate

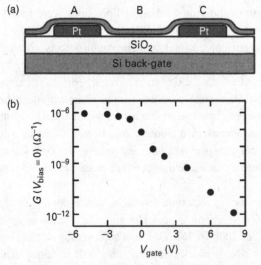

Fig. 6.16   (a) A schematic side view of the 'TUBEFET' device constructed by Tans *et al.* (6.78). A single semiconducting nanotube is contacted by two Pt electrodes. The Si substrate, which is covered by a 300 nm layer of SiO_2, acts as a back-gate. (b) Conductance of the nanotube at $V_{bias} = 0$ as a function of the gate voltage V_{gate}.

voltage is applied the conductance saturates at $10^{-6}\ \Omega^{-1}$, while positive gate voltages produce a rapid fall in conductance. Overall, the conductance can be varied over at least six orders of magnitude. A short time after the Dekker work, similar devices were described by Phaedon Avouris and colleagues, this time using both single- and multi-walled tubes (6.26).

The first devices produced by the Avouris and Dekker groups had high contact resistances (maximum conductance $\sim 10^{-6}\ \Omega^{-1}$, contact resistance $\sim 1\ \mathrm{M}\ \Omega$). Techniques have subsequently been developed for achieving much reduced contact resistances. Avouris *et al.* used methods involving 'end-bonded' rather than 'side-bonded' tubes (6.79), while Hongjie Dai's group used nanotubes grown *in situ* on substrates, as discussed in Section 3.4.4 (6.80).

In 2001, Avouris and colleagues described the fabrication of arrays of nanoscale FETs, utilizing ropes of pure semiconducting SWNTs (6.81). They were able to achieve this by selectively 'burning' away the metallic tubes within the ropes, as described Chapter 4 (p. 99). To fabricate the arrays, the SWNT ropes were deposited on an oxidized Si wafer, which also served as the back gate. An array of source, drain and side-gate electrodes was then fabricated lithographically on top of the permanently modified SWNT ropes. The back gate was used to deplete the semiconducting tubes, followed by the application of a voltage to destroy the metallic tubes. The resulting devices were shown to have reasonable FET characteristics. This approach has been taken up by other groups. For example, in 2004 Robert Seidel from Infineon Technologies in Germany and colleagues produced SWNT-based devices which could be used to control macroscopic devices such as light emitting diodes and electric motors (6.82). An alternative approach to fabricating nanotube devices is self-assembly. In an example of this, Israeli researchers exploited the interaction of proteins and DNA to assemble nanotubes into a field-effect transistor which operated at room temperature (6.83), as previously mentioned in Chapter 4.

In the studies of nanotube FETs described here, the transistors are 'on' for a negative gate voltage, and are therefore *p*-type. This is despite the fact that the tubes were not intentionally doped. Although not understood in the early days of research in this area, this can now be explained in terms of the sensitivity of nanotubes' electronic properties to oxygen. The Zettl group showed in 2000 that exposure to air or oxygen dramatically influences the electrical resistance, thermoelectric power and local density of states of carbon nanotubes (6.84). Several explanations have been proposed for this behaviour, such as doping by oxygen during the synthesis and handling of the nanotubes, or charge transfer from the metal electrodes. Whatever the mechanism, it is likely that oxygen exposure was responsible for the *p*-type behaviour, since the early FETs were fabricated in air. This oxygen sensitivity is clearly a potential problem when constructing electronic devices from nanotubes, but it can be used in a positive way in chemical gas sensors, as discussed in Chapter 11 (p. 280).

6.4.3 Logic circuits

In order to produce useful logic circuits, both *n*- and *p*-type FETs are needed. We saw in the previous section that as-prepared nanotube transistors are invariably *p*-type. Several

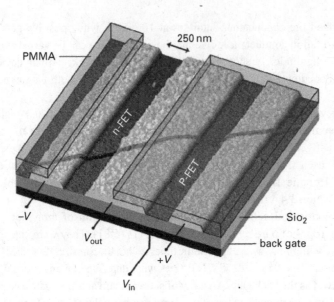

250 nm

PMMA

n-FET

P-FET

−V

V_{out}

V_{in}

+V

Sio₂

back gate

Fig. 6.17   An intramolecular logic gate based on a single nanotube bundle (6.86).

groups have therefore explored ways of inducing n-type behaviour. In 2000, workers from Berkeley and Rice Universities (6.85) described experiments on SWNT ropes, in which intercalation with alkali metals changed the characteristics of the ropes from p- to n-type. A simpler way of producing this effect was reported by the Avouris group in 2001 (6.86). It was shown that simply annealing an SWNT-based FET in a vacuum produced n-type behaviour. In the same paper, Avouris and colleagues used both the doping approach and the annealing method to induce n-type behaviour in short sections of nanotube bundles. In this way they were able to build the first nanotube-based logic gate, namely a voltage inverter, or 'NOT' gate. Figure 6.17 illustrates how they used doping to construct the gate. A single nanotube bundle was positioned over gold electrodes supported on SiO_2, producing two p-type CNTFETs in series. The device was covered by PMMA and a window was produced by e-beam lithography to expose part of the nanotube. Potassium was then evaporated through this window to produce an n-CNTFET, while the other CNTFET remains p-type. It was demonstrated that this device operated as a NOT gate.

Since this pioneering work, groups have fabricated nanotube-based circuits that exhibit a range of digital logic operations. These include inverters, random-access memory cells, and ring oscillator circuits (6.87–6.89), For reviews of this fast-moving field, refs 6.36 and 6.37 can be recommended.

6.5   Magnetic properties of nanotubes

The magnetic properties of graphite are dominated by the presence of ring currents, i.e. electron orbits circulating above and below the hexagonal lattice planes which include

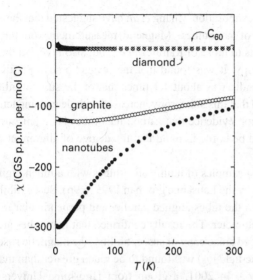

Fig. 6.18 The magnetic susceptibilities of carbon allotropes as a function of temperature, from the work by Ramirez and colleagues (6.94).

several atoms within their radius (6.90). These result in a relatively large negative susceptibility which is highly anisotropic. Thus, when the field is oriented perpendicular to the layer planes the susceptibility, χ_c, is 22×10^{-6} emu g^{-1}, while the susceptibility with the field parallel to the planes, χ_{ab} is 0.5×10^{-6} emu g^{-1}. On a very simple model, one might assume that the magnetic properties of a carbon nanotube would approximate to those of a rolled-up graphene sheet. The susceptibility of tubes aligned perpendicular to the field, $\chi_{\perp}$, would therefore approximate to $(\chi_c + \chi_{ab})/2$ and the susceptibility of tubes aligned parallel to the field, $\chi_{\parallel}$, would approximate to χ_{ab}. Since $\chi_c \gg \chi_{ab}$, this would suggest that $\chi_{\perp} \gg \chi_{\parallel}$ for nanotubes. Detailed theoretical work has tended to confirm this simple-minded model (6.91–6.93).

Some early studies of the magnetic susceptibility of nanotubes were carried out by Arthur Ramirez and colleagues from Bell Labs, in collaboration with Smalley's group from Rice University (6.94). These workers studied a variety of carbons, including an unpurified sample of nanotubes, over a range of temperatures from absolute zero to room temperature, using a SQUID magnetometer. The results are shown in Fig. 6.18. It can be seen that the nanotube sample displays quite different behaviour to the other carbons, having a large diamagnetic susceptibility (i.e. negative χ) which increases with decreasing temperature. The results clearly indicate that nanotubes have a greater susceptibility than graphite, although it is not possible to say whether this susceptibility lies parallel or perpendicular to the tube axis since the tubes in the sample were randomly oriented. Ramirez *et al.* speculated that the large susceptibility of nanotubes might result from ring currents flowing around the tube circumferences.

The first attempts to measure the magnetic properties of aligned nanotube samples were made by Robert Chang and colleagues from Northwestern University (6.95), who carried out their measurements on the columnar deposits which are sometimes produced on the cathode following arc-evaporation. Chang *et al.* have suggested that these deposits consist of aligned bundles of nanotubes. Magnetic measurements on this material produced rather similar results to those of Ramirez *et al.*, but also revealed the presence of a small degree of anisotropy. It was found that the magnetic susceptibility with *H* parallel to the axis of the bundle was about 1.1 times that of the susceptibility with *H* perpendicular to the axis, and that this anisotropy increased with decreasing temperature. These results represent the first evidence that carbon nanotubes have anisotropic magnetic properties, but it should be borne in mind that the degree of alignment within the bundles is probably not high.

Magnetic measurements on samples of multiwalled tubes with a much higher degree of alignment were carried out by the Lausanne group in 1995 (6.96). Susceptibilities were measured from 4–300 K, with the tubes aligned parallel and perpendicular to the field, again using a SQUID magnetometer. The results confirmed that nanotubes are diamagnetic, and showed a pronounced anisotropy of susceptibility. The magnetic susceptibility of tubes aligned parallel to the field ($\chi_{\parallel}$) was found to be much greater than that of tubes perpendicular to the field ($\chi_{\perp}$). In 2001, a group from Hiroshima University (6.97) reported magnetic measurements on individual MWNTs, and again found that the susceptibility parallel to the tube axis was greater than that perpendicular to the axis. These results are the reverse of the theoretical predictions noted above, and the reason for this discrepancy is not clear.

6.6   Nanotube field emitters

Work on the field emission properties of carbon nanotubes began in about 1995 (6.98), and has developed into a major field of research. This work has primarily been driven by the prospect of using arrays of field emitting nanotubes in flat screen displays, but there is also interest in field emission from individual tubes. Before the advent of carbon nanotubes, most of the research in this area focused on the use of metal cones deposited onto silicon substrates using photolithographic techniques. However, a conical shape is not ideal for field emission. In 1991, Takao Utsumi discussed the merits of variously shaped field emitters, and assigned each a 'figure of merit', according to their emission properties (6.99). The best structure was an extended narrow pillar, a geometry that approximates closely to a carbon nanotube grown perpendicularly from a flat substrate. The high conductivity and stability of nanotubes at high temperatures are further reasons for their suitability as field emission sources.

Probably the earliest experiments on carbon as a field emission source were described in 1972 by a group from the UK Ministry of Defence (6.100), who reported that under some conditions, conventional carbon fibres showed superior performance to metals. More recently, the field emission properties of a range of carbon materials have been investigated, including chemical vapour deposited (CVD) diamond (6.101), amorphous diamond-like carbon (6.102) and nanostructured carbon (6.103). However, none of these carbons have the combination of characteristics that make carbon nanotubes such ideal field emitters.

Some of the first experiments on field emission from nanotubes were carried out by Walt de Heer and colleagues in 1995 (6.104). They used MWNTs prepared by arc-evaporation, which were purified and then deposited onto a plastic surface. Partial alignment of the tubes could be achieved by lightly rubbing the surface with a Teflon or aluminium foil. A 3 mm copper electron microscopy grid was held above the film of aligned nanotubes, at a distance of approximately 20 μm. The anode was situated about 1 cm above the copper grid, and current densities of up to $0.1 \, A \, cm^{-1}$ were reported, at a voltage of approximately 700 V. This level of emission should be sufficient to produce an image on a phosphor-coated display. Other groups have used 'post-synthesis' methods to produce nanotube films for field emission. For example, Robert Chang's group demonstrated field emission from a flat panel display employing a nanotube-epoxy composite as the source (6.105). However, the degree of nanotube alignment achieved in post-synthesis processing is usually limited, so that field emission tends to be dominated by a small number of tubes. For this reason, most studies of field emission from nanotubes have employed catalytically-grown arrays of nanotubes, and this work will now be summarized.

The growth of aligned MWNTs on substrates using chemical vapour deposition (CVD) and plasma enhanced chemical vapour deposition (PECVD) was described in Section 3.2.1. Using this process, it is possible to grow aligned arrays sufficiently large to serve as flat-panel displays. The first studies of field emission from arrays of tubes grown in this way appeared in 1999–2000 (6.106–6.108). These early studies showed that close packed arrays of nanotubes are not ideal for field emission applications as the close packing of the tubes screens the applied field, effectively reducing the field enhancement of the high aspect ratio tubes. It has been shown (6.109) that the optimum emitted current density would be achieved with an array of individual vertically aligned tubes spaced apart by twice their height. As a result, techniques have been developed for growing patterned arrays of well-separated tubes, as shown in Fig. 3.5. A number of other factors affect the field emission characteristics of nanotubes. High crystallinity results in high conductivity, which gives improved field emission properties, as does a good contact between the tubes and the substrate. Milne and co-workers have shown that both crystallinity and the quality of the contact can be improved by rapid thermal annealing of the nanotube arrays in high vacuum (6.110). Several groups have compared emission from open and closed nanotubes, with conflicting results. Thus, early work by Smalley and colleagues showed that that field emission was enhanced when the tubes were opened by laser vaporization or oxidation (6.111), while more recent work found that emission from closed tips was more efficient (6.110). The consensus seems to be that open tips, being 'sharp', require less voltage to turn on, but tend to be unstable in terms of emission current, while closed caps are very stable, with a smooth emission pattern, but require more voltage to turn on (6.112).

At the time of writing, it appears that no nanotube-based field emission displays are commercially available. However, a number of prototypes have been demonstrated. For example, in 2005, Motorola displayed a 5-inch colour video display which utilized an array of nanotubes grown on glass (Fig. 6.19). Other companies who claim to have produced working prototypes include Samsung and Applied Nanotech Inc. Commercial products may not be far behind.

Fig. 6.19   A colour field emission display using carbon nanotubes, developed by Motorola Labs.

In addition to the work on arrays of nanotubes, field emission experiments have also been carried out on individual tubes and on clusters of tubes. The earliest work in this area arose out of studies by Smalley's group into the use of single nanotubes as probes in scanning tunnelling or atomic force microscopy (see Chapter 11). During these studies they developed a method for mounting nanotubes on the tips of conventional carbon fibres, and this enabled them to determine the field emission characteristics of individual tubes (6.111). As mentioned above, they found that opened tubes produced the best results. Field emission from individual MWNTs has also been studied by Yahachi Saito and colleagues, who have used closed and open tubes as the sources in a field emission microscope (6.113). The motivation behind these studies is to explore the possibility of using individual nanotube field emitters in cathode ray tubes or as guns in electron microscopes. There is also interest in nanotube-based light sources. In 2004, a team from Lausanne described a luminescent tube of 40 cm length based on carbon nanotube field emission (6.114). Nanotubes were grown on a metal wire, which was placed inside a glass tube coated with a phosphor and a conductive layer (to conduct away the electrons). It was demonstrated that the device performed well in comparison to conventional fluorescent tubes. Whether the costs can be reduced to make the product commercially viable is of course another question.

A number of excellent reviews of the application of carbon nanotubes as field emitters are available (6.35, 6.109, 6.115).

6.7   Conclusions

Since the mid 1990s, huge strides have been made in understanding the electronic properties of nanotubes and in constructing nanotube-based electronic devices. Among the landmarks in this field are the direct correlation of electronic properties with SWNT structure, by Dekker and co-workers in 1998; the demonstration in 1997 of quantum transport in SWNTs by Dekker, Smalley and colleagues and in MWNTs one year later by de Heer's group; and the construction in 1998 of SWNT field effect transistor and in 2001 of SWNT-based single-electron transistors which operated at room temperature, both by the Dekker group. Other highlights include the first logic gate based on a single nanotube bundle, constructed by the Avouris group in 2001. The observation of the Aharonov–Bohm effect in both multi- and single-walled tubes is also worthy of note.

Despite these impressive achievements, a commercial nanotube-based personal computer or iPod still seems a very distant prospect, for reasons which have become familiar throughout this book. We still do not have a way of preparing nanotubes with a defined structure, or of reliably arranging them in a defined manner in order to construct the kind of complex circuit that will be needed for realistic applications. As far as the controlled arrangement of nanotubes is concerned, some progress has been made, both using the directed growth of SWNTs, as discussed in Chapter 3 (p. 61) and using methods for positioning ready-made tubes as outlined in Chapter 4 (p. 92). However, these processes will be of little avail until the more fundamental problem of making nanotubes with a known structure is solved.

One area in which commercial electronic products based on nanotubes may soon become a reality is field emission. A carbon nanotube has almost the perfect geometry for a field emitter, and large arrays of nanotubes can now readily be grown on flat substrates. As noted above, working field emission displays using nanotubes have been demonstrated by several companies. Nanotube-based field emitters may also find applications as light or X-ray sources and in a range of other areas.

References

(6.1) P. R. Wallace, 'The band theory of graphite', *Phys. Rev.*, **71**, 622 (1947).

(6.2) I. L. Spain, 'Electronic transport properties of graphite, carbons and related materials', *Chem. Phys. Carbon*, **16**, 119 (1981).

(6.3) J. C. Slonczewski and P. R. Weiss, 'Band structure of graphite', *Phys. Rev.*, **109**, 272 (1958).

(6.4) J. W. McClure, 'Band structure of graphite and de Haas–van Alphen effect', *Phys. Rev.*, **108**, 612 (1957).

(6.5) R. R. Haering and S. Mrozowski, 'Band structure and electronic properties of graphite crystals', *Prog. Semicond.*, **5**, 273 (1960).

(6.6) I. L. Spain, 'The electronic properties of graphite', *Chem. Phys. Carbon*, **8**, 1 (1973).

(6.7) B. T. Kelly, *The Physics of Graphite*, Applied Science Publishers, London and New Jersey, 1981.

(6.8) R. Saito, M. Fujita, G. Dresselhaus *et al.*, 'Electronic structure of graphene tubules based on C_{60}', *Phys. Rev. B*, **46**, 1804 (1992).

(6.9) M. S. Dresselhaus, G. Dresselhaus and R. Saito, 'Physics of carbon nanotubes', *Carbon*, **33**, 883 (1995).

(6.10) M. S. Dresselhaus, G. Dresselhaus and P. C. Eklund, *Science of Fullerenes and Carbon Nanotubes*, Academic Press, San Diego, 1996.

(6.11) N. Hamada, S. Sawada and A. Oshiyama, 'New one-dimensional conductors: graphitic microtubules', *Phys. Rev. Lett.*, **68**, 1579 (1992).

(6.12) J. W. Mintmire, B. I. Dunlap and C. T. White, 'Are fullerene tubules metallic?', *Phys. Rev. Lett.*, **68**, 631 (1992).

(6.13) C. Kittel, *Introduction to Solid State Physics*, 5th edn, John Wiley, New York, 1976, p. 133.

(6.14) R. Saito, M. Fujita, G. Dresselhaus *et al.*, 'Electronic structure of chiral graphene tubules', *Appl. Phys. Lett.*, **60**, 2204 (1992).

(6.15) S. Reich, C. Thomsen and J. Maultzsch, *Carbon Nanotubes*, Wiley-VCH, Weinheim, 2004.

(6.16) X. Blase, L. X. Benedict, E. L. Shirley *et al.*, 'Hybridization effects and metallicity in small radius carbon nanotubes', *Phys. Rev. Lett.*, **72**, 1878 (1994).

(6.17) C. L. Kane and E. J. Mele, 'Size, shape, and low energy electronic structure of carbon nanotubes', *Phys. Rev. Lett.*, **78**, 1932 (1997).

(6.18) P. Delaney, H. J. Choi, J. Ihm *et al.*, 'Broken symmetry and pseudogaps in ropes of carbon nanotubes', *Nature*, **391**, 466 (1998).

(6.19) P. Delaney, H. J. Choi, J. Ihm *et al.*, 'Broken symmetry and pseudogaps in ropes of carbon nanotubes', *Phys. Rev. B.*, **60**, 7899 (1999).

(6.20) Y. K. Kwon, S. Saito and D. Tománek, 'Effect of intertube coupling on the electronic structure of carbon nanotube ropes', *Phys. Rev. B.*, **58**, 13 314 (1998).

(6.21) S. Datta, *Electronic Transport Properties in Mesoscopic Systems*, Cambridge University Press, Cambridge, 1995.

(6.22) W. D. Tian and S. Datta, 'Aharonov–Bohm-type effect in graphene tubules – a Landauer approach', *Phys. Rev. B*, **49**, 5097 (1994).

(6.23) M. F. Lin and K. W.-K. Shung, 'Magnetoconductance of carbon nanotubes', *Phys. Rev. B*, **51**, 7592 (1995).

(6.24) L. Chico, L. X. Benedict, S. G. Louie *et al.*, 'Quantum conductance of carbon nanotubes with defects', *Phys. Rev. B*, **54**, 2600 (1996).

(6.25) S. Sanvito, Y.-K. Kwon, D. Tománek *et al.*, 'Fractional quantum conductance in carbon nanotubes', *Phys. Rev. Lett.*, **84**, 1974 (2000).

(6.26) R. Martel, T. Schmidt, H. R. Shea *et al.*, 'Single and multi-wall carbon nanotube field-effect transistors', *Appl. Phys. Lett.*, **73**, 2447 (1998).

(6.27) T. Dürkop, S. A. Getty, E. Cobas *et al.*, 'Extraordinary mobility in semiconducting carbon nanotubes', *Nano Lett.*, **4**, 35 (2004).

(6.28) P. L. McEuen, M. Bockrath, D. H. Cobden *et al.*, 'Disorder, pseudospins, and backscattering in carbon nanotubes', *Phys. Rev. Lett.*, **83**, 5098 (1999).

(6.29) P. L. McEuen and J. Y. Park, 'Electron transport in single-walled carbon nanotubes', *MRS Bulletin*, **29**, 272 (2004).

(6.30) G. Pennington and N. Goldsman, 'Low-field semiclassical carrier transport in semiconducting carbon nanotubes', *Phys. Rev. B*, **71**, 205318 (2005).

(6.31) H. Ajiki and T. Ando, 'Electronic states of carbon nanotubes', *J. Phys. Soc. Jpn.*, **62**, 1255 (1993).

(6.32) H. Ajiki and T. Ando, 'Energy bands of carbon nanotubes in magnetic fields', *J. Phys. Soc. Jpn.*, **65**, 505 (1996).

(6.33) T. Ando, 'Theory of electronic states and transport in carbon nanotubes', *J. Phys. Soc. Jpn.*, **74**, 777 (2005).

(6.34) M. Ouyang, J. L. Huang, and C. M. Lieber, 'Fundamental electronic properties and applications of single-walled carbon nanotubes', *Acc. Chem. Res.*, **35**, 1018 (2002).

(6.35) M. P. Anantram and F. Léonard, 'Physics of carbon nanotube electronic devices', *Rep. Prog. Phys.*, **69**, 507 (2006).

(6.36) J. C. Charlier, X. Blase and S. Roche, 'Electronic and transport properties of nanotubes', *Rev. Mod. Phys.*, **79**, 677 (2007).

(6.37) P. Avouris, 'Electronics with carbon nanotubes', *Phys. World*, **20**, 40 (2007).

(6.38) L. Langer, V. Bayot, E. Grivei *et al.*, 'Quantum transport in a multiwalled carbon nanotube', *Phys. Rev. Lett.*, **76**, 479 (1996).

(6.39) T. W. Ebbesen, H. J. Lezec, H. Hiura *et al.*, 'Electrical conductivity of individual carbon nanotubes', *Nature*, **382**, 54 (1996).

(6.40) P. G. Collins, M. Hersam, M. Arnold *et al.*, 'Current saturation and electrical breakdown in multiwalled carbon nanotubes', *Phys. Rev. Lett.*, **86**, 3128 (2001).

(6.41) J. W. G. Wildöer, L. C. Venema, A. G. Rinzler *et al.*, 'Electronic structure of atomically resolved carbon nanotubes', *Nature*, **391**, 59 (1998).

(6.42) L. C. Venema, V. Meunier, P. Lambin *et al.*, 'Atomic structure of carbon nanotubes from scanning tunneling microscopy', *Phys. Rev. B*, **61**, 2991 (2000).

(6.43) P. Kim, T. W. Odom, J.-L. Huang *et al.*, 'Electronic density of states of atomically-resolved single-walled carbon nanotubes: van Hove singularities and end states', *Phys. Rev. Lett.*, **82**, 1225 (1999).

(6.44) T. W. Odom, J. L. Huang, P. Kim *et al.*, 'Structure and electronic properties of carbon nanotubes', *J. Phys. Chem. B*, **104**, 2794 (2000).

(6.45) M. Ouyang, J. L. Huang, C. L. Cheung *et al.*, 'Energy gaps in "metallic" single-walled carbon nanotubes', *Science*, **292**, 702 (2001).

(6.46) R. Saito, T. Takeya, T. Kimura *et al.*, 'Raman intensity of single-wall carbon nanotubes', *Phys. Rev. B*, **57**, 4145 (1998).

(6.47) A. Kasuya, M. Sugano, T. Maeda *et al.*, 'Resonant Raman scattering and the zone-folded electronic structure in single-wall nanotubes', *Phys. Rev. B*, **57**, 4999 (1998).

(6.48) M. A. Pimenta, A. Marucci, S. A. Empedocles *et al.*, 'Raman modes of metallic carbon nanotubes', *Phys. Rev. B*, **58**, R16016 (1998).

(6.49) A. Jorio, A. G. Souza, G. Dresselhaus *et al.*, 'G-band resonant Raman study of 62 isolated single wall carbon nanotubes', *Phys. Rev. B*, **65**, 155412 (2002).

(6.50) C. Fantini, A. Jorio, M. Souza *et al.*, 'Optical transition energies for carbon nanotubes from resonant Raman spectroscopy: environment and temperature effects', *Phys. Rev. Lett.*, **93**, 147406 (2004).

(6.51) S. Frank, P. Poncharal, Z. L. Wang *et al.*, 'Carbon nanotube quantum resistors', *Science*, **280**, 1744 (1998).

(6.52) P. Poncharal, S. Frank, Z. L. Wang *et al.*, 'Conductance quantization in multiwalled carbon nanotubes', *Eur. Phys. J. D*, **9**, 77 (1999).

(6.53) P. Poncharal, C. Berger, Y. Yi *et al.*, 'Room temperature ballistic conduction in carbon nanotubes', *J. Phys. Chem. B*, **106**, 12 104 (2002).

(6.54) M. Bockrath, D. H. Cobden, P. L. McEuen *et al.*, 'Single-electron transport in ropes of carbon nanotubes', *Science*, **275**, 1922 (1997).

(6.55) S. J. Tans, M. H. Devoret, H. J. Dai *et al.*, 'Individual single-wall carbon nanotubes as quantum wires', *Nature*, **386**, 474 (1997).

(6.56) L. J. Geerligs, D. V. Averin and J. E. Mooij, 'Observation of macroscopic quantum tunneling through the coulomb energy barrier', *Phys. Rev. Lett.*, **65**, 3037 (1990).

(6.57) H. W. C. Postma, T. Teepen, Z. Yao *et al.*, 'Carbon nanotube single-electron transistors at room temperature', *Science*, **293**, 76 (2001).

(6.58) D. Mann, A. Javey, J. Kong *et al.*, 'Ballistic transport in metallic nanotubes with reliable Pd ohmic contacts', *Nano Lett.*, **3**, 1541 (2003).

(6.59) M. Ahlskog, R. Tarkiainen, L. Roschier *et al.*, 'Single-electron transistor made of two crossing multiwalled carbon nanotubes and its noise properties', *Appl. Phys. Lett.*, **77**, 4037 (2000).

(6.60) A. Kanda, Y. Ootuka, K. Tsukagoshi *et al.*, 'Electron transport in metal/multiwall carbon nanotube/metal structures (metal = Ti or Pt/Au)', *Appl. Phys. Lett.*, **79**, 1354 (2001).

(6.61) N. Yoneya, E. Watanabe, K. Tsukagoshi *et al.*, 'Coulomb blockade in multiwalled carbon nanotube island with nanotube leads', *Appl. Phys. Lett.*, **79**, 1465 (2001).

(6.62) E. Brown, L. Hao, J. C. Gallop *et al.*, 'Ballistic thermal and electrical conductance measurements on individual multiwall carbon nanotubes', *Appl. Phys. Lett.*, **87**, 023107 (2005).

(6.63) A. Bachtold, C. Strunk, J. P. Salvetat *et al.*, 'Aharonov–Bohm oscillations in carbon nanotubes', *Nature*, **397**, 673 (1999).

(6.64) U. C. Coskun, T. C. Wei, S. Vishveshwara *et al.*, 'h/e magnetic flux modulation of the energy gap in nanotube quantum dots', *Science*, **304**, 1132 (2004).

(6.65) B. Lassagne, J. P. Cleuziou, S. Nanot *et al.*, 'Aharonov–Bohm conductance modulation in ballistic carbon nanotubes', *Phys. Rev. Lett.* **98**, 176802 (2007).

(6.66) S. Zaric, G. N. Ostojic, J. Kono *et al.*, 'Optical signatures of the Aharonov–Bohm phase in single-walled carbon nanotubes', *Science*, **304**, 1129 (2004).

(6.67) J. Cao, Q. Wang, M. Rolandi *et al.*, 'Aharonov–Bohm interference and beating in single-walled carbon-nanotube interferometers', *Phys. Rev. Lett.* **93**, 216803 (2004).

(6.68) N. Wang, Z. K. Tang, G. D. Li *et al.*, 'Single-walled 4 Å carbon nanotube arrays', *Nature*, **408**, 50 (2000).

(6.69) Z. K. Tang, L. Y. Zhang and N. Wang, 'Superconductivity in 4 angstrom single-walled carbon nanotubes', *Science*, **292**, 2462 (2001).

(6.70) M. Kociak, A. Y. Kasumov, S. Guéron *et al.*, 'Superconductivity in ropes of single-walled tubes', *Phys. Rev. Lett.* **86**, 2416 (2001).

(6.71) I. Takesue, J. Haruyama, N. Kobayashi *et al.*, 'Superconductivity in entirely end-bonded multiwalled carbon nanotubes', *Phys. Rev. Lett.* **96**, 057001 (2006).

(6.72) P. Lambin, J. P. Vigneron, A. Fonseca *et al.*, 'Atomic structure and electronic properties of bent carbon nanotubes', *Synth. Met.*, **77**, 249 (1996).

(6.73) L. Chico, V. H. Crespi, L. X. Benedict *et al.*, 'Pure carbon nanoscale devices: nanotube heterojunctions', *Phys. Rev. Lett.*, **76**, 971 (1996).

(6.74) P. G. Collins, A. Zettl, H. Bando *et al.*, 'Nanotube nanodevice', *Science*, **278**, 100 (1997).

(6.75) Z. Yao, H. W. C. Postma, L. Balents *et al.*, 'Carbon nanotube intramolecular junctions', *Nature*, **402**, 273 (1999).

(6.76) C. Zhou, J. Kong, E. Yenilmez *et al.*, 'Modulated chemical doping of individual carbon nanotubes', *Science*, **290**, 1552 (2000).

(6.77) J. U. Lee, P. P. Gipp and C. M. Heller, 'Carbon nanotube p–n junction diodes', *Appl. Phys. Lett.*, **85**, 145 (2004).

(6.78) S. J. Tans, A. R. M. Verschueren and C. Dekker, 'Room-temperature transistor based on a single carbon nanotube', *Nature*, **393**, 49 (1998).

(6.79) R. Martel, H. S. P. Wong, K. Chan *et al.*, 'Carbon nanotube field-effect transistors for logic applications', *Proc. IEDM*, 2001, p. 159.

(6.80) C. Zhou, J. Kong and H. J. Dai, 'Electrical measurements of individual semiconducting single-walled nanotubes of various diameters', *Appl. Phys. Lett.*, **76**, 1597 (2000).

(6.81) P. G. Collins, M. S. Arnold and P. Avouris, 'Engineering carbon nanotubes and nanotube circuits using electrical breakdown', *Science*, **292**, 706 (2001).

(6.82) R. Seidel, A. P. Graham, E. Unger *et al.*, 'High-current nanotube transistors', *Nano Lett.*, **4**, 831 (2004).

(6.83) K. Keren, R. S. Berman, E. Buchstab *et al.*, 'DNA-templated carbon nanotube field-effect transistor', *Science*, **302**, 1380 (2003).

(6.84) P. G. Collins, K. Bradley, M. Ishigami *et al.*, 'Extreme oxygen sensitivity of electronic properties of carbon nanotubes', *Science*, **287**, 1801 (2000).

(6.85) M. Bockrath, J. Hone, A. Zettl *et al.*, 'Chemical doping of individual carbon nanotube ropes', *Phys. Rev. B*, **61**, R10606 (2000).

(6.86) V. Derycke, R. Martel, J. Appenzeller *et al.*, 'Carbon nanotube inter- and intramolecular logic gates', *Nano Lett.*, **1**, 453 (2001).

(6.87) A. Bachtold, P. Hadley, T. Nakanishi *et al.*, 'Logic circuits with carbon nanotube transistors', *Science*, **294**, 1317 (2001).

(6.88) A. Bachtold, P. Hadley, T. Nakanishi *et al.*, 'Logic circuits based on carbon nanotubes', *Physica E*, **16**, 42 (2003).

(6.89) Z. Chen, J. Appenzeller, Y.-M. Lin *et al.*, 'An integrated logic circuit assembled on a single carbon nanotube', *Science*, **311**, 1735 (2006).

(6.90) R. C. Haddon, 'Magnetism of the carbon allotropes', *Nature*, **378**, 249 (1995).

(6.91) H. Ajiki and T. Ando, 'Magnetic properties of carbon nanotubes', *J. Phys. Soc. Jpn.* **62**, 2470 (1993).

(6.92) H. Ajiki and T. Ando, 'Magnetic properties of ensembles of carbon nanotubes', *J. Phys. Soc. Jpn.* **64**, 4382 (1995).

(6.93) M. F. Lin and W. K. Shung, 'Magnetization of graphene tubules', *Phys. Rev. B*, **52**, 8423 (1995).

(6.94) A. P. Ramirez, R. C. Haddon, O. Zhou *et al.*, 'Magnetic susceptibility of molecular carbon: Nanotubes and fullerite', *Science*, **265**, 84 (1994).

(6.95) X. K. Wang, R. P. H. Chang, A. Patashinski *et al.*, 'Magnetic susceptibility of buckytubes', *J. Mater. Res.*, **9**, 1578 (1994).

(6.96) O. Chauvet, L. Forro, W. Bacsa *et al.*, 'Magnetic anisotropies of aligned carbon nanotubes', *Phys. Rev. B*, **52**, R6963 (1995).

(6.97) M. Fujiwara, E. Oki, M. Hamada *et al.*, 'Magnetic orientation and magnetic properties of a single carbon nanotube', *J. Phys. Chem. A*, **105**, 4383 (2001).

(6.98) R. F. Service, 'Nanotubes show image-display talent', *Science*, **270**, 1119 (1995).

(6.99) T. Utsumi, 'Vacuum microelectronics – what's new and exciting', *IEEE Trans. Electron Devices*, **38**, 2276 (1991).

(6.100) F. S. Baker, A. R. Osborn and J. Williams, 'Field-emission from carbon fibres – new electron source', *Nature*, **239**, 96 (1972).

(6.101) W. Zhu, G. P. Kochanski, S. Jin *et al.*, 'Defect-enhanced electron field-emission from chemical-vapor-deposited diamond', *J. Appl. Phys*, **78**, 2707 (1995).

(6.102) G. A. J. Amaratunga and S. R. P. Silva, 'Nitrogen containing hydrogenated amorphous carbon for thin-film field emission cathodes', *Appl. Phys. Lett.*, 1996, **68**, 2529.

(6.103) B. S. Satyanarayana, J. Robertson and W. I. Milne, 'Low threshold field emission from nanoclustered carbon grown by cathodic arc', *J. Appl. Phys.*, **87**, 3126 (2000).

(6.104) W. A. de Heer, A. Châtelain and D. Ugarte, 'A carbon nanotube field-emission electron source', *Science*, **270**, 1179 (1995).

(6.105) Q. H. Wang, A. A. Setlur, J. M. Lauerhaas *et al.*, 'A nanotube-based field-emission flat panel display', *Appl. Phys. Lett.*, **72**, 2912 (1998).

(6.106) J. M. Bonard, J. P Salvetat, T. Stöckli *et al.*, 'Field emission from carbon nanotubes: perspectives for applications and clues to the emission mechanism', *Appl. Phys. A*, **69**, 245 (1999).

(6.107) S. S. Fan, M. G. Chapline, N. R. Franklin *et al.*, 'Self-oriented regular arrays of carbon nanotubes and their field emission properties', *Science*, **283**, 512 (1999).

(6.108) L. Nilsson, O. Groening, C. Emmenegger *et al.*, 'Scanning field emission from patterned carbon nanotube films', *Appl. Phys. Lett.*, **76**, 2071 (2000).

(6.109) W. I. Milne, K. B. K. Teo, G. A. J. Amaratunga *et al.*, 'Carbon nanotubes as field emission sources', *J. Mater. Chem.*, **14**, 933 (2004).

(6.110) E. Minoux, O. Groening, K. B. K. Teo *et al.*, 'Achieving high-current carbon nanotube emitters', *Nano Lett.*, **5**, 2135 (2005).

(6.111) A. G. Rinzler, J. H. Hafner, P. Nikolaev *et al.*, 'Unravelling nanotubes: field emission from an atomic wire', *Science*, **269**, 1550 (1995).

(6.112) S. M. Yoon, J. Chae and J. S. Suh, 'Comparison of the field emissions between highly ordered carbon nanotubes with closed and open tips', *Appl. Phys. Lett.*, **84**, 825 (2004).

(6.113) Y. Saito, K. Hamaguchi, K. Hata *et al.*, 'Conical beams from open nanotubes', *Nature*, **389**, 554 (1997).

(6.114) M. Croci, I. Arfaoui, T. Stöckli *et al.*, 'A fully sealed luminescent tube based on carbon nanotube field emission', *Microelectron. J.*, **35**, 329 (2004).

(6.115) W. I. Milne, K. B. K. Teo, M. Mann *et al.* "Carbon nanotubes as electron sources", *Physica Status Solidi A*, **203**, 1058 (2006).

7 Physical properties II: mechanical, optical and thermal

The mechanical properties of carbon nanotubes have attracted just as much interest as their electronic properties, and with good reason. It is now well established that nanotubes are the stiffest and strongest fibres ever produced. Thus, the Young's modulus of the best nanotubes can be as high as 1000 GPa, approximately five times higher than steel, while their tensile strength can be up to 63 GPa, around 50 times higher than steel. These properties, coupled with their low density, give nanotubes huge potential in a whole range of structural applications. In many cases, exploiting these properties involves incorporating the tubes into composite materials, and this is discussed in Chapter 9. The present chapter covers the mechanical properties of carbon nanotubes in detail, beginning with a review of the theoretical predictions and then giving a summary of experimental measurements on multi- and single-walled nanotubes. Optical properties of carbon nanotubes are considered next, and the application of various forms of spectroscopy to nanotubes is described. This is followed by a brief discussion of the thermal properties of nanotubes, and finally some comments on the physical stability of nanotubes.

7.1 Mechanical properties of carbon nanotubes

7.1.1 Theoretical predictions

Before discussing detailed calculations of the mechanical properties of nanotubes, we can carry out some simple calculations to illustrate the relationship between the diameter of a nanotube and its stiffness. Consider first a tube with an inner diameter of 1 nm. We assume a wall thickness of 0.34 nm, so the outer diameter is 1.68 nm and the cross-sectional area is 1.43×10^{-18} m^2. If we now apply a tensile load of 100 nN to the tube, this results in a stress of $\sim 7 \times 10^{10}$ Nm^{-2}. The corresponding strain, assuming a Young's modulus of 1060 GPa, is approximately 6.6%. Now consider a tube with an inner diameter of 10.0 nm and an outer diameter of 10.68 nm. In this case a tensile load of 100 nN results in a stress of 9.05×10^9 Nm^{-2}, and a strain of about 0.85%. These figures clearly demonstrate the way in which stiffness increases with tube diameter. This is consistent with the observation that single-walled nanotubes, with diameters typically of the order of 1 nm, are usually curly, while multiwall tubes tend to be straight.

In this simple calculation, it was assumed that the Young's modulus for nanotubes will be equal to that for a graphene sheet, i.e. 1060 GPa. This can only be an approximation, and a number of groups have attempted to calculate the modulus for nanotubes with various diameters and structures, using a variety of approaches. Some of the first detailed calculations were carried out by Jian Ping Lu of the University of North Carolina at Chapel Hill, who carried out tight-binding calculations on single-walled tubes with diameters from 0.34 to 13.5 nm (7.1), and found a Young's modulus of 970 GPa. This is close to the modulus for a graphene sheet, and was found to be independent of tube structure or diameter. A short time later, Angel Rubio of the University of Valladolid and colleagues found slightly higher values for the Young's moduli (typically 1240 GPa) of tubes with a range of structures and diameters (7.2). Unlike Lu, Rubio *et al.* found that the moduli depended on both tube diameter and structure. Subsequent work has generally supported the view that modulus depends on tube diameter and structure (e.g. 7.3–7.8). A review of computational studies (7.7) found that the predicted values of Y are in the range 0.5–5.5 TPa. This rather wide range of predicted values is partly due to the fact that different authors chose different values for the wall thickness of a single-walled tube. Most authors assume a wall thickness equal to the separation between adjacent walls in a multiwalled tube, i.e. 0.34 nm. The high values of 5 TPa and above were found in studies assuming much smaller values for the wall thickness. It is important to appreciate that while the *modulus* can vary with tube diameter, the *stiffness* of small tubes will always tend to be less than for larger tubes, as illustrated in the simple-minded calculation above.

The behaviour of nanotubes under compression has been studied by workers from North Carolina State University (NCSU) (e.g. 7.9, 7.10). Some of their results are illustrated in Fig. 7.1. Here, the effect of axial compression on an armchair (7, 7) nanotube has been simulated using molecular dynamics. Fig. 7.1(a) shows a plot of strain energy vs. longitudinal strain, ε. This shows that at small strains, the strain energy varies with ε^2, as expected from Hooke's Law, but at higher strains a series of discontinuities are observed, and the strain energy curve becomes approximately linear. The discontinuities labelled b–e correspond to the four buckled configurations shown in the simulations. Nanotube bending was also simulated, as illustrated in Fig. 7.2 for the case of a (13, 0) zigzag tube. Again a kink is observed in the strain energy curve, corresponding to the buckled structure shown in Fig. 7.2(b). A number of other groups have also modelled the mechanism of nanotube buckling (7.11–7.14).

The behaviour of nanotubes under high tensile loads, which leads to eventual fracture, has been widely studied theoretically (e.g. 7.15–7.19). Again the NCSU group led the way. Their calculations have shown that the maximum theoretical tensile strain, i.e. elongation, of a single-walled tube is almost 20% (7.15, 7.16). Many researchers believe that the early stages of the fracture mechanism may involve the formation of Stone–Wales defects (see p. 115). The brittle fracture of nanotubes was simulated by workers from Nanyang Technological University, Singapore (7.17). They found that a (5-7-7-5) defect is formed at a strain of 0.24. As the strain increases, more (5-7-7-5) defects are generated, and when the strain reaches 0.256, two bonds are broken leading to two holes, as shown in Fig. 7.3(a). With increasing strain, more bonds are broken and the holes become larger until the tube fractures (Fig. 7.3b–d). As discussed on p. 137, (5-7-7-5) defects have been directly imaged in single-walled tubes using HRTEM.

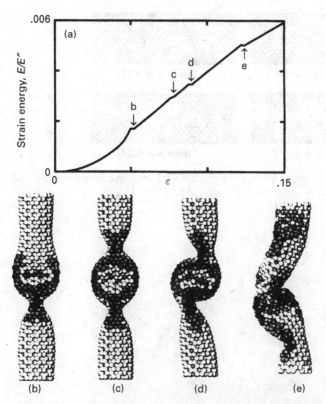

Fig. 7.1 Molecular dynamics simulation of armchair (7, 7) nanotube under axial compression, from the work of Yakobson *et al.* (7.9). (a) Plot of strain energy vs. longitudinal strain, ε. (b)–(e) Morphological changes corresponding to singularities in strain energy curves.

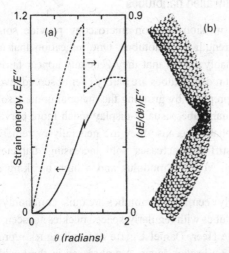

Fig. 7.2 The simulated bending of a (13, 0) zigzag tube. (b) The morphology of a tube beyond buckling point (7.9).

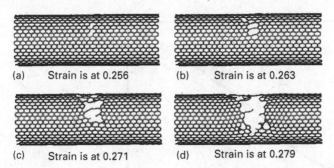

(a)   Strain is at 0.256   (b)   Strain is at 0.263

(c)   Strain is at 0.271   (d)   Strain is at 0.279

Fig. 7.3   A simulation of the brittle fracture of a single-walled nanotube, from the work of Liew *et al.* (7.17).

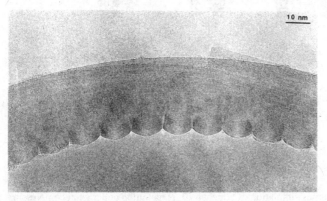

Fig. 7.4   A TEM image of a bent nanotube (radius of curvature ~400 nm), showing wavelike distortion (7.23).

7.1.2   Experimental observations: multiwalled nanotubes

General observations using transmission electron microscopy provide some useful insights into the stiffness and strength of nanotubes. One indication that arc-grown multiwalled nanotubes are reasonably stiff is that they generally appear fairly straight in TEM images. Completely fractured nanotubes are hardly ever observed, even though samples for TEM are sometimes prepared by grinding the material under a solvent in a pestle and mortar. Single-walled nanotubes usually display much more curvature than multiwalled tubes, but this is probably because they are generally extremely thin (as noted in the previous section, stiffness increases with increasing diameter). These observations indicate both a high Young's modulus and a high breaking stress for nanotubes.

Although broken tubes are rarely seen, bent nanotubes are quite commonly observed (e.g. 7.20–7.23). Frequently, bent tubes with regularly-spaced buckles are seen, as shown in the beautiful image by Walt de Heer, Daniel Ugarte and colleagues reproduced in Fig. 7.4 (7.23). It is striking that no broken layers are observed in the buckled areas, demonstrating the great flexibility of graphene layers. It has been demonstrated that when

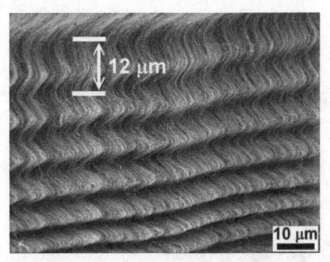

Fig. 7.5   A SEM micrograph showing a buckled array of aligned MWNTs under compression (7.25).

the stress constraining a tube is released, the tube can return to its original straight form (7.24). This behaviour sets nanotubes apart from conventional carbon fibres, and all other currently known fibres, which are much more susceptible to fracture when subject to stress beyond the elastic limit. It also raises the fascinating possibility that composites could be produced which would snap back into place following deformation. Ajayan's group and collaborators have studied the effect of compression on aligned MWNT films (7.25). They found that the tubes collectively formed zigzag buckles, as shown in Fig. 7.5. Like bent nanotubes, these compressed tubes could fully unfold to their original length upon load release. Buckled single-walled tubes have also been described by Iijima and colleagues (7.22), some of whose work is shown in Fig. 7.6.

The first quantitative TEM measurements of the mechanical properties of nanotubes were carried out by Treacy, Ebbesen and Gibson in 1996 (7.26). Clusters of nanotubes were deposited on TEM grids such that isolated tubes extended for a considerable distance into empty space. The specimens were then placed in a special holder which enabled *in situ* heating to be carried out in the TEM. Images were then recorded of a number of individual, freely vibrating, nanotubes at temperatures up to 800 °C. Examples are shown in Fig. 7.7. By analysing the mean-square amplitude as a function of temperature it was possible to obtain estimates for the Young's modulus. These ranged from 410 GPa to 4.15 TPa, with an average of 1.8 TPa. The large spread in values results from uncertainties in estimating the lengths of the anchored tubes, and from the presence of defects in the tube structures.

Charles Lieber and colleagues from Harvard were probably the first to use scanning probe microscopy to probe the mechanical properties of nanotubes. In 1997 they described a method of fixing nanofibres at one end, and then determining the bending force of the fibres as a function of displacement (7.27). The fibres were firstly dispersed on a single crystal MoS substrate, and then pinned to this surface by depositing square

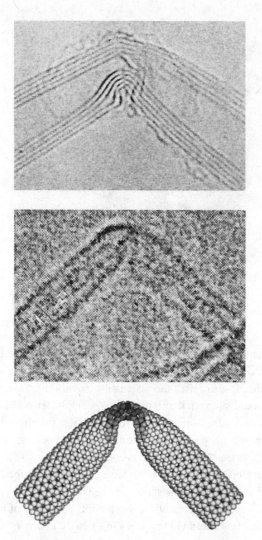

Fig. 7.6   Images by Iijima of buckled multiwalled and single-walled nanotubes, and a simulation of structure (7.22).

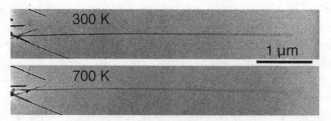

Fig. 7.7   TEM images of free-standing multiwalled nanotubes showing blurring of the tips due to thermal vibration (from 300 to 600 K) (7.26).

pads of SiO through a mask. This left many fibres with one end trapped under the pads and one end free. The samples were then imaged by AFM, enabling the free nanofibres to be located. Repeated scans of individual fibres then enabled lateral force vs. displacement (*F–d*) data to be recorded. Scanning could be carried out in such a way that the tip rode over the fibre when a certain applied force was reached, so that the fibre sprang back to its equilibrium position, enabling further scans to be made. Measurements on arc-produced multiwalled nanotubes produced linear *F–d* curves for small deflections, and the results implied a value of ~1.28 TPa for the elastic modulus. It was found that the tubes could accommodate large deflections without breaking. However, for deflections larger than 10° an abrupt change in the slope of the *F–d* curve was seen. This was attributed to elastic buckling of the kind seen in TEM studies.

In a paper published shortly after the Harvard work, Richard Superfine and colleagues from the University of North Carolina at Chapel Hill also described AFM studies of arc-synthesized MWNTs subjected to large bending stresses (7.28). For these experiments the tubes were not fixed at one end, but were supported on mica. In many cases the friction was sufficient to pin the tube in a strained configuration for imaging. Again it was found that the tubes could be bent repeatedly through large angles without fracturing. Fig. 7.8 shows an individual tube that has been manipulated into a severely deformed configuration. Despite repeated bending, the nanotubes exposed to this treatment showed no sign of plastic damage. As in the TEM studies, regularly spaced buckles were observed in the bent tubes, which disappeared when the tubes were straightened. These experiments provide further evidence of the extraordinary resilience of carbon nanotubes.

Since this early work, the mechanical properties of multiwalled nanotubes have been widely studied using AFM (7.29–7.34), TEM (7.35, 7.36) and SEM (7.37–7.39). All of these studies have confirmed that MWNTs, at least those produced by arc-evaporation, have exceptional stiffness and strength. Some of the most important studies will now be summarized. For further information, a number of reviews are available (7.40–7.43).

László Forró and colleagues at Lausanne described detailed studies of arc-produced MWNTs using AFM in 1999 (7.30). In this work the tubes were deposited on a polished alumina ultra-filtration membrane with 200 nm pores, and an AFM tip was used to apply a load to a tube which was suspended across a pore. The average value of the Young's modulus for 11 individual arc-grown carbon nanotubes was found to be 810 GPa.

Manipulating carbon nanotubes inside an electron microscope poses an even greater challenge than doing so using AFM, but a few groups have achieved this. In 2000, John Cumings and Alex Zettl described *in situ* experiments inside a TEM in which the reversible telescopic extension of arc-grown multiwalled tubes was demonstrated (7.35). There appeared to be almost no friction between the layers, and no evidence of wear or fatigue was seen, suggesting that the use of MWNTs in nanomechanical systems (7.44) may be a practical possibility. In a subsequent study, the Zettl group used the same *in situ* system to carry out a series of pulling and bending tests on individual MWNTs (7.36). In some of these tests they were able to break the tubes, and therefore calculated the tensile strengths: values of approximately of 150 GPa were found. From bending studies, the Young's modulus of the tubes was estimated to be 900 GPa.

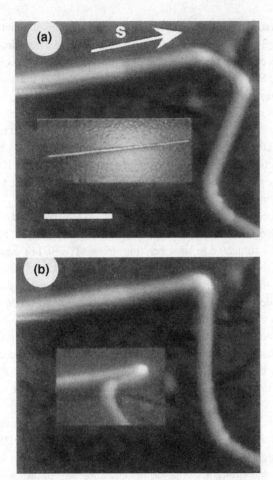

Fig. 7.8   (a), (b) Atomic force microscope images of MWNTs on mica, showing bending and buckling induced by the tip (7.28). The scale bar in (a) is 300 nm.

Exceptionally skilful work on the mechanical properties of nanotubes has been described by Rodney Ruoff and his team (7.37–7.39). This involved the use of a micromanipulator inside an SEM to attach individual arc-grown MWNTs to the opposing tips of AFM cantilever probes, as shown in Fig. 7.9. In order to attach the tube ends to the probes, the electron beam was used to deposit carbonaceous material on the point of contact. When a tube had been successfully attached at both ends, one of the cantilevers was driven away from the other, to apply a tensile load. Analysis of the stress–strain curves for individual MWNTs enabled the Young's moduli of the outermost layer to be determined. Quite a wide range of values were found, varying from 270 to 950 GPa. The most likely explanation for this variation is the presence of

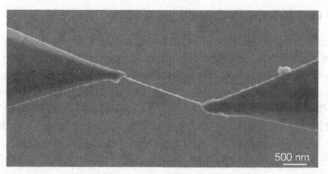

Fig. 7.9   A SEM image of individual MWNT mounted between two opposing AFM tips, for mechanical testing (7.37)

defects. It should also be recognized that only the outer layers of the tubes were being stressed in these experiments, unlike some of the other methods where the moduli of whole tubes were determined. When sufficient stress was applied, the outer layers fractured, leaving the inner ones intact, a failure mode known as 'sword-in-sheath'. The measured tensile strength of the outer layers ranged from 11 to 63 GPa. The phenomenon of sliding shells in multiwalled tubes was also observed by the Ruoff group.

All of the studies discussed so far have concerned multiwalled nanotubes produced by arc-evaporation. Forró and colleagues have also measured the mechanical properties of catalytically-produced MWNTs, and generally found them to be much inferior to those of the arc-grown ones. In the first study, published in 1999 they used the AFM method described above to directly compare arc-grown and catalytically-produced MWNTs (7.31). The average Young's modulus for the arc-grown tubes was found to be 870 GPa, very similar to the value found in the previous study (7.30). The average figure for catalytically-produced tubes, however, was just 27 GPa. In a later study (7.32) it was shown that the elastic moduli of the CVD tubes were not significantly improved by heating at temperatures up to 2400 °C. The best value for the modulus, even after heat treatment, was still below 100 GPa. This is a very important result, which demonstrates the inherent structural superiority of arc-produced MWNTs. Later work using AFM (7.33) produced the slightly higher figure of 350 ± 110 GPa, but this is still far short of the values for arc-grown tubes.

Clearly most catalytically-grown tubes contain defects that it is almost impossible to remove. This may not be true of all catalytically-produced tubes however. The Forró group have shown that catalytically-produced MWNTs *with a small number of layers* can have excellent mechanical properties (7.34). Thus, moduli as high as 1 TPa were seen for tubes with 2 or 3 walls. This indicates a much lower defect density than in many-walled CVD tubes, and may indicate that 'few-walled' nanotubes grow by a different mechanism, as already noted (p. 71).

7.1.3   Experimental observations: single-walled nanotubes

Measuring the elastic properties of single-walled tubes presents an even greater challenge than for MWNTs, and there have been very few measurements on individual SWNTs. Ebbesen, Treacy and others were the first to report such measurements, in 1998 (7.45). Using methods similar to those first used for MWNTs, i.e. observing room temperature vibrations of the tubes in a TEM, they found an average Young's modulus of 1.25 TPa. This is rather similar to the typical values found for arc-grown MWNTs.

While experiments on individual SWNTs are extraordinarily difficult, the mechanical properties of SWNT 'ropes' are rather easier to study. Ruoff and co-workers used the 'nanostressing stage' described in the previous section to measure the breaking strength and Young's modulus of SWNTs produced by laser ablation (7.46). From these measurements they were able to determine values for individual tubes, and found an average strength of 30 GPa and average modulus of 1.0 TPa. The maximum elongation of a single-walled tube under tensile strain was found to be 6%, much lower than the 20% predicted by theory (7.15, 7.16).

A group led by Jean-Paul Salvetat and László Forró used scanning probe microscopy to measure the mechanical properties of SWNT 'ropes' prepared by arc-discharge in 1999 (7.47). They used the AFM method described in the previous section, with ropes suspended across 200 nm pores. In this way they found elastic and shear moduli of the ropes were to be of the order of 1 TPa and 1 GPa, respectively. A study of CVD-grown SWNTs was conducted by Thomas Tombler of Stanford and colleagues a short time later (7.48). Although the main purpose of this work was to look at the electromechanical properties of the tubes, the Young's modulus was also measured and a value of 1.2 TPa was found. A group from Rice used AFM to measure the tensile strength of single-walled nanotube ropes produced by laser vaporization (7.49). The tensile strength was found to be approximately 45 GPa, with a maximum strain of 5.8%. Taken together, these studies show that SWNTs have exceptional mechanical properties, independent of the preparation method.

7.2   Optical properties of nanotubes

The optical properties of carbon materials vary from the sparkling transparency of diamonds to the deep black of soot. Diamonds owe their transparency to the fact that all of the electrons are associated with the sp^3 bonds, which absorb only IR light, so that visible light passes straight through. Their sparkle results from an exceptionally high refractive index, and the ability to disperse visible light into its spectral components. Soot, on the other hand, like all finely divided forms of sp^2 carbon, has an abundance of π-electrons capable of absorbing light at a wide range of wavelengths, giving it a black colour. Graphite does not have the matt black colour of more finely divided sp^2 carbons, as can be seen by comparing the mark made by a pencil with that made by a charcoal crayon. This is because the metallic character of graphite gives it reflective properties; these are particularly evident in the extremely crystalline from of graphite known as highly-oriented pyrolytic graphite.

Samples of nanotubes usually appear black. However, as we saw in Chapter 4 (p. 91 and p. 96), very thin layers of tubes can be transparent. This is because the penetration depth for nanotubes is rather large. Penetration depth is the distance electromagnetic radiation of a specific wavelength can penetrate into the material. For visible light the penetration depth varies inversely with the free carrier density and since nanotubes are a low carrier density system they have a large penetration depth.

In the following two sections we consider the use of optical spectroscopy techniques in the study of carbon nanotubes.

7.2.1   Optical absorption spectroscopy

Optical absorption spectroscopy has not been widely used to study carbon nanotubes, as it is not enormously informative. For example, absorption spectra cannot be used to determine nanotube structure. The relatively few studies that have been carried out have almost exclusively involved single-walled tubes. One of the most important of these was published by Hiromichi Kataura, then at Tokyo Metropolitan University, and colleagues in 1999 (7.50). Figure 7.10, taken from this work, shows absorption spectra of as-prepared and purified SWNTs. Three large absorption peaks at approximately 0.68, 1.2 and 1.7 eV can be seen, superimposed on the broad absorption due to the π plasmon. It was also shown that the positions of these peaks varied slightly in spectra from SWNTs with different diameter distributions. This is to be expected, since the absorption bands are due to transitions between spikes in the densities of states in the electronic structure of the tubes. As we saw in the previous chapter, the positions of these singularities in the densities of states depend on the structure and diameter of the tube. Kataura *et al.* determined the theoretical gap energies between mirror-image spikes in the densities of

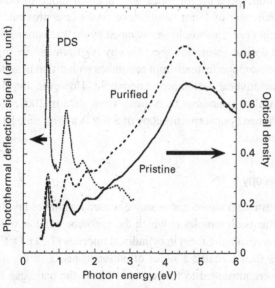

Fig. 7.10   Optical absorption spectra of SWNTs prepared by the arc method, adapted from the work of Kataura and colleagues (7.50).

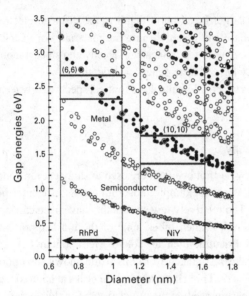

Fig. 7.11   The Kataura plot (7.50), showing calculated gap energies between mirror-image spikes in the density of states for single-walled nanotubes with different diameters. Solid circles indicate metallic SWNTs and open circles semiconducting ones. Double circles indicate armchair SWNTs. Arrows show the diameter distributions for tubes produced by arc-evaporation or laser vaporization using RhPd and Ni/Y 'catalysts'. Two horizontal lines in each catalyst area show a 'metallic window' in which only the optical transitions of metallic tubes would be observed.

states for a large number of single-walled tubes. The resulting graph, shown in Fig. 7.11, has come to be known as the Kataura plot. The plot shows where peaks should be observed in an absorption spectrum for tubes with a given range of diameters. For example, tubes with diameters of 1 nm would have peaks at approximately 0.8, 1.6 and 2.3 eV (peaks at higher energies would be swamped by the π plasmon). The Kataura plot shows why simple optical absorption spectroscopy is of limited use in identifying nanotube structure: the absorption features from nanotubes with different structures often overlap, making an unequivocal assignment impossible. However, the discovery of fluorescence in single-walled nanotubes by a group from Rice in 2002 (7.51) opened the way to structure-assigned optical spectroscopy of SWNTs a short time later (7.52), as discussed in the next section.

7.2.2   Fluorescence spectroscopy

The first study to demonstrate fluorescence in single-walled nanotubes involved making spectroscopic measurements on samples in which the nanotube bundles were separated into isolated tubes by encapsulating them in cylindrical micelles (7.51). Optical spectroscopy of these samples then revealed a series of emission peaks in the near infrared (~800–1600 nm) that were attributed to fluorescence across the band gap of semiconducting nanotubes. The fluorescence can be understood by reference to Fig. 7.12, which shows schematically the density of states for a semiconducting nanotube. Fluorescence

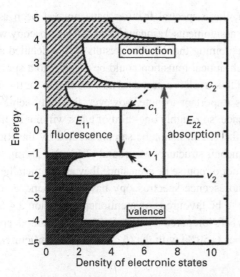

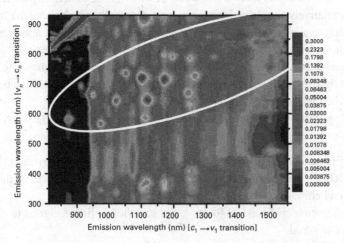

Fig. 7.12   A schematic of the density of electronic states for a single nanotube. The solid arrows depict the optical excitation and emission transitions of interest; the dashed arrows denote non-radiative relaxation of the electron (in the conduction band) and hole (in the valence band) before emission (7.52).

Fig. 7.13   A contour plot of fluorescence intensity versus excitation and emission wavelengths for SWNTs (7.52).

occurs when light absorption at photon energy E_{22} is followed by fluorescence emission near E_{11}. The values of E_{11} and E_{22} depend on tube structure.

The Kataura plot shows that nanotubes with different structures can have similar E_{11} values, so it is necessary to consider both the excitation wavelengths and the emission wavelengths in order to separate out the spectral features of particular nanotubes. The Rice group represented the excitation and emission wavelengths using 2D plots of the kind shown in Fig. 7.13. Here the dark spots within the oval represent transitions from individual nanotubes. However, these spectral features have not yet been assigned to (n, m) structures. To make these assignments, resonance Raman spectroscopy was carried out on the same samples. By combining these Raman results with a detailed analysis of the spectrofluorimetric data, each optical transition could be mapped to a specific (n, m) nanotube structure. The result of this work is a powerful new technique for the determination of nanotube structure. It is important to note, however, that only semiconducting nanotubes will exhibit fluorescence, so roughly one-third of tubes will not be detected by this form of spectroscopy. For the HiPco nanotube sample studied by the Rice team, a significant bias towards near-armchair structures was found (7.53). No zigzag tubes were detected; armchair tubes would not of course be seen since they are all metallic.

Subsequent studies using fluorescence spectroscopy have also found evidence that near-armchair structures appear to be favoured. As mentioned in Section 3.4.5, Resasco and colleagues found that SWNTs prepared by the CoMoCAT method appeared to contain a high proportion of tubes (~50%) with the (6, 5) and (7, 5) structures.

7.3   Raman spectroscopy

Raman spectroscopy has proved to be an extremely valuable tool for the detection of nanotubes in bulk samples and for the study of nanotube structure. Experimentally the technique is relatively simple, at least when applied to bulk samples, and the instrumentation is widely available. Spectra can be recorded at room temperature and pressure, and the technique is quick, non-destructive and sensitive. Raman can also be applied to individual nanotubes, although this is experimentally much more demanding. The discussion that follows refers mainly to single-walled nanotubes; the technique has been of less value when applied to multiwalled tubes.

A typical spectrum of a sample of single-walled tubes is shown in Fig. 7.14. The main features are as follows.

(1) A low-frequency peak ($< 200 \, cm^{-1}$), assigned to the A_{1g} symmetry radial breathing mode (RBM). The frequency of the peak depends on the diameter of the tube, as discussed below.
(2) A strong feature at around $1340 \, cm^{-1}$, the so-called D line, assigned to disordered graphitic material.
(3) A group of peaks in the approximate range $1550–1600 \, cm^{-1}$ labelled the G band. In graphite, the G band exhibits a single peak at $1582 \, cm^{-1}$ related to the tangential mode vibrations of the C atoms. Although not clear in Fig. 7.14, the G-band for SWNTs is composed of two features due the confinement of the vibrational wave vectors in the circumferential direction.
(4) A line at around $2600 \, cm^{-1}$, the second order harmonic of the D mode, labelled G'.
(5) Some second order modes between about 1700 and $1800 \, cm^{-1}$; these are not of great value in characterizing nanotubes.

The vibrations responsible for the RBM and G-band vibrations are shown in Fig. 7.15. The radial breathing feature occurring just below $200 \, cm^{-1}$ is a unique phonon mode,

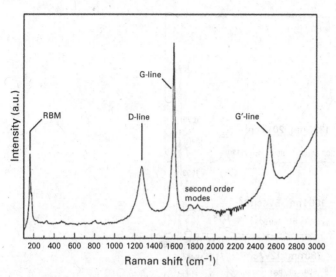

Fig. 7.14 Raman spectrum from a SWNT sample, from the work of Belin and Epron (7.54).

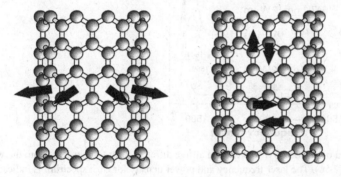

Fig. 7.15 Atomic displacements associated with the RBM and *G*-band normal-mode vibrations (7.55).

appearing only in spectra from single-walled carbon nanotubes, and is therefore very useful in confirming that bulk samples contain SWNTs. Importantly, it can also provide information concerning tube diameter, and can be used to infer structure. This feature was first described in detail in a classic paper published in *Science* in 1997 (7.56). This paper, a collaborative effort by workers from five different centres in the USA and Japan, described a Raman study of purified nanotube 'rope' samples. A number of new nanotube-related features were identified, with the RBM peak being much the strongest. Strikingly, it was demonstrated that both the strength and position of this peak varied according to the laser excitation frequency, as shown in Fig. 7.16. These results were interpreted in terms of a resonant Raman scattering process, and provide evidence that tubes with different diameters couple with different efficiencies to the laser field. Resonant Raman scattering occurs when the energy of an incident photon matches the energy of strong optical absorption electronic transitions. This greatly increases the intensity of the observed Raman effect. The allowed optical transitions have been

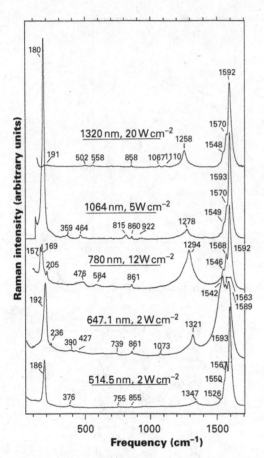

Fig. 7.16   Raman spectra of purified SWNTs excited at five different laser frequencies, from the work of Rao *et al.* (7.56). The laser frequency and power density for each spectrum is indicated.

calculated for a large number of tube structures (see Fig. 7.11), so the laser energy required to maximize the signal from nanotubes with a given diameter can be determined.

Theory has shown that the frequency of the radial breathing mode, ω_{RBM}, varies inversely with diameter, d, at least for small tubes (7.57):

$$\omega_{RBM} = \alpha/d$$

where α is a factor that depends on the nature of the sample. This has been confirmed in a number of experimental studies (e.g. 7.58, 7.59). For tube diameters larger than about 2 nm, the character of the electronic states becomes essentially independent of tube diameter, and therefore approximates to that of a graphene sheet. It has also been established that ω_{RBM} is independent of the (n, m) indices of a tube (7.60, 7.61). However, the (n, m) values can be inferred, as discussed below.

A very large number of studies have now been published which exploit the resonant Raman effect to study single-walled nanotubes (e.g. 7.62–7.67). When studying samples in solution, a dispersant such as sodium dodecyl sulphate (SDS) is often used to separate

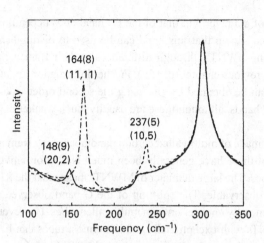

Fig. 7.17 Raman spectra from individual SWNTs on an SiO_2 substrate (7.68). The frequencies and linewidths (in brackets) are given, together with the (n, m) assignments. The 303 cm^{-1} feature comes from the substrate and is used for calibration.

bundles of tubes, and avoid complicating inter-tube effects. Especially noteworthy has been the application of the technique to the study of *individual* nanotubes. The first such experiments were reported by the Dresselhaus group in 2001 (7.68). In this work, isolated SWNTs were prepared directly by CVD on a Si substrate containing Fe catalyst particles. The Si substrate was then oxidized to give a thin SiO_2 layer. Atomic force microscopy showed that the tubes were generally well separated from each other, with an average diameter of about 2 nm. Resonant Raman spectra were obtained from individual SWNTs using a micro-Raman spectrometer with a 1 µm laser spot. Figure 7.17 shows the radial breathing mode region of spectra recorded from three different isolated SWNTs. The diameters of the tubes can be found directly, using the relation between ω_{RBM} and d given above. In order to determine the (n, m) values of the tubes, we should ideally know the electronic transition energy, E_{ii}, for each tube. However, in many cases an $(n; m)$ determination can be made since there is generally only one (n, m) value that has both a diameter close to the measured d value, determined from ω_{RBM}, while also being within the resonant window of the laser excitation energy. In this way, the Dresselhaus group were able to make (n, m) assignments for 25 individual tubes without determining E_{ii} (7.68). Of these, 21 were chiral, two zigzag and two armchair. In subsequent work they showed that values of E_{ii} for individual tubes can be obtained by a careful analysis of the laser energy dependence of Stokes and anti-Stokes Raman processes (7.69). By combining measurements of ω_{RBM} and E_{ii}, (n, m) assignments were made for 46 HiPco nanotubes wrapped in sodium dodecyl sulphate (SDS) and dispersed in an aqueous solution (7.70). They found that 40 were chiral, three zigzag and three armchair.

In addition to the RBM line, the G band can also be used to determine nanotube diameter. As already mentioned, the G-band for SWNTs is composed of two features, one at 1590 cm^{-1} labelled G^+ and the other at about 1570 cm^{-1} named G^-. Work by Dresselhaus *et al.* on individual SWNTs has shown that the frequency of the G^+ band is essentially independent of diameter, but that of the G^- band does depend on d (7.65, 7.71, 7.72). It has also been shown that this band can be used to distinguish between metallic and semiconducting SWNTs, through differences in their Raman lineshapes. This was discussed in the previous chapter (p. 159). Further information about structure and electronic properties can be obtained by analysing the second order features in the spectrum, i.e. the D and G' bands, although these are usually much weaker than the first order features (e.g. 7.62).

A few studies have been made of multiwalled carbon nanotubes using Raman spectroscopy (e.g. 7.73–7.76), but these have generally been much less informative than the studies of single-walled tubes. The large diameter of MWNTs means that the RBM signal is usually too weak to be observable. The splitting of the G band observed in single-walled nanotubes is also generally unobservable in multiwalled tubes. However, as noted in Chapter 2 (p. 16) MWNTs with exceptionally thin innermost tubes can be made by carrying out arc-evaporation in hydrogen. In this case both the RBM line and the splitting of the G band can be observed (7.77, 7.78).

Several reviews of Raman spectroscopy of nanotubes have been given (7.55, 7.79–7.82), and the book by Reich *et al.* also gives an excellent discussion of the topic (7.83).

7.4   Thermal properties of nanotubes

Crystalline carbons display the highest measured thermal conductivities of all known materials. For pure diamond the thermal conductivity, k, is 2000–2500 W m^{-1} K^{-1}, while for graphite, the in-plane conductivity at room temperature can reach 2000 W m^{-1} K^{-1}. For carbon nanotubes, even higher values are predicted: calculations by David Tománek and colleagues produced a value of 6600 W m^{-1} K^{-1} for an isolated (10, 10) nanotube at room temperature (7.84).

Early experimental work on the thermal properties of nanotubes was carried out on ropes or bundles of tubes. Alex Zettl's group at Berkeley measured the thermal conductivity of SWNT mats, made up of tangled bundles of ropes (7.85). Initially, the values obtained for room temperature thermal conductivity of the mats were not high, ranging from about 2–35 W m^{-1} K^{-1}. However, these values did not take into account the highly tangled nature of the ropes. An attempt to translate these numbers into thermal conductivities for individual ropes produced values in the range 1750–5800 W m^{-1} K^{-1}. Thermal conductivity measurements of SWNT mats were also made at temperatures down to 7 K. It was found that the temperature dependence of k differed markedly from that of graphite. In high-quality graphite, the in-plane thermal conductivity, which is dominated by acoustic phonons, varies as T^{2-3} up to 150 K (7.86). The low-temperature thermal conductance of the SWNT samples studied by Zettl and colleagues was found to vary linearly with temperature and extrapolated to zero at $T = 0$. This result is consistent with quantization of thermal conductance of SWNTs (see below).

It appears that there have as yet been no measurements of the thermal conductivity of individual single-walled tubes. Certainly such experiments would pose a major experimental challenge. Measurements have been made on individual multiwalled nanotubes, however, and again very high conductivities have been found, at least for arc-grown tubes. Paul McEuen's group at Berkeley have used a microfabricated device to measure the thermal conductivity of MWNTs produced by arc-evaporation (7.87). Values of greater than $3000 \, W \, m^{-1} \, K^{-1}$ at room temperature were found. The thermal conductivity of catalytically-produced MWNTs appears to be much lower, however. Workers from Singapore grew arrays of tubes on Si substrates and, using a method called pulsed photothermal reflectance, found values of approximately $200 \, W \, m^{-1} \, K^{-1}$ (7.88).

Like electronic conduction, heat conduction in nanoscale structures can become quantized. This is because at temperatures close to absolute zero, only a limited number of phonons remain active in small devices (7.89, 7.90). Quantized thermal conductance was first observed experimentally in 2000 by Schwab *et al.* using a device containing Si_3N_4 'phonon wave guides' (7.91). With their extremely small diameters and high stiffness, SWNTs are good candidates for the measurement of quantized thermal conductance. They have a large phonon mean-free-path, of the order of $1 \, \mu m$. A theoretical analysis of quantized thermal conductance in carbon nanotubes has been given by Takahiro Yamamoto and colleagues (7.92). It was shown that the phonon-derived thermal conductance of semiconducting SWNTs exhibits a universal quantization in the low-temperature limit, independent of the radius or atomic geometry. This has not yet been confirmed experimentally. However, some studies have produced results which reflect quantum effects in one dimension. The work of the Zettl's group on SWNT mats has already been mentioned (7.85). As noted in the previous chapter, researchers from the National Physical Laboratory carried out simultaneous thermal and electrical measurements on individual SWNTs using a temperature sensing scanning probe microscope and found some evidence for ballistic transport of phonons (7.93).

As well as carrying out detailed studies of the thermal properties of carbon nanotubes, Alex Zettl's group have used nanotubes to fabricate a thermal rectifier, a device that directs the flow of heat (7.94). This was achieved by loading one end of a tube with a high-mass-density material, specifically, trimethyl cyclopentadienyl platinum. It was shown that the system exhibited asymmetric thermal conductance with greater heat flow in the direction of decreasing mass density. Such a 'thermal diode' might be useful in preventing overheating in microelectronic devices.

7.5 The physical stability of nanotubes

The earliest theoretical work on the stability of carbon tubules as a function of diameter was carried out by Gary Tibbetts in 1983 (7.95). Using a continuum model, Tibbetts found that the strain energy of a thin graphitic tube varies with 1/(diameter). This implies that the strain energy per atom varies with 1/(diameter)^2. Following the discovery of fullerene-related nanotubes, a number of groups have carried out more detailed theoretical studies of nanotube stability (e.g. 7.96–7.98). John Mintmire and colleagues from the Naval Research Laboratory in Washington used empirical potentials to calculate the strain energies per carbon atom for all possible tubes with diameters less than 1.8 nm (7.96). They also found that, to a good approximation, the strain energy per atom varied with $1/(\text{diameter})^2$. For tubes with diameters larger than about 1.6 nm, the strain energy becomes very close to that in planar graphite. It is interesting to note that this diameter is approximately the same as the smallest observed experimentally in multiwalled nanotubes. Mintmire and colleagues found that strain energy was independent of tube structure.

Other workers have taken a slightly different approach to determining the stability of nanotubes (7.97, 7.98). This involves considering a nanotube as a rolled-up graphene strip, and balancing the energy gained due to the elimination of edge atoms against the energy cost of bond-bending. In calculations of this kind, Sawada and Hamada found that the critical diameter above which tubes are more stable than strips is approximately 0.4–0.6 nm (7.97). Independent calculations by Lucas, Lambin and Smalley produced a similar result (7.98).

The energetics of multiwalled carbon tubes have been considered by Charlier and Michenaud (7.99). They found that the energy gained by adding a new cylindrical layer to a central one was of the same order as the one in graphite bilayering. The optimum interlayer distance between an inner (5, 5) nanotube and an outer (10, 10) tube was found to be 0.339 nm. This is somewhat smaller than the 0.344 nm {002} spacing found in turbostratic graphite. Experimental measurements show that the interlayer spacings in nanotubes can vary quite considerably, but are typically around 0.34 nm (see Section 5.3.1).

The extreme thermal stability of multiwalled carbon nanotubes was demonstrated experimentally by Zettl's group in 2007 (7.100). Using a specially designed thermal platform compatible with extreme temperature operation and real-time TEM, they found atomic-scale stability to temperatures approaching 3000 °C. This suggests that carbon nanotubes are more robust than either graphite or diamond.

7.6   Discussion

A list of the physical properties of carbon nanotubes would be packed with superlatives. The mechanical properties, for example, of the best quality nanotubes are superior to those of any other known fibre. As we have seen, a whole series of studies have found that arc-grown multiwalled tubes have Young's moduli of around 1 TPa, with tensile strengths of the order of 50 GPa (although higher values for tensile strength have been recorded in some studies). Similar figures were found for single-walled tubes. The only other materials with properties which remotely approach these are conventional carbon fibres, produced by pyrolysing organic precursors. However carbon fibres do not possess the combination of properties that are found in nanotubes: they are either high modulus or high strength, but not both. Carbon nanotubes also have extremely high thermal stability, as noted in the preceding section. This stability partially explains the exceptionally high electric current-carrying capability of the tubes, which was mentioned in the previous chapter. The thermal conductivity of carbon nanotubes also seems to be extraordinary. Values of greater than 3000 W m^{-1} K^{-1} at room temperature have been found for arc-grown MWNTs. Considering these properties it is easy to understand the excitement which has surrounded carbon nanotubes. However it is important to recognize that not *all* nanotubes possess such outstanding characteristics. In particular, catalytically-produced MWNTs fall short in a number of areas, reflecting the relatively high concentration of defects in these tubes. Thus, the studies of László Forró and colleagues found an average figure for the elastic modulus of CVD-grown tubes of just 27 GPa (7.31). It might be thought that the stiffness could be improved by annealing out the defects. However, the Forró group found that even after heating to 2400 °C, the highest moduli were still below 100 GPa (7.32). Similarly the thermal conductivities of catalytically-produced MWNTs are much lower than those of the arc-grown tubes (7.88). It appears, therefore, that catalytically-grown MWNTs should not be used when the best properties are required, and that there is still a need to develop methods for making arc-quality tubes in large quantities.

As well as mechanical and thermal properties, the optical properties of nanotubes have also been discussed in this chapter. These do not lend themselves so readily to applications, but the application of fluorescence spectroscopy is proving to be a valuable method for identifying nanotube structure (although it only applies to semiconducting tubes). Raman spectroscopy has also become an extremely important and widely-used way of characterizing both bulk samples of nanotubes and individual tubes.

References

(7.1) J. P. Lu, 'Elastic properties of carbon nanotubes and nanoropes', *Phys. Rev. Lett.*, **79**, 1297 (1997).

(7.2) E. Hernández, C. Goze, P. Bernier *et al.*, 'Elastic properties of C and $B_xC_yN_z$ composite nanotubes', *Phys. Rev. Lett.*, **80**, 4502 (1998).

(7.3) N. Yao and V. Lordi, 'Young's modulus of single-walled carbon nanotubes', *J. Appl. Phys.*, **84**, 1939 (1998).

(7.4) S. Govindjee and J. L. Sackman, 'On the use of continuum mechanics to estimate the properties of nanotubes', *Solid State Commun.*, **110**, 227 (1999).

(7.5) T. Ozaki, Y. Iwasa and T. Mitani, 'Stiffness of single-walled carbon nanotubes under large strain', *Phys. Rev. Lett.*, **84**, 1712 (2000).

(7.6) G. Van Lier, C. Van Alsenoy, V. Van Doren *et al.*, '*Ab initio* study of the elastic properties of single-walled carbon nanotubes and graphene', *Chem. Phys. Lett.*, **326**, 181 (2000).

(7.7) A. Sears and R. C. Batra, 'Macroscopic properties of carbon nanotubes from molecular-mechanics simulations', *Phys. Rev. B*, **69**, 235406 (2004).

(7.8) X. Zhou, J. J. Zhou and Z. C. Ou-Yang, 'Strain energy and Young's modulus of single-wall carbon nanotubes calculated from electronic energy-band theory', *Phys. Rev. B*, **62**, 13 692 (2000).

(7.9) B. I. Yakobson, C. J. Brabec and J. Bernholc, 'Nanomechanics of carbon tubes: instabilities beyond linear response', *Phys. Rev. Lett.*, **76**, 2511 (1996).

(7.10) B. I. Yakobson, M. P. Campbell, C. J. Brabec *et al.*, 'High strain rate fracture and C-chain unraveling in carbon nanotubes', *Comp. Mat. Sci.*, **8**, 341 (1997).

(7.11) T. Kuzumaki, T. Hayashi, H. Ichinose *et al.*, '*In-situ* observed deformation of carbon nanotubes', *Phil. Mag. A*, **77**, 1461 (1998).

(7.12) T. Vodenitcharova and L. C. Zhang, 'Mechanism of bending with kinking of a single-walled carbon nanotube', *Phys. Rev. B*, **69**, 115410 (2004).

(7.13) A. Kutana and K. P. Giapis, 'Transient deformation regime in bending of single-walled carbon nanotubes', *Phys. Rev. Lett.*, **97**, 245501 (2006).

(7.14) T. C. Chang and J. Hou, 'Molecular dynamics simulations on buckling of multiwalled carbon nanotubes under bending', *J. Appl. Phys.*, **100**, 114327 (2006).

(7.15) M. B. Nardelli, B. I. Yakobson and J. Bernholc, 'Mechanism of strain release in carbon nanotubes', *Phys. Rev. B*, **57**, R4277 (1998).

(7.16) Q. Z. Zhao, M. B. Nardelli and J. Bernholc, 'Ultimate strength of carbon nanotubes: a theoretical study', *Phys. Rev. B*, **65**, 144105 (2002).

(7.17) K. M. Liew, X. Q. He and C. H. Wong, 'On the study of elastic and plastic properties of multi-walled carbon nanotubes under axial tension using molecular dynamics simulation', *Acta Mater.*, **52**, 2521 (2004).

(7.18) Q. Wang, K. M. Liew, X. Q. He *et al.*, 'Local buckling of carbon nanotubes under bending', *Appl. Phys. Lett.*, **91**, 093128 (2007).

(7.19) T. Dumitrica, M. Hua and B. I. Yakobson, 'Symmetry-, time-, and temperature-dependent strength of carbon nanotubes', *Proc. Natl. Acad. Sci. USA*, **103**, 6105 (2006).

(7.20) J. F. Despres, E. Daguerre and K. Lafdi, 'Flexibility of graphene layers in carbon nanotubes', *Carbon*, **33**, 87 (1995).

(7.21) K. L. Lu, R. M. Lago, Y. K. Chen *et al.*, 'Mechanical damage of carbon nanotubes by ultrasound', *Carbon*, **34**, 814 (1996).

(7.22) S. Iijima, C. Brabec, A. Maiti *et al.*, 'Structural flexibility of carbon nanotubes', *J. Chem. Phys.*, **104**, 2089 (1996).

(7.23) P. Poncharal, Z. L. Wang, D. Ugarte *et al.*, 'Electrostatic deflections and electromechanical resonances of carbon nanotubes', *Science*, **283**, 1513 (1999).

(7.24) B. I. Yakobson and P. Avouris, 'Mechanical properties of carbon nanotubes', *Topics Appl. Phys.*, **80**, 287 (2001).

(7.25) A. Cao, P. L. Dickrell, W. G. Sawyer *et al.*, 'Super-compressible foamlike carbon nanotube films', *Science*, **310**, 1307 (2005).

(7.26) M. M. J. Treacy, T. W. Ebbesen and J. M. Gibson, 'Exceptionally high Young's modulus observed for individual carbon nanotubes', *Nature*, **381**, 678 (1996).

(7.27) E. W. Wong, P. E. Sheehan and C. M. Lieber, 'Nanobeam mechanics: elasticity, strength, and toughness of nanorods and nanotubes', *Science*, **277**, 1971 (1997).

(7.28) M. R. Falvo, G. J. Clary, R. M. Taylor *et al.*, 'Bending and buckling of carbon nanotubes under large strain', *Nature*, **389**, 582 (1997).

(7.29) T. Hertel, R. Martel and P. Avouris, 'Manipulation of individual carbon nanotubes and their interaction with surfaces', *J. Phys. Chem. B*, **102**, 910 (1998).

(7.30) J. P. Salvetat, J. M. Bonard, N. H. Thomson *et al.*, 'Mechanical properties of carbon nanotubes', *Appl. Phys. A*, **69**, 255 (1999).

(7.31) J. P. Salvetat, A. J. Kulik, J. M. Bonard *et al.*, 'Elastic modulus of ordered and disordered multiwalled carbon nanotubes', *Adv. Mater.*, **11**, 161 (1999).

(7.32) B. Lukić, J. W. Seo, E. Couteau *et al.*, 'Elastic modulus of multi-walled carbon nanotubes produced by catalytic chemical vapour deposition', *Appl. Phys. A*, **80**, 695 (2005).

(7.33) G. Guhados, W. K. Wan, X. L. Sun *et al.*, 'Simultaneous measurement of Young's and shear moduli of multiwalled carbon nanotubes using atomic force microscopy', *J. Appl. Phys.*, **101**, 033514 (2007).

(7.34) B. Lukić, J. W. Seo, R. R. Bacsa *et al.*, 'Catalytically grown carbon nanotubes of small diameter have a high Young's modulus', *Nano Lett.*, **5**, 2074 (2005).

(7.35) J. Cumings and A. Zettl, 'Low-friction nanoscale linear bearing realized from multiwall carbon nanotubes', *Science*, **289**, 602 (2000).

(7.36) B. G. Demczyk, Y. M. Wang, J. Cumings *et al.*, 'Direct mechanical measurement of the tensile strength and elastic modulus of multiwalled carbon nanotubes', *Mater. Sci. Eng. A*, **334**, 173 (2002).

(7.37) M. F. Yu, O. Lourie, M. J. Dyer *et al.*, 'Strength and breaking mechanism of multiwalled carbon nanotubes under tensile load', *Science*, **287**, 637 (2000).

(7.38) M. F. Yu, B. I. Yakobson and R. S. Ruoff, 'Controlled sliding and pullout of nested shells in individual multiwalled carbon nanotubes', *J. Phys. Chem. B*, **104**, 8764 (2000).

(7.39) R. S. Ruoff, 'Time, temperature, and load: the flaws of carbon nanotubes', *Proc. Natl. Acad. Sci. USA*, **103**, 6779 (2006).

(7.40) R. S. Ruoff, D. Qian and W. K. Liu, 'Mechanical properties of carbon nanotubes: theoretical predictions and experimental measurements', *Comptes Rendus Phys.*, **4**, 993 (2003).

(7.41) J. P. Salvetat, S. Bhattacharyya and R. B. Pipes, 'Progress on mechanics of carbon nanotubes and derived materials', *J. Nanosci. Nanotech.*, **6**, 1857 (2006).

(7.42) T. Hayashi, Y. A. Kim, T. Natsuki *et al.*, 'Mechanical properties of carbon nanomaterials', *Chem. Phys. Chem.*, **8**, 999 (2007).

(7.43) A. V. Eletskii, 'Mechanical properties of carbon nanostructures and related materials', *Phys. Usp.*, **50**, 225 (2007).

(7.44) R. E. Tuzun, D. W. Noid and B. G. Sumpter, 'The dynamics of molecular bearings', *Nanotechnology*, **6**, 64 (1995).

(7.45) A. Krishnan, E. Dujardin, T. W. Ebbesen *et al.*, 'Young's modulus of single-walled nanotubes', *Phys. Rev. B*, **58**, 14 013 (1998).

(7.46) M. F. Yu, B. S. Files, S. Arepalli *et al.*, 'Tensile loading of ropes of single wall carbon nanotubes and their mechanical properties', *Phys. Rev. Lett.*, **84**, 5552 (2000).

(7.47) J. P. Salvetat, G. A. D. Briggs, J. M. Bonard *et al.*, 'Elastic and shear moduli of single-walled carbon nanotube ropes', *Phys. Rev. Lett.*, **82**, 944 (1999).

(7.48) T. W. Tombler, C. W. Zhou, L. Alexseyev *et al.*, 'Reversible electromechanical characteristics of carbon nanotubes under local-probe manipulation', *Nature*, **405**, 769 (2000).

(7.49) D. A. Walters, L. M. Ericson, M. J. Casavant *et al.*, 'Elastic strain of freely suspended single-wall carbon nanotube ropes', *Appl. Phys. Lett.*, **74**, 3803 (1999).

(7.50) H. Kataura, Y. Kumazawa, Y. Maniwa *et al.*, 'Optical properties of single-wall carbon nanotubes', *Synthetic Metals*, **103**, 2555 (1999).

(7.51) M. J. O'Connell, S. M. Bachilo, C. B. Huffman *et al.*, 'Band gap fluorescence from individual single-walled carbon nanotubes', *Science*, **297**, 593 (2002).

(7.52) S. M. Bachilo, M. S. Strano, C. Kittrell *et al.*, 'Structure-assigned optical spectra of single walled carbon nanotubes', *Science*, **298**, 2361 (2002).

(7.53) S. M. Bachilo, L. Balzano, J. E. Herrera *et al.*, 'Narrow (n, m)-distribution of single-walled carbon nanotubes grown using a solid supported catalyst', *J. Amer. Chem. Soc.*, **125**, 11 186 (2003).

(7.54) T. Belin and F. Epron, 'Characterization methods of carbon nanotubes: a review', *Mater. Sci. Eng. B*, **119**, 105 (2005).

(7.55) A. Jorio, R. Saito, G. Dresselhaus *et al.*, 'Determination of nanotubes properties by Raman spectroscopy', *Phil. Trans. Roy. Soc. A*, **362**, 2311 (2004).

(7.56) A. M. Rao, E. Richter, S. Bandow *et al.* 'Diameter-selective Raman scattering from vibrational modes in carbon nanotubes', *Science*, **275**, 187 (1997).

(7.57) S. Bandow, S. Asaka, Y. Saito *et al.*, 'Effect of the growth temperature on the diameter distribution and chirality of single-wall carbon nanotubes', *Phys. Rev. Lett.*, **80**, 3779 (1998).

(7.58) A. M. Rao, J. Chen, E. Richter *et al.*, 'Effect of van der Waals interactions on the Raman modes in single walled carbon nanotubes', *Phys. Rev. Lett.*, **86**, 3895 (2001).

(7.59) M. Hulman, R. Pfeiffer and H. Kuzmany, 'Raman spectroscopy of small-diameter nanotubes', *New J. Phys.*, **6**, 1 (2004).

(7.60) R. A. Jishi, L. Venkataraman, M. S. Dresselhaus *et al.*, 'Phonon modes in carbon nanotubules', *Chem. Phys. Lett.*, **209**, 77 (1993).

(7.61) D. Sanchez-Portal, E. Artacho, J. M. Soler *et al.*, '*Ab initio* structural, elastic, and vibrational properties of carbon nanotubes', *Phys. Rev. B*, **59**, 12 678 (1999).

(7.62) S. D. M. Brown, P. Corio, A. Marucci *et al.*, 'Second-order resonant Raman spectra of single-walled carbon nanotubes', *Phys. Rev. B*, **61**, 7734 (2000).

(7.63) L. Alvarez, A. Righi, T. Guillard *et al.*, 'Resonant Raman study of the structure and electronic properties of SWNTs', *Chem. Phys. Lett.*, **316**, 186 (2000).

(7.64) P. Corio, M. L. A. Temperini, P. S. Santos *et al.*, 'Resonant Raman scattering characterization of carbon nanotubes grown with different catalysts', *Chem. Phys. Lett.*, **350**, 373 (2001).

(7.65) A. Jorio, M. A. Pimenta, A. G. Souza Filho *et al.*, 'Resonance Raman spectra of carbon nanotubes by cross-polarized light', *Phys. Rev. Lett.*, **90**, 107403 (2003).

(7.66) A. Grüneis, R. Saito, J. Jiang *et al.*, 'Resonance Raman spectra of carbon nanotube bundles observed by perpendicularly polarized light', *Chem. Phys. Lett.*, **387**, 301 (2004).

(7.67) M. A. Pimenta, A. P. Gomes, C. Fantini *et al.* 'Optical studies of carbon nanotubes and nanographites', *Physica E*, **37**, 88 (2007).

(7.68) A. Jorio, R. Saito, J. H. Hafner *et al.*, 'Structural (n, m) determination of isolated single-wall carbon nanotubes by resonant Raman scattering', *Phys. Rev. Lett.*, **86**, 1118 (2001).

(7.69) A. G. Souza Filho, A. Jorio, J. H. Hafner *et al.*, 'Electronic transition energy E_{ii} for an isolated (n, m) single-wall carbon nanotube obtained by anti-Stokes/Stokes resonant Raman intensity ratio', *Phys. Rev. B*, **63**, 241404R (2001).

(7.70) C. Fantini, A. Jorio, M. Souza *et al.*, 'Optical transition energies for carbon nanotubes from resonant Raman spectroscopy: environment and temperature effects', *Phys. Rev. Lett.*, **93**, 147406 (2004).

(7.71) V. W. Brar, G. G. Samsonidze, M. S. Dresselhaus *et al.*, 'Second-order harmonic and combination modes in graphite single-wall carbon nanotube bundles and isolated single-wall carbon nanotubes', *Phys. Rev. B*, **66**, 155418 (2002).

(7.72) M. Souza, A. Jorio, C. Fantini *et al.*, 'Single and double resonance Raman *G*-band processes in carbon nanotubes', *Phys. Rev. B*, **69**, 241403 (2004).

(7.73) H. Hiura, T. W. Ebbesen, K. Tanigaki *et al.*, 'Raman studies of carbon nanotubes', *Chem. Phys. Lett.*, **202**, 509 (1993).

(7.74) N. Chandrabhas, A. K. Sood, D. Sundararaman *et al.*, 'Structure and vibrational properties of carbon tubules', *Pramana-J. Phys.*, **42**, 375 (1994).

(7.75) J. Kastner, T. Pichler, H. Kuzmany *et al.*, 'Resonance Raman and infrared spectroscopy of carbon nanotubes', *Chem. Phys. Lett.*, **221**, 53 (1994).

(7.76) J. M. Benoit, J. P. Buisson, O. Chauvet *et al.*, 'Low-frequency Raman studies of multiwalled carbon nanotubes: experiments and theory', *Phys. Rev. B*, **66**, 073417 (2002).

(7.77) X. Zhao, Y. Ando, L.-C. Qin *et al.*, 'Radial breathing modes of multiwalled carbon nanotubes', *Chem. Phys. Lett.*, **361**, 169 (2002).

(7.78) X. Zhao, Y. Ando, L.-C. Qin *et al.*, 'Multiple splitting of *G*-band modes from individual multiwalled carbon nanotubes', *Appl. Phys. Lett.*, **81**, 2550 (2002).

(7.79) M. S. Dresselhaus, G. Dresselhaus, A. Jorio *et al.*, 'Science and applications of single nanotube Raman spectroscopy', *J. Nanosci. Nanotech.*, **3**, 19 (2003).

(7.80) M. S. Dresselhaus, G. Dresselhaus, R. Saito *et al.*, 'Raman spectroscopy of carbon nanotubes', *Phys. Rep.*, **409**, 47 (2005).

(7.81) M. S. Dresselhaus, F. Villalpando-Paez, G. G. Samsonidze *et al.*, 'Raman scattering from one-dimensional carbon systems', *Physica E*, **37**, 81 (2007).

(7.82) C. Thomsen, S. Reich and J. Maultzsch, 'Resonant Raman spectroscopy of nanotubes', *Phil. Trans. Roy. Soc. A*, **362**, 2337 (2004).

(7.83) S. Reich, C. Thomsen and J. Maultzsch, *Carbon Nanotubes*, Wiley-VCH, Weinheim, 2004.

(7.84) S. Berber, Y. K. Kwon, and D. Tománek, 'Unusually high thermal conductivity of carbon nanotubes', *Phys. Rev. Lett.*, **84**, 4613 (2000).

(7.85) J. Hone, M. Whitney, C. Piskoti *et al.*, 'Thermal conductivity of single-walled carbon nanotubes', *Phys. Rev. B*, **59**, R2514 (1999).

(7.86) B. T. Kelly, *Physics of Graphite*, Applied Science Publishers, London, 1981.

(7.87) P. Kim, L. Shi, A. Majumdar *et al.*, 'Thermal transport measurements of individual multiwalled nanotubes', *Phys. Rev. Lett.*, **87**, 215502 (2001).

(7.88) D. J. Yang, Q. Zhang, G. Chen *et al.*, 'Thermal conductivity of multiwalled carbon nanotubes', *Phys. Rev. B*, **66**, 165440 (2002).

(7.89) L. G. C. Rego and G. Kirczenow, 'Quantized thermal conductance of dielectric quantum wires', *Phys. Rev. Lett.*, **81**, 232 (1998).

(7.90) D. E. Angelescu, M. C. Cross and M. L. Roukes, 'Heat transport in mesoscopic systems', *Superlatt. Microstruct.*, **23**, 673 (1998).

(7.91) K. Schwab, E. A. Henrlksen, J. M. Worlock *et al.*, 'Measurement of the quantum of thermal conductance', *Nature*, **404**, 974 (2000).

(7.92) T. Yamamoto, S. Watanabe and K. Watanabe, 'Universal features of quantized thermal conductance of carbon nanotubes', *Phys. Rev. Lett.*, **92**, 075502 (2004).

(7.93) E. Brown, L. Hao, J. C. Gallop *et al.*, 'Ballistic thermal and electrical conductance measurements on individual multiwall carbon nanotubes', *Appl. Phys. Lett.*, **87**, 023107 (2005).

(7.94) C. W. Chang, D. Okawa, A. Majumdar *et al.*, 'Solid-state thermal rectifier', *Science*, **314**, 1121 (2006).

(7.95) G. G. Tibbetts, 'Why are carbon filaments tubular?', *J. Cryst. Growth*, **66**, 632 (1983).

(7.96) D. H. Robertson, D. W. Brenner and J. W. Mintmire, 'Energetics of nanoscale graphitic tubules', *Phys. Rev. B*, **45**, 12 592 (1992).

(7.97) S. Sawada and N. Hamada, 'Energetics of carbon nanotubes', *Solid State Commun.*, **83**, 917 (1992).

(7.98) A. A. Lucas, P. Lambin and R. E. Smalley, 'On the energetics of tubular fullerenes', *J. Phys. Chem. Solids*, **54**, 587 (1993).

(7.99) J. C. Charlier and J. P. Michenaud, 'Energetics of multilayered carbon tubules', *Phys. Rev. Lett.*, **70**, 1858 (1993).

(7.100) G. E. Begtrup, K. G. Ray, B. M. Kessler *et al.*, 'Probing nanoscale solids at thermal extremes', *Phys. Rev. Lett.*, **99**, 155901 (2007).

8 Chemistry and biology of nanotubes

The first insights into the chemistry of carbon nanotubes grew out of efforts to open and fill the tubes in the early 1990s. As described in Chapter 10, this work clearly demonstrated one very basic aspect of nanotube chemistry: they are most reactive at the tips. Indeed, the reaction of nanotubes with acids and other reagents could be exquisitely selective, with attack occurring only at defective regions. Since this early work, interest in the functionalization of carbon nanotubes has grown rapidly. In many studies the aim of functionalization has been simply to solubilize the tubes, by attaching hydrophilic species to these normally hydrophobic structures. However, as will be seen, the chemical modification of nanotubes is proving to be valuable in a wide range of areas, from the preparation of carbon nanotube composites to the production of sensors. As well as covalent functionalization, there is great interest in 'non-covalent functionalization', i.e. connecting molecules to nanotubes without actually forming chemical bonds. The advantage of this approach is that it avoids disrupting the structure of the tubes, enabling their full properties to be retained. Non-covalent functionalization can be achieved by forming van der Waals bonds between planar groups and the tube walls, or by wrapping molecules helically round the tubes. This chapter begins with an overview of the methods that can be used to chemically functionalize carbon nanotubes. This is followed by a discussion of functionalization with biomolecules. Interest in this area was greatly stimulated by the demonstration that functionalized nanotubes can cross cell membranes, suggesting that tubes could be used to ferry therapeutic agents into cells. Of course, before any such applications can be considered it is essential to assess the potential toxicity of carbon nanotubes, and a brief review of the work that has been carried out on this subject is given at the end of the chapter.

8.1 Covalent functionalization

An exhaustive review of the rapidly growing field of nanotube chemistry would not be possible here; instead the aim is to illustrate the various different ways in which nanotubes can be functionalized through selected examples. For more comprehensive treatments, a number of reviews are available (e.g. 8.1–8.6), of which the 2006 article 'Chemistry of carbon nanotubes' by Maurizio Prato and colleagues (8.5) can be particularly recommended.

8.1.1 Functionalization of nanotube ends and defects

As already noted, functionalization of nanotube ends and defects occurs when acids are used in nanotube purification and opening. Among the acids and acid mixtures that have been used for this are HNO_3 (8.7), $HNO_3 + H_2SO_4$ (8.8) and $H_2SO_4 + KMnO_4$ (8.9). Such treatments result in the formation of carboxyl and other groups at the sites attacked by the acid. Further functionalization can be achieved by reaction with these groups. Robert Haddon and colleagues from the University of Kentucky were among the first to use this approach. In 1998 they described a method for condensing long-chain alkylamines with carboxyl groups attached to the ends of shortened SWNTs (8.10). This was achieved by activating the carboxyl groups with thionychloride and then reacting with octadecyla-mine, as in Fig. 8.1. The tubes had substantial solubility in chloroform, dichloromethane, aromatic solvents (benzene, toluene, chlorobenzene, 1,2-dichlorobenzene) and CS_2. Having successfully solubilized shortened SWNTs by 'end-functionalization' Haddon and collea-gues then demonstrated the solubilization of full-length SWNTs (8.11). They achieved this by carrying out a treatment that simultaneously broke up the SWNT ropes and formed an octadecylammonium SWNT-carboxylate zwitterion.

Also in 1998, the Smalley group demonstrated that end-functionalization could be used to connect together single-walled nanotubes (8.12). This work, previously discussed in Chapter 4 (Section 4.4.1), involved firstly cutting the tubes into short lengths by sonicating in acid. The shortened tubes were treated with thionyl chloride and then

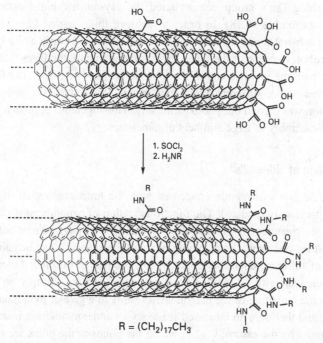

$$R = (CH_2)_{17}CH_3$$

Fig. 8.1   Defect-group functionalization of SWNTs with octadecylamine to produce soluble tubes (8.2, 8.10).

with NH_2–$(CH_2)_{11}$–SH to produce an amide link between the alkanethiol and the nanotube ends. Gold particles were then anchored to the thiol-derivitized tips of the tubes, and used as links for connecting tubes together.

In another notable early study, Charles Lieber and colleagues from Harvard described the use of functionalized nanotubes as tips for atomic force microscopy (8.13, 8.14). Multiwalled tubes were opened and shortened by treatment in an oxidizing environment. Various functional groups were then attached to the carboxyl groups at the open tip ends, and the tubes were then used to sense specific interactions with functional groups on substrates. The use of nanotubes as tips for AFM is discussed further in Chapter 11.

Researchers from Clemson University, South Carolina, have grafted polymers with amino terminal groups, such as poly(propionylethylenimine-*co*-ethylenimine) (PPEI- EI), onto oxidized nanotubes through amide formation (8.15). The resulting polymer-grafted SWNTs and MWNTs are highly soluble in most common organic solvents and water. The aqueous suspensions had useful optical limiting properties. The same reaction was also used to covalently attach polymers such as poly(vinyl acetate-co-vinyl alcohol) (PVA-VA) onto oxidized carbon nanotubes. Karl Coleman and colleagues have prepared iodinated SWNTs by treating oxidized tubes with elemental iodine and iodosobenzene diacetate (8.16).

Adding silyl moieties to the walls of nanotubes is of interest in connection with incorporating the tubes into composite materials, and possibly adjusting their electrical properties. Masami Aizawa and Milo Shaffer described the silylation of multiwalled tubes by reaction of various silyl sources with polar groups introduced onto the tubes surface by acid oxidation (8.17).

In 2004, Liming Dai's group demonstrated the asymmetric end-functionalization of multiwalled nanotubes (8.18). In order to achieve this, aligned films of MWNTs were grown on substrates and then removed and placed on the surface of a liquid such as water or alcohol. Because of the tubes' hydrophobic nature, the films floated on the liquid surface, enabling just one end of the tubes to be functionalized. The films could then be inverted and a different functional group attached to the opposite ends. Functionalization was carried out photochemically. This ingenious approach has potential for the self-assembly of large numbers of nanotubes.

8.1.2   Functionalization of sidewalls

The preceding section was mainly concerned with the functionalization of nanotubes that had been previously oxidized. The functionalization of nanotube sidewalls usually involves the direct treatment of 'pristine' tubes. As Prato *et al.* have pointed out (8.4, 8.5), some important lessons about the reactivity of nanotube sidewalls can be gained from a consideration of fullerene chemistry (8.19, 8.20). Studies of different fullerenes have shown that their reactivity in addition reactions depends very strongly on curvature. An increase in the curvature of the carbon shell results in a greater pyramidalization of the sp^2 atoms, and therefore an increased tendency to undergo addition reactions. This effect is illustrated by the case of C_{70}, in which the bonds at the poles are much more reactive than those around the flatter equatorial region. Addition to the equatorial atoms requires very reactive species, such as arynes, carbenes or halogens. In the case of nanotubes, the curvature is considerably less than that of small fullerenes like C_{70}, so functionalization of the sidewalls will only occur if a highly reactive reagent is used. It also follows that small diameter nanotubes are likely to display enhanced reactivity relative to larger-diameter tubes. As mentioned in Chapter 4 (Section 4.5.3) it has been suggested that metallic SWNTs may be more reactive than semiconducting tubes, although there is little clear evidence for this. The most important methods employed for sidewall functionalization are now outlined.

Fluorination

One of the first attempts to functionalize the sidewalls of SWNTs involved fluorination and was described by workers from Rice in 1998 (8.21). The starting material for these experiments was purified buckypaper, and this was treated with elemental fluorine at temperatures between 150–600 °C. For tubes treated at 600 °C, the degree of fluorination was estimated to be 0.1–1. Treatment of the functionalized tubes with hydrazine at room temperature caused defluorination, apparently restoring the tubes to their original structures. Subsequent work revealed that the fluorinated SWNTs could be dissolved in alcohols by ultrasonication (8.22). Elemental fluorine has subsequently been used by a number of groups to fluorinate nanotubes (e.g. 8.23), and the use of XeF_2 has also been explored (8.24). Fluorination has been used as the basis for further functionalization of nanotubes, for example using alkyl lithium reagents.

Addition of carbenes

At about the same time as the Rice work on fluorination, Haddon and colleagues demonstrated the functionalization of nanotube sidewalls using a carbene. It is well established that 1,1-dichlorocarbene can attack C=C bonds connecting two adjacent six-membered carbon rings, as shown in Fig. 8.2. Haddon and colleagues used this reaction to derivatize firstly insoluble SWNTs, and then shortened SWNTs dissolved in toluene (8.25). The dichlorocarbene was generated from the mercury complex $PhCCl_2HgBr$.

Although carbenes normally react with carbon double bonds by forming a cyclopropane ring, there are cases where just a single covalent bond is formed between the substrate and a zwitterionic adduct. A carbene that reacts in this way is the dipyridyl imidazolium carbene. This carbene has been used to functionalize fullerenes, and in 2001 was used by Andreas Hirsch of the University of Erlangen and colleagues to functionalize SWNTs (8.26), as shown in Fig. 8.3. The derivitized nanotubes were quite soluble in dimethyl sulphoxide (DMSO).

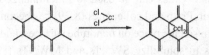

Fig. 8.2 The addition of dichlorocarbene to a double bond in a hexagonal network.

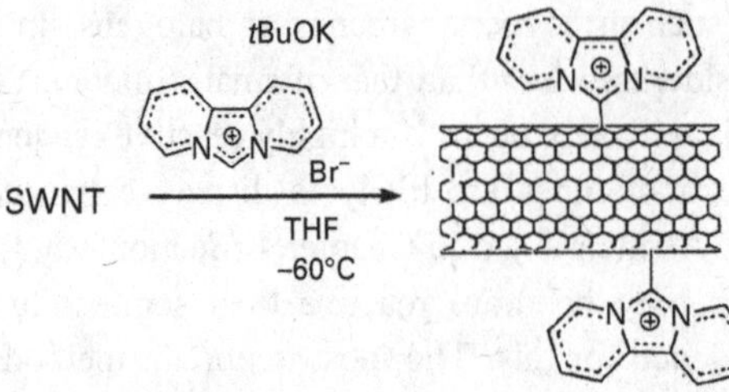

Fig. 8.3   Sidewall functionalization by the addition of nucleophilic carbenes (8.4).

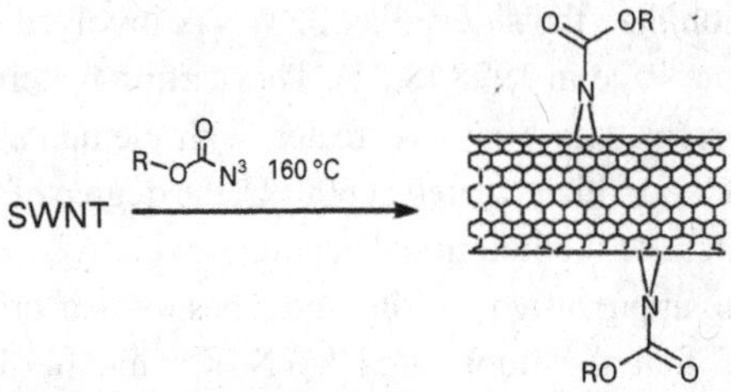

Fig. 8.4   Sidewall functionalization by the cycloaddition of nitrenes (8.27).

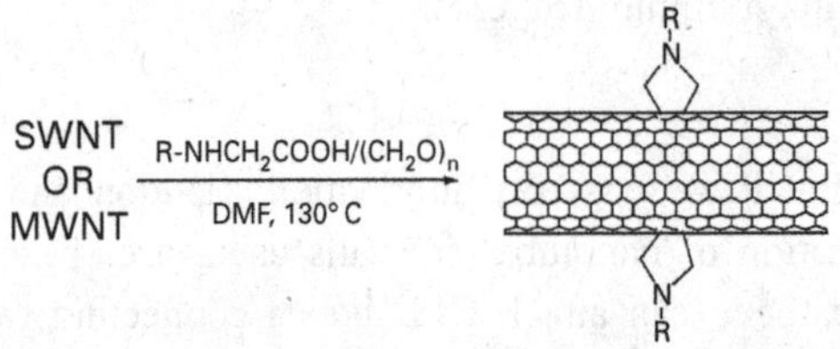

Fig. 8.5   Functionalization by the 1,3-Dipolar cycloaddition of azomethine ylides (8.5, 8.30).

Addition of nitrenes

Hirsch and colleagues also demonstrated the functionalization of nanotube sidewalls using nitrenes, as shown in Fig 8.4 (8.27). Alkyl azides were used as the nitrene precursors, and the resulting functionalized SWNTs were again found to be soluble in DMSO. Maxine McCall and colleagues from CSIRO Molecular Science, Australia, have used azide photolysis to functionalize MWNTs (8.28), and then to attach DNA oligonucleotides to the tubes, as discussed further below (Section 8.4.2).

Modification via 1,3-dipolar cycloaddition of azomethine ylides

1,3-dipolar cycloaddition of azomethine ylides can be used to functionalize C_{60}. This reaction, first demonstrated by Maurizio Prato and colleagues (8.29) has become known as the Prato reaction. The reaction can also take place on nanotube surfaces, and has been used by Prato, Hirsch and colleagues to solubilize SWNTs and MWNTs (8.30). The ylides were generated by condensation of an α-amino acid and an aldehyde, as in Fig 8.5.

The triethylene glycol group was chosen as the N-substituent group of the α-amino acid due to its high-solubilizing power, and various aldehydes were employed. Single-walled nanotubes modified in this way showed a high solubility in chloroform (up to 50 g l^{-1}), and other organic solvents and even in water.

Prato and co-workers showed that the functionalization of SWNTs with azomethine ylides could be used to purify SWNTs produced by the HiPco process (8.31). In subsequent work, the same group (8.32) demonstrated that SWNTs functionalized in this way were able to associate with plasmid DNA through electrostatic interactions, as discussed in Section 8.4.2 below.

In 2006 French researchers described an alternative method for the covalent functionalization of SWNTs by azomethine ylides (8.33). Their approach was based on the generation of the 1,3-dipole by double deprotonation of the corresponding trialkyl-N-oxide. They claimed that their approach resulted in the selective functionalization of semiconducting tubes, and hence provided a method for separating these from metallic tubes.

Solution-phase ozonolysis

Sarbajit Banerjee and Stanislaus Wong of the State University of New York, Stony Brook have used ozonolysis to functionalize SWNT sidewalls (8.34). Ozonolysis was carried out at low temperature in the solution phase. Ozone (O_3) adds to the double bonds of the sidewalls through a 1,3-dipolar cycloaddition similar to the addition of azomethine ylides. This forms a rather unstable ozonide, which can be cleaved to yield a range of functional groups, including aldehydic, ketonic, alcoholic and carboxyl groups. In subsequent work they showed that smaller tubes reacted more readily in the ozonolysis reaction than larger tubes (8.35). As well as leading to functionalized tubes, ozonolysis also purifies the nanotube samples, as described in Chapter 4 (Section 4.2.2).

Addition of radicals

Radicals have been quite widely used for nanotube functionalization. In some cases this has been achieved electrochemically, as discussed below, and in other cases thermal and photochemical methods have been used. In a study published in 2003, the Rice group generated radicals by thermally decomposing alkyl or aryl peroxides and used them to produce phenyl and undecyl sidewall functionalized SWNTs (8.36). László Forró and colleagues described a similar method at about the same time (8.37). The photoinduced addition of perfluorinated alkyl radicals to tube walls has been described by Hirsch et al. (8.2). This was accomplished by illuminating a mixture of SWNTs and an alkyl iodide with a mercury lamp. As already mentioned, McCall et al. used azide photochemistry to functionalize MWNT sidewalls (8.28).

Silylation

The work of Aizawa and Shaffer on silylation of MWNTs (8.17) was mentioned above. In this work, silylation was achieved by reacting silyl sources with acid-oxidized tubes. In 2006, Hemraj-Benny and Wong described the silylation of pristine single-walled tubes (8.38). Two different organosilanes were used, trimethoxysilane and hexaphenyldisilane, and the reactions were activated by UV irradiation. Examination of the functionalized tubes by electron microscopy and various types of spectroscopy suggested that they had not been structurally damaged by the chemical treatment.

Electrochemical reactions

The electrochemical approach is a potentially valuable method for the chemical alteration of nanotubes, as the extent of reaction can be directly adjusted by an applied potential. Workers from Rice University have derivatized SWNTs via electrochemical reduction of a variety of aryl diazonium salts (8.39). This resulted in the formation of a bond between the nanotube surface and a benzene ring. Spectroscopic characterization of functionalized tubes suggested that the tubes' electronic properties had been drastically altered. In subsequent work, a group from Stuttgart (8.40), demonstrated that using single-walled nanotubes as either the anode or the cathode in an electrochemical cell enabled oxidation or reduction of small molecules to occur at the nanotube surface, forming radical species that could then attack the carbon lattice to form covalent bonds. Reductive coupling resulted in C–C bond formation, while oxidative coupling resulted in the attachment of amines through an NH group, as in Fig. 8.6. The growth of polymerized layers up to 6 nm thick was demonstrated.

In a unique study published in 2007, Philip Collins and colleagues from the University of California, Irvine, used electrochemistry to control and monitor covalent functionalization of individual SWNTs (8.41). The nanotubes were grown by CVD on oxidized Si wafers and connected to metal electrodes on the wafers. The devices were then placed in an electrochemical cell with an acidic electrolyte solution. By applying a potential, an electrooxidation reaction could be induced, resulting in large jumps in conductance, which could be attributed to individual oxidation events. The potential could be switched off with microsecond accuracy, thus stopping the reaction after just a single chemical event. This approach has great potential for studying functionalization, and possibly for constructing molecular circuits.

Attachment of Polymers

There has been interest in covalently attaching polymers to nanotubes, with the specific aim of improving bonding in nanotube/polymer composites (see also Section 9.1). In 2003,

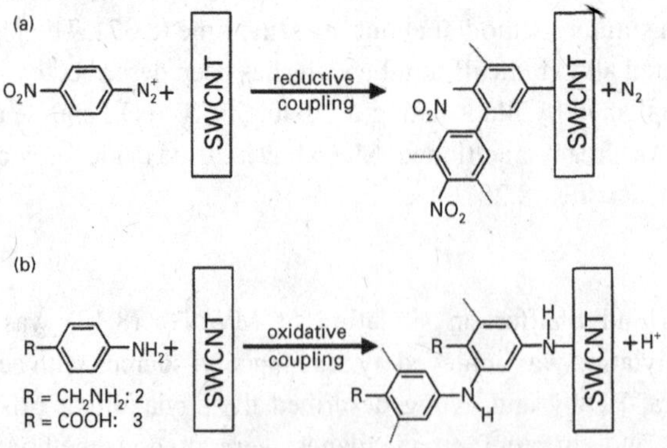

Fig. 8.6   Electrochemical modification of SWNTs. (a) Reductive coupling, (b) oxidative coupling (8.40).

Ajayan's group reported a method for preparing polystyrene-grafted single-walled nanotubes by firstly treating the tubes with sec-butyl lithium and then using these as initiators for the anionic polymerization of styrene (8.42). A short time later Jonathan Coleman and co-workers described a similar technique, this time with halogenated polymers (8.43).

8.2 Non-covalent functionalization

The use of commercial surfactants to enhance the solubility of nanotubes can be thought of as a kind of non-covalent functionalization. The ionic surfactant sodium dodecyl sulphate (SDS), widely employed in household products such as shampoo and shaving foam, was found to be of great value in solubilizing both MWNTs and SWNTs. Its use in the purification of MWNTs was mentioned in Chapter 4, while the use of SDS in suspending single-walled tubes for fluorescence spectroscopy was discussed in Chapter 7. Typically, about 1% of SDS is added to the aqueous suspension, and ultrasonication is frequently used to facilitate dispersion. In some cases the surfactant molecules form ordered layers on the nanotube surfaces (8.44). Ionic surfactants work by transferring a charge to the surfaces of the nanotubes. The nanotubes are then dispersed by electrostatic forces, and the behaviour of the dispersion will show a strong dependence on pH. Arjun Yodh and colleagues from the University of Pennsylvania have showed that sodium dodecyl benzene sulphonate (SDBS) can be more effective than SDS in dispersing SWNTs in water, because it contains a benzyl ring (8.45). Non-ionic surfactants (e.g. triton X-100) have also been used to solubilize nanotubes, for example in the production of composites (8.46). These operate using a combination of hydrogen bonding and/or steric dispersion forces. At high concentrations the surfactants form micelles, with the nanotubes held in the hydrophobic core of the micelles.

Apart from the use of surfactants, there are essentially two different approaches to the non-covalent functionalization of nanotubes. The first is to employ relatively small molecules containing planar groups that irreversibly adsorb to the nanotube surfaces by π-stacking forces, while the second involves wrapping larger polymeric molecules around the tubes. The first approach was pioneered by Hongjie Dai's group in 2001 (8.47). The molecule they used possessed a planar pyrenyl group to form the van der Waals bonds, as shown in Fig. 8.7, and a 'tail' consisting of a succinimidyl ester group. It was shown that biomolecules could be immobilized on the nanotube sidewalls by

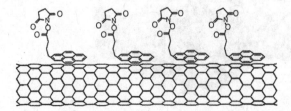

Fig. 8.7   Non-covalent functionalization of SWNT with 1-pyrenebutanoic acid, succinimidyl ester (8.47).

forming bonds with this group, as discussed in Section 8.4.1. Following this pioneering work, pyrenes have been quite widely used to solubilize nanotubes (8.48, 8.49) and to attach species such as magnetic nanoparticles (8.50) and fullerenes (8.51) to nanotubes.

Heterocyclic polyaromatic molecules such as porphyrins and phthalocyanines have also been used for the non-covalent functionalization and solubilization of nanotubes (e.g. 8.52–8.54). The interaction of these molecules with SWNTs is believed to be essentially van der Waals in nature, like that of pyrenes. A detailed discussion on the interactions between single-walled nanotubes and pyrenes, porphyrins and other molecules has been given by Dirk Guldi, Maurizio Prato and colleagues (8.55).

The second approach to the non-covalent functionalization of SWNTs involves helically wrapping the tubes with a polymer. Of course, the interaction of nanotubes with polymers has been intensively studied in connection with the production of composite materials, and this is covered in the next chapter. Here we are primarily concerned with the use of polymeric systems to solubilize nanotubes.

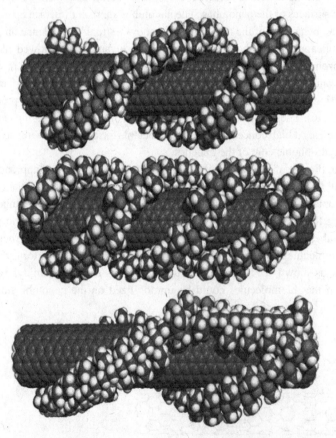

Fig. 8.8   Models of wrapping arrangements of a PVP polymer around an (8, 8) nanotube (8.56).

In 2001 the Smalley group described the solubilization of SWNTs by non-covalently associating them with linear polymers such as polyvinyl pyrrolidone (PVP) and polystyrene sulphonate (PSS) (8.56). The polymers were found to wrap around the tubes. Some possible wrapping arrangements are shown in Fig. 8.8. The nanotubes could be unwrapped by changing the solvent system. At about the same time, Fraser Stoddart and James Heath of the University of California, Los Angeles, and their colleagues reported similar experiments using the conjugated polymer poly(m-phenylenevinylene) substituted with octyloxy chains. By adding SWNTs to a solution of this polymer and sonication the mixture, they were able to produce a stable nanotube suspension (8.57). In further work (8.58) they used related polymers to wrap both individual nanotubes and small nanotube bundles. They believe that this approach could eventually be used to assemble nanoscale electronic devices. The UCLA group have also used stilbenoid dendrimers for the non-covalent functionalization of SWNTs (8.59, 8.60), as illustrated in the molecular model shown in Fig. 8.9. The authors believe that these relatively rigid polymers, with their well-defined structures, should be particularly useful in breaking up bundles of nanotubes and stabilizing individual tubes.

A number of other studies of the non-covalent functionalization of nanotubes using polymers have been described, including the use of crosslinked, amphiphilic copolymer micelles (8.61), pyrene-containing poly(phenylacetylene) chains (8.62) and poly(phenyleneethynylene) (8.63). A useful review has been given by Peng Liu (8.64).

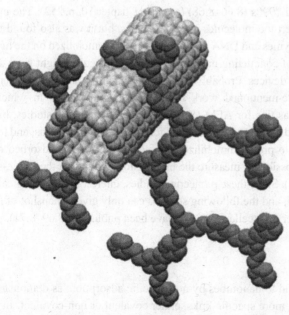

Fig. 8.9   Model of stilbenoid dendrimer wrapped around a (10, 10) nanotube (8.60).

8.3   Characterizing chemically functionalized nanotubes

Preceding sections have been illustrated with various reaction schemes showing the formation of covalent and non-covalent bonds with carbon nanotubes. In fact, achieving clear evidence that such bonds have been formed is often difficult. There are a number of reasons for this. To begin with, most samples are contaminated with non-nanotube carbon, and this material can also become functionalized. There is also the problem that tubes with a range of different structures and sizes are always present, and this can complicate the analysis of functionalized samples. Finally, when spectroscopic techniques are used, the interpretation of spectral features themselves is not always straightforward. Despite these difficulties, a number of different forms of spectroscopy have been used to analyse functionalized nanotubes, including X-ray photoelectron spectroscopy (8.16, 8.40, 8.65), infrared (8.21, 8.40), Raman (8.16, 8.21, 8.30, 8.36, 8.37) and ultraviolet–visible spectroscopy (8.30, 8.36). Direct imaging of functional groups on nanotubes sidewalls by transmission electron microscopy has not yet been demonstrated, but TEM is useful in establishing whether the structure of the tubes has been affected by functionalization. Energy dispersive X-ray microanalysis in the TEM can also be used to confirm the presence of non-carbon species attached to tubes.

8.4   Biological functionalization

Probably the earliest work on the interaction of biological molecules with carbon nanotubes was that carried by Malcolm Green and Edman Tsang from Oxford and their colleagues in the mid 1990s (8.66–8.68) (see also Chapter 10, p. 252). The initial aim of this work was to insert the molecules inside the tubes, but it was also found that biomolecules including enzymes and DNA oligomers could be immobilized on the tube surfaces. Such combinations of conducting nanotubes with biomolecules might have applications in nanoscale sensing devices. Probably the earliest actual use of bio-functionalized nanotubes was the above-mentioned work by Lieber and colleagues in which modified MWNTs were used as tips for AFM (8.13, 8.14). In one of these studies, biotin ligand was covalently linked to nanotube tips by the formation of amide bonds, and the modified tips were then used to probe immobilized streptavidin molecules adsorbed on mica. In this way it proved possible to measure the binding forces between the biotin-streptavidin pairs (see also p. 278). Since these pioneering studies, interest in the biological functionalization has exploded, and the following sections can only give a snapshot of this subject. Once again, a number of excellent reviews have been published (8.69–8.71).

8.4.1   Proteins

Proteins can be bound to nanotubes by non-specific adsorption, as demonstrated by the Oxford group, or by more specific links, either covalent or non-covalent. In the case of non-specific adsorption, the observed strong bonding of proteins may be partly a result of the amino affinity of carbon nanotubes. Covalent or non-covalent bonding of a range of proteins has been achieved via one of the various functionalization strategies discussed in previous sections.

In an interesting study of the non-specific adsorption of proteins, French researchers (8.72) reported in 1999 that streptavidin and HupR (a transcriptional regulator from a photosynthetic bacterium) were adsorbed onto MWNT surfaces in an ordered way, forming a helical structure. It was suggested that crystallizing proteins on nanotubes could be useful in structural investigations by TEM. Hongjie Dai and co-workers studied the interaction of streptavidin with SWNTs (8.73). They found that, like MWNTs, SWNTs could be readily coated with the protein. Single-walled tubes can also be solubilized by sonication in the presence of streptavidin, as shown by a joint US–Italian group (8.74). In some applications, such as in the development of specific biosensors, the nanotube surfaces need to be receptive to some proteins while rejecting others. Dai *et al.* demonstrated (8.75) that this could be achieved by coating the tubes with a poly-(ethylene oxide)-containing surfactant, and then conjugating specific receptors to the coated tubes. The adsorption of specific proteins to the functionalized tubes could then be detected electronically, as also discussed in Chapter 11 (p. 282). Both the Oxford (8.76) and Clemson (8.77) groups reported strong adsorption of ferritin onto SWNTs in an aqueous environment. The Clemson workers reported that ferritin adsorption was in fact so significant that it resulted in the solubilization of the SWNTs in water.

As discussed in Section 8.2, Hongjie Dai and colleagues have demonstrated the non-covalent binding of a molecule possessing a planar pyrenyl group to the surfaces of SWNTs (8.47). The 'tail' of this molecule consisted of a succinimidyl ester group. These workers demonstrated that proteins could be immobilized on the nanotubes by using a reaction that involved the nucleophilic substitution of the succinimide by an amine group on the protein, resulting in the formation of an amide bond. The immobilization of other biomolecules, including ferritin and streptavidin was also demonstrated. Figure 8.10 shows the immobilization of ferritin on SWNTs supported on a SiO_2 substrate.

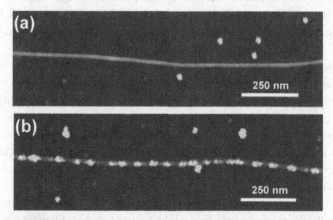

Fig. 8.10 Immobilization of ferritin on SWNTs, from the work of Dai *et al.* (8.47). (a) AFM image showing SWNT bundle before adsorption of ferritin, (b) image showing ferritin molecules adsorbed on SWNTs.

A great deal of work has also been carried out on the binding of enzymes to nanotubes, partly due to possible applications in sensors. The early work of the Oxford group on the immobilization of enzymes on both the inner and outer surfaces of MWNTs has already been mentioned (8.66, 8.68). It was demonstrated in this work that the enzymes retained moderate bioactivity while attached to the tubes. In 2004, workers from the Rensselaer Polytechnic Institute, New York, studied the activities of two enzymes, R-chymotrypsin (CT) and soybean peroxidase (SBP), adsorbed onto SWNTs (8.78). The soybean peroxidase retained up to 30% of its native activity upon adsorption, while the adsorbed R-chymotrypsin retained only 1% of its native activity. Analysis using FT–IR spectroscopy showed that both enzymes underwent structural changes upon adsorption. A group from Arizona State University showed that enzyme multilayers on MWNTs exhibited remarkably sensitive electrochemical detection of proteins and nucleic acids (8.79), while Australian researchers have attached the enzyme glucose oxidase (GOx) to vertically aligned arrays of SWNTs on gold, for possible use as a sensing device (8.80). Cees Dekker's group (8.81) used a planar pyrenyl group linking molecule to immobilize glucose oxidase on semiconducting SWNTs grown on a silicon wafer. They used the GOx-decorated SWNTs as sensors to measure enzyme activity as discussed further in Chapter 11 (p. 283).

Much of the work on the bio-functionalization of nanotubes has been aimed at possible therapeutic applications. In this connection, a very important development has been the discovery that functionalized tubes can pass through cell membranes. It appears that the first demonstration of this was given in 2004 by a European team led by Alberto Bianco, Kostas Kostarelos and Maurizio Prato (8.32). This work, which involved the transport of plasmid DNA into mammalian cells is discussed further below. A short time later Nadine Kam and Hongjie Dai reported that protein-functionalized nanotubes could cross cell membranes (8.82). The mechanism of this process was believed to be endocytosis, in which cells absorb external bodies by engulfing them with protrusions of their cell membranes. In a further study Dai's group showed that nanotubes could be targeted to cancer cells growing in culture and used to kill the cell with high temperature (8.83). This was accomplished by attaching folic acid to the surface of the nanotubes. Folic acid binds to a folic acid receptor protein found in abundance on the surfaces of many types of cancer cells.

A more direct way of targeting cancer cells would be to use antibodies specific to antigens on the surfaces of cancer cells. Antibody-functionalized nanotubes might also be useful in biosensors. With these kinds of applications in mind, several groups have explored the binding of antibodies to nanotubes. James Rusling of the University of Connecticut prepared aligned arrays of SWNTs using a Langmuir–Blodgett method, and then attached antibodies to the ends of the tubes (8.84). These were then used in an electrochemical sensor which could detect low levels of the prostate specific antigen secondary antibodies.

Balaji Panchapakesan of the University of Delaware and his co-workers have shown how antibody-functionalized nanotubes might be used in photodynamic therapy for breast cancer (8.85). Monoclonal antibodies specific to breast cancer cells were attached to SWNTs using a succinimidyl ester link. The complexes were attached to specific cancer cells, and then excited with near-infrared radiation using lasers. This generated enough heat to destroy the cells. Antibody-functionalized nanotubes have also been used to deliver radioactive therapeutic agents to tumours (8.86).

8.4.2   Nucleic acids

Attaching DNA and other nucleic acids to nanotubes is potentially interesting for a number of reasons. At a fairly simple level, functionalizing nanotubes with nucleic acids might provide a useful method for solubilizing the tubes. Nucleic acid-modified nanotubes might also be useful in sensors, and there is the possibility that DNA-guided assembly could be used to build nanotube networks. Perhaps most excitingly, nanotubes could be used to deliver therapeutic DNA into cells. Work on the preparation and applications of DNA-nanotube complexes is now briefly reviewed.

Ming Zheng and colleagues from DuPont, with colleagues from MIT and the University of Illinois at Urbana-Champaign were among the first to explore the solubilization of nanotubes with DNA (8.87–8.89), as already discussed in Chapter 4 (p. 101). They showed that bundled SWNTs could be dispersed in water by sonication in the presence of single-stranded DNA (ssDNA). Optical absorption and fluorescence spectroscopy and AFM measurements provided evidence for individually dispersed carbon nanotubes. Molecular modelling suggested that the ssDNA was helical wrapped around the tubes, with the bases innermost, bound to the surface by π–π interactions, and the hydrophilic sugar–phosphate backbone pointing outwards and providing the solubility in water. They also demonstrated that DNA-coated carbon nanotubes could be separated into fractions with different electronic structures by ion-exchange chromatography. A number of other groups have used DNA to solubilize nanotubes (e.g. 8.90–8.92).

Zheng and colleagues have explored the use of DNA-wrapped single-walled tubes as sensing devices. In an initial study, they showed that SWNTs could detect subtle rearrangements in the structure of adsorbed DNA induced by metal ions (8.93). They began by wrapping DNA molecules around the tubes, and then introduced various ions (e.g. Ca^{2+}, Hg^{2+}, Na^+), which induced a conformational change in the DNA. This reduced the contact area between the DNA and the tubes, resulting in a perturbation of the electronic structure, which produced changes in the IR spectra. The system therefore acts as a sensor, and Zheng's group demonstrated that it could detect Hg^{2+} ions in whole blood, ink and living mammalian cells and tissues. In a subsequent paper (8.94) they described using a similar method to detect specific DNA sequences. This was accomplished by wrapping single strands of DNA around the tubes, and then using this to detect the binding of complementary DNA strands. This kind of sensor could be used in the detection of cancer-related genes or in identifying infectious organisms.

As well as studies of non-specific interactions of DNA with nanotubes, there has been a large amount of work on the covalent bonding of DNA to tubes. One of the first demonstrations was given by Chris Dwyer and colleagues of the University of North Carolina at Chapel Hill (8.95). Single-walled nanotubes were firstly purified in HNO_3 and then oxidized in a HNO_3/H_2SO_4 mixture to produce a solution of open-ended tubes with terminal carboxyl groups. A condensation reaction was then carried out between these carboxyl groups and amino-terminated DNA strands, to form amide bonds between the tube ends and the DNA. As mentioned earlier, Australian workers have used azide photolysis to functionalize MWNTs (8.28, 8.96), and then to attach DNA oligonucleotides to the tubes. The sites of DNA attachment were visualized by binding gold nanoparticles modified with DNA of complementary sequence to the functionalized tubes.

Several groups have shown that carbon nanotubes functionalized with DNA can be used as sensitive DNA sensors (e.g. 8.97–8.100). As an example, Liming Dai and colleagues covalently attached specific DNA sequences to nanotubes and used these to sense complementary DNA and DNA chains of specific sequences with a high sensitivity and selectivity (8.99).

DNA has also been used to assemble nanotube networks. In one of the first examples of this, Israeli researchers showed in 2003 that the interactions of proteins and DNA could be used to assemble nanotubes into a field-effect transistor which operated at room temperature (8.101) (see also p. 94 and p. 167). At about the same time, Huijun Xin and Adam Woolley from Brigham Young University described the use of aligned DNA molecules on a Si substrate to position single-walled tubes (8.102). Pyrenemethylamine (PMA) was used as a bridging compound between the DNA and the tubes. The amine group of this molecule is attracted electrostatically to the negatively charged phosphate backbone of DNA, while its pyrenyl group interacts with SWNT surfaces through stacking forces. AFM images showed that over 60% of all the deposited SWNTs on substrates were aligned on DNA fragments. The use of DNA to manufacture networks of nanotubes and other nanostructures is still at an early stage of development. Prospects for the field have been reviewed (8.103–8.105).

The use of nanotubes to deliver DNA into cells for possible gene therapy is also a very new area of research. As mentioned above the first demonstration of this seems to be the 2004 work by Bianco, Kostarelos and Prato (8.32). In this work, plasmid DNA was attached to ammonium-functionalized nanotube walls. The functionalized tubes were then brought into contact with a culture of mammalian cells, and observed to enter the cells. This pioneering work stimulated great interest, and there are undoubtedly many groups around the world attempting to use nanotubes to ferry nucleic acids into cells. To date, however, there have been relatively few published studies. One example is a 2007 paper by Hongjie Dai's group which describes a way in which RNA-functionalized nanotubes could be used to block HIV infection (8.106). The short RNA fragments were attached to SWNT sidewalls via amine groups and the nanotube–RNA hybrids were then introduced into human T-cells and primary blood cells. The RNA fragments 'switched off' the genes for HIV-specific receptors cell surfaces, thus inhibiting attack by HIV viruses.

As well as the review paper already cited, there are some general reviews on carbon nanotubes and DNA (8.107, 8.108).

8.5   Toxicity of carbon nanotubes

Research into the interaction of carbon nanotubes with biological systems, and into the toxicity of nanotubes, is still at an early stage, and a consensus has yet to emerge on the safety and biocompatibility of nanotubes. The aim of this section, therefore, is merely to highlight some of the studies which have been done in this field, without making any attempt to draw definitive conclusions.

To begin, it is worth considering the biocompatibility of some well-established forms of carbon. As pointed out by Darren Martin and colleagues in a useful review (8.109), carbon materials, including pyrolytic carbon and diamond-like carbon, are already quite widely used in medicine. On the other hand, high-purity carbon black has been reported to increase oxidative stress in human lung cells in vitro (8.110) and pulmonary tumours in rats (8.111). Significantly, the toxicity of small carbon black particles was found to be greater than that of larger ones, which may suggest that special care needs to be taken with highly dispersed carbon materials.

One of the first attempts to assess the health hazards of carbon nanotubes was carried out by Andrzej Huczko and colleagues from the University of Warsaw in 2001 (8.112). This group looked at the effect of nanotube-containing soot on the pulmonary function of guinea pigs, and found no evidence of any abnormalities. However, in subsequent studies, the same group reported that exposure to nanotubes could induce measurable pulmonary pathology in guinea pigs (8.113, 8.114). Other groups have confirmed that nanotubes can induce inflammation of the lung tissue of mice (8.115, 8.116). In a highly-cited study published in 2004, David Warheit and co-workers described the effects of instilling the lungs of rats with SWNTs (8.117). It was found that 15% of the instilled rats died within 24 h, but this was attributed to mechanical blockage of the large airways by aggregates of tubes, and not toxicity *per se*. This kind of blockage might not occur if the nanotubes were breathed in, rather than deliberately instilled. Warheit has therefore recommended that an inhalation toxicity study should be carried out using aerosols of tubes (8.118).

There have been several studies of the cytotoxicity of carbon nanotubes, i.e. their toxicity to individual cells. One of the first was reported in 2003 by Anna Shvedova of the US National Institute for Occupational Safety and Health and colleagues who showed that exposure to nanotubes resulted in damage to human keratinocyte (skin) cells (8.119). On the other hand, Irish workers found that SWNTs had very low toxicity to human lung cells (8.120). In an interesting study, workers from Rice found that the cytotoxicity of sidewall functionalized SWNTs was significantly less than that of non-functionalized tubes (8.121).

In their 2006 review, Martin and colleagues concluded that 'there is still much work to be done in establishing the toxicity and biocompatibility of carbon nanotubes' (8.109). Vicki Colvin of Rice put the situation more starkly in November 2007: 'The bad news is that we have way over five different opinions about carbon nanotube toxicity right now' (8.122). In the light of the continuing uncertainty in this area, it is obviously wise to err on the side of caution when preparing and handling these materials. Although it appears that nanotubes are less likely to become airborne than previously thought (8.123), measures should be taken to avoid inhalation of nanotube-containing material. The wearing of a face mask is therefore recommended. As far as contact with the skin is concerned, there is little concrete data available here, but wearing gloves when handling nanotube material would seem to be a sensible precaution. For further information on the interaction of carbon nanotubes with biological systems, several reviews are available (8.109, 8.118, 8.124–8.127).

8.6   Discussion

The chemistry and biology of nanotubes has advanced rapidly in recent years. Until about 1999, when the earlier version of this book was published, the chemical modification of nanotubes was largely limited to the formation of carboxyl and other groups at the tips by acid treatment. Subsequent work has shown that nanotube sidewalls can be functionalized, using reactive species such as halogens, arynes or carbenes. In this way, a wide range of molecules, including biomolecules, can be chemically attached to the sidewalls. As well as covalent functionalization, great advances have also been made in the non-covalent functionalization of nanotubes, either by π-bonding small molecules to nanotube walls, or by coiling polymeric molecules around the tubes. Proteins, nucleic acids and other biomolecules can also be non-covalently bonded to nanotubes.

The benefits of these advances are being felt across the whole of nanotube science. The ability to solubilize and separate nanotubes has greatly facilitated both purification or processing of the tubes. As discussed in the next chapter, functionalization is often used in the preparation of carbon nanotube composites in order to improve the bonding between the nanotubes and the matrix material. Several groups have also shown that functionalization can be used as a means of arranging nanotubes into defined arrays or networks. The production of both chemical and biological sensors using functionalized nanotubes is another rapidly growing area, and the possibility of using nanotubes in medical applications, including gene therapy, is creating much excitement.

Despite these important developments, there is still much to be done. We need a much better understanding of the interaction of carbon nanotubes with biological systems, and of their toxicity, before any medical applications can be envisaged. As far as carbon nanotube chemistry is concerned, it could be argued that this will not reach maturity until we have a totally reliable way of making tubes with a known structure. Only then will it be possible to fully characterize chemically modified nanotubes using spectroscopic and other methods.

References

(8.1) J. L. Bahr and J. M. Tour, 'Covalent chemistry of single-wall carbon nanotubes', *J. Mater. Chem.*, **12**, 1952 (2002).

(8.2) A. Hirsch, 'Functionalization of single-walled carbon nanotubes', *Angew. Chem.-Int. Ed.*, **41**, 1853 (2002).

(8.3) A. Hirsch and O. Vostrowsky, 'Functionalization of carbon nanotubes', *Topics Current Chem.*, **245**, 193 (2005).

(8.4) D. Tasis, N. Tagmatarchis, V. Georgakilas *et al.*, 'Soluble carbon nanotubes', *Chem. Eur. J.*, **9**, 4001 (2003).

(8.5) D. Tasis, N. Tagmatarchis, A. Bianco *et al.*, 'Chemistry of carbon nanotubes', *Chem. Rev.*, **106**, 1105 (2006).

(8.6) M. Prato, K. Kostarelos and A. Bianco, 'Functionalized carbon nanotubes in drug design and discovery', *Acc. Chem. Res.*, **41**, 60 (2008).

(8.7) S. C. Tsang, Y. K. Chen, P. J. F. Harris *et al.*, 'A simple chemical method of opening carbon nanotubes', *Nature*, **372**, 159 (1994).

(8.8) E. Dujardin, T. W. Ebbesen, A. Krishnan *et al.*, 'Purification of single-shell nanotubes', *Adv. Mater.*, **10**, 611 (1998).

(8.9) H. Hiura, T. W. Ebbesen and K. Tanigaki, 'Opening and purification of carbon nanotubes in high yields', *Adv. Mater.*, **7**, 275 (1995).

(8.10) J. Chen, M. A. Hamon, H. Hu *et al.*, 'Solution properties of single-walled carbon nanotubes', *Science*, **282**, 95 (1998).

(8.11) J. Chen, A. M. Rao, S. Lyuksyutov *et al.*, 'Dissolution of full-length single-walled carbon nanotubes', *J. Phys. Chem. B*, **105**, 2525 (2001).

(8.12) J. Liu, A. G. Rinzler, H. J. Dai *et al.*, 'Fullerene pipes', *Science*, **280**, 1253 (1998).

(8.13) S. S. Wong, E. Joselevich, A. T. Woolley *et al.*, 'Covalently functionalized nanotubes as nanometre-sized probes in chemistry and biology', *Nature*, **394**, 52 (1998).

(8.14) S. S. Wong, A. T. Woolley, E. Joselevich *et al.*, 'Covalently-functionalized single-walled carbon nanotube probe tips for chemical force microscopy', *J. Amer. Chem. Soc.*, **120**, 8557 (1998).

(8.15) J. E. Riggs, Z. X. Guo, D. L. Carroll *et al.*, 'Strong luminescence of solubilized carbon nanotubes', *J. Amer. Chem. Soc.*, **122**, 5879 (2000).

(8.16) K. S. Coleman, A. K. Chakraborty, S. R. Bailey *et al.*, 'Iodination of single-walled carbon nanotubes', *Chem. Mater.*, **19**, 1076 (2007).

(8.17) M. Aizawa and M. S. P. Shaffer, 'Silylation of multi-walled carbon nanotubes', *Chem. Phys. Lett.*, **368**, 121 (2003).

(8.18) K. M. Lee, L. C. Li and L. M. Dai, 'Asymmetric end-functionalization of multi-walled carbon nanotubes', *J. Amer. Chem. Soc.*, **127**, 4122 (2005).

(8.19) R. Taylor and D. R. M. Walton, 'The chemistry of fullerenes', *Nature*, **363**, 685 (1993).

(8.20) C. Thilgen and F. Diederich, 'The higher fullerenes: covalent chemistry and chirality', *Topics Current Chem.*, **199**, 135 (1999).

(8.21) E. T. Mickelson, C. B. Huffman, A. G. Rinzler *et al.*, 'Fluorination of single-wall carbon nanotubes', *Chem. Phys. Lett.*, **296**, 188 (1998).

(8.22) E. T. Mickelson, I. W. Chiang, J. L. Zimmerman *et al.*, 'Solvation of fluorinated single-wall carbon nanotubes in alcohol solvents', *J. Phys. Chem. B*, **103**, 4318 (1999).

(8.23) S. Kawasaki, K. Komatsu, F. Okino *et al.*, 'Fluorination of open- and closed-end single-walled carbon nanotubes', *Phys. Chem. Chem. Phys.*, **6**, 1769 (2004).

(8.24) E. Unger, M. Liebau, G. S. Duesberg *et al.*, 'Fluorination of carbon nanotubes with xenon difluoride', *Chem. Phys. Lett.*, **399**, 280 (2004).

(8.25) Y. Chen, R. C. Haddon, S. Fang *et al.*, 'Chemical attachment of organic functional groups to single-walled carbon nanotube material', *J. Mater. Res.*, **13**, 2423 (1998).

(8.26) M. Holzinger, O. Vostrowsky, A. Hirsch *et al.*, 'Sidewall functionalization of carbon nanotubes', *Angew. Chem. Int. Ed.*, **40**, 4002 (2001).

(8.27) M. Holzinger, J. Abraham, P. Whelan *et al.*, 'Functionalization of single-walled carbon nanotubes with (R-)oxycarbonyl nitrenes', *J. Amer. Chem. Soc.*, **125**, 8566 (2003).

(8.28) M. J. Moghaddam, S. Taylor, M. Gao *et al.*, 'Highly efficient binding of DNA on the sidewalls and tips of carbon nanotubes using photochemistry', *Nano Lett.*, **4**, 89 (2004).

(8.29) M. Maggini, G. Scorrano and M. Prato, 'Addition of azomethine ylides to C_{60}: synthesis, characterization, and functionalization of fullerene pyrrolidines'. *J. Amer. Chem. Soc.*, **115**, 9798 (1993).

(8.30) V. Georgakilas, K. Kordatos, M. Prato *et al.*, 'Organic functionalization of carbon nanotubes', *J. Amer. Chem. Soc.*, **124**, 760 (2002).

(8.31) V. Georgakilas, D. Voulgaris, E. Vazquez *et al.*, 'Purification of HiPco carbon nanotubes via organic functionalization', *J. Amer. Chem. Soc.*, **124**, 14 318 (2002).

(8.32) D. Pantarotto, R. Singh, D. McCarthy *et al.*, 'Functionalized carbon nanotubes for plasmid DNA gene delivery', *Angew. Chem. Int. Ed.*, **43**, 5242 (2004).

(8.33) C. Menard-Moyon, N. Izard, E. Doris *et al.*, 'Separation of semiconducting from metallic carbon nanotubes by selective functionalization with azomethine ylides', *J. Amer. Chem. Soc.*, **128**, 6552 (2006).

(8.34) S. Banerjee and S. S. Wong, 'Rational sidewall functionalization and purification of single-walled carbon nanotubes by solution-phase ozonolysis', *J. Phys. Chem. B*, **106**, 12 144 (2002).

(8.35) S. Banerjee and S. S. Wong, 'Demonstration of diameter-selective reactivity in the sidewall ozonation of SWNTs by resonance Raman spectroscopy', *Nano Lett.*, **4**, 1445 (2004).

(8.36) H. Peng, P. Reverdy, V. N. Khabashesku *et al.*, 'Sidewall functionalization of single-walled carbon nanotubes with organic peroxides', *Chem. Commun.*, 362 (2003).

(8.37) P. Umek, J. W. Seo, K. Hernadi *et al.*, 'Addition of carbon radicals generated from organic peroxides to single wall carbon nanotubes', *Chem. Mater.*, **15**, 4751 (2003).

(8.38) T. Hemraj-Benny and S. S. Wong, 'Silylation of single-walled carbon nanotubes', *Chem. Mater.*, **18**, 4827 (2006).

(8.39) J. L. Bahr, J. P. Yang, D. V. Kosynkin *et al.*, 'Functionalization of carbon nanotubes by electrochemical reduction of aryl diazonium salts: a bucky paper electrode', *J. Amer. Chem. Soc.*, **123**, 6536 (2001).

(8.40) S. E. Kooi, U. Schlecht, M. Burghard *et al.*, 'Electrochemical modification of single carbon nanotubes', *Angew. Chem. Int. Ed.*, **41**, 1353 (2002).

(8.41) B. R. Goldsmith, J. G. Coroneus, V. R. Khalap, *et al.*, 'Conductance-controlled point functionalization of single-walled carbon nanotubes', *Science*, **315**, 77 (2007).

(8.42) G. Viswanathan, N. Chakrapani, H. Yang *et al.*, 'Single-step *in situ* synthesis of polymer-grafted single-wall nanotube composites', *J. Amer. Chem. Soc.*, **125**, 9258 (2003).

(8.43) R. Blake, Y. K. Gun'ko, J. Coleman *et al.*, 'A generic organometallic approach toward ultra-strong carbon nanotube polymer composites', *J. Amer. Chem. Soc.*, **126**, 10 226 (2004).

(8.44) C. Richard, F. Balavoine, P. Schultz *et al.*, 'Supramolecular self-assembly of lipid derivatives on carbon nanotubes', *Science*, **300**, 775 (2003).

(8.45) M. F. Islam, E. Rojas, D. M. Bergey *et al.*, 'High weight fraction surfactant solubilization of single-wall carbon nanotubes in water', *Nano Lett.*, **3**, 269 (2003).

(8.46) M. Kang, S. J. Myung and H. J. Jin, 'Nylon 610 and carbon nanotube composite by *in situ* interfacial polymerization', *Polymer*, **47**, 3961 (2006).

(8.47) R. J. Chen, Y. G. Zhan, D. W Wang *et al.*, 'Noncovalent sidewall functionalization of single-walled carbon nanotubes for protein immobilization', *J. Amer. Chem. Soc.*, **123**, 3838 (2001).

(8.48) N. Nakashima, Y. Tomonari and H. Murakami, 'Water-soluble single-walled carbon nanotubes via noncovalent sidewall-functionalization with a pyrene-carrying ammonium ion', *Chem. Lett.*, **31**, 638 (2002).

(8.49) H. Murakami and N. Nakashima, 'Soluble carbon nanotubes and their applications', *J. Nanosci. Nanotech.*, **6**, 16 (2006).

(8.50) V. Georgakilas, V. Tzitzios, D. Gournis *et al.*, 'Attachment of magnetic nanoparticles on carbon nanotubes and their soluble derivatives', *Chem. Mater.*, **17**, 1613 (2005).

(8.51) D. M. Guldi, 'Biomimetic assemblies of carbon nanostructures for photochemical energy conversion', *J. Phys. Chem. B*, **109**, 11 432 (2005).

(8.52) J. Y. Chen and C. P. Collier, 'Noncovalent functionalization of single-walled carbon nanotubes with water-soluble porphyrins', *J. Phys. Chem. B*, **109**, 7605 (2005).

(8.53) H. Murakami, G. Nakamura, T. Nomura *et al.*, 'Noncovalent porphyrin-functionalized single-walled carbon nanotubes: solubilization and spectral behaviors', *J. Porphyrins Phthalocyanines*, **11**, 418 (2007).

(8.54) A. M. Ma, J. Lu, S. H. Yang *et al.*, 'Quantitative non-covalent functionalization of carbon nanotubes', *J. Cluster Sci.*, **17**, 599 (2006).

(8.55) C. Ehli, G. M. A. Rahman, N. Jux *et al.* 'Interactions in single wall carbon nanotubes/pyrene/porphyrin nanohybrids', *J. Amer. Chem. Soc.*, **128**, 11 222 (2006).

(8.56) M. J. O'Connell, P. Boul, L. M. Ericson *et al.*, 'Reversible water-solubilization of single-walled carbon nanotubes by polymer wrapping', *Chem. Phys. Lett.*, **342**, 265 (2001).

(8.57) A. Star, J. F. Stoddart, D. Steuerman *et al.*, 'Preparation and properties of polymer-wrapped single-walled carbon nanotubes', *Angew. Chem. Int. Ed.*, **40**, 1721 (2001).

(8.58) A. Star, Y. Liu, K. Grant *et al.*, 'Non-covalent side-wall functionalization of single-walled carbon nanotubes', *Macromolecules*, **36**, 553 (2003).

(8.59) D. W. Steuerman, A. Star, R. Narizzano *et al.*, 'Interactions between conjugated polymers and single-walled carbon nanotubes', *J. Phys. Chem. B*, **106**, 3124 (2002).

(8.60) A. Star and J. F. Stoddart, 'Dispersion and solubilization of single-walled carbon nanotubes with a hyperbranched polymer', *Macromolecules*, **35**, 7516 (2002).

(8.61) Y. J. Kang and T. A. Taton, 'Micelle-encapsulated carbon nanotubes: a route to nanotube composites', *J Amer. Chem. Soc.*, **125**, 5650 (2003).

(8.62) W. Z. Yuan, J. Z. Sun, Y. Dong *et al.*, 'Wrapping carbon nanotubes in pyrene-containing poly (phenylacetylene) chains: solubility, stability, light emission, and surface photovoltaic properties', *Macromolecules*, **39**, 8011 (2006).

(8.63) J. Mao, Q. Liu, X. Lv *et al.*, 'A water-soluble hybrid material of single-walled carbon nanotubes with an amphiphilic poly(phenyleneethynylene): preparation, characterization, and photovoltaic properties', *J. Nanosci. Nanotech.*, **7**, 2709 (2007).

(8.64) P. Liu, 'Modifications of carbon nanotubes with polymers', *Eur. Polymer J.*, **41**, 2693 (2005).

(8.65) C. A. Dyke, M. P. Stewart, F. Maya, *et al.* 'Diazonium-based functionalization of carbon nanotubes: XPS and GC-MS analysis and mechanistic implications', *Synlett*, **155** (2004).

(8.66) S. C. Tsang, J. J. Davis, M. L. H. Green *et al.*, 'Immobilisation of small proteins in carbon nanotubes: high resolution transmission electron microscopy study and catalytic activity', *Chem. Commun.*, 1803 (1995).

(8.67) S. C. Tsang, Z. Guo, Y. K. Chen *et al.*, 'Immobilisation of platinised and iodinated oligo-nucleotides on carbon nanotubes', *Angew. Chem. Int. Ed.*, **36**, 2198 (1997).

(8.68) J. J. Davis, M. L. H. Green, H. A. O. Hill *et al.*, 'The immobilisation of proteins in carbon nanotubes', *Inorganica Chimica Acta*, **272**, 261 (1998).

(8.69) Y. Lin, S. Taylor, H. P. Li *et al.*, 'Advances toward bioapplications of carbon nanotubes', *J. Mater. Chem.*, **14**, 527 (2004).

(8.70) W. R. Yang, P. Thordarson, J. J. Gooding *et al.*, 'Carbon nanotubes for biological and biomedical applications', *Nanotechnology*, **18**, 412001 (2007).

(8.71) D. X. Cui, 'Advances and prospects on biomolecules functionalized carbon nanotubes', *J. Nanosci. Nanotech.*, **7**, 1298 (2007).

(8.72) F. Balavoine, P. Schultz, C. Richard *et al.*, 'Helical crystallization of proteins on carbon nanotubes: a first step towards the development of new biosensors', *Angew. Chem. Int. Ed.*, **38**, 1912 (1999).

(8.73) M. Shim, N. W. S. Kam, R. J. Chen *et al.*, 'Functionalization of carbon nanotubes for biocompatibility and biomolecular recognition', *Nano Lett.*, **2**, 285 (2002).

(8.74) M. Bottini, F. Cerignoli, L. Tautz *et al.*, 'Adsorption of streptavidin onto single-walled carbon nanotubes: application in fluorescent supramolecular nanoassemblies', *J. Nanosci. Nanotech.*, **6**, 3693 (2006).

(8.75) R. J. Chen, S. Bangsaruntip, K. A. Drouvalakis *et al.*, 'Non-covalent functionalization of carbon nanotubes for highly specific electronic biosensors', *Proc. Nat. Acad. Sci. USA*, **100**, 4984 (2003).

(8.76) B. R. Azamian, J. J. Davis, K. S. Coleman *et al.*, 'Bioelectrochemical single-walled carbon nanotubes', *J. Amer. Chem. Soc.*, **124**, 12 664 (2002).

(8.77) Y. Lin, L. F. Allard and Y. P. Sun, 'Protein-affinity of single-walled carbon nanotubes in water', *J. Phys. Chem. B*, **108**, 3760 (2004).

(8.78) S. S. Karajanagi, A. A. Vertegel, R. S. Kane *et al.*, 'Structure and function of enzymes adsorbed onto single-walled carbon nanotubes', *Langmuir*, **20**, 11 594 (2004).

(8.79) B. Munge, G. Liu, G. Collins *et al.*, 'Multiple enzyme layers on carbon nanotubes for electrochemical detection down to 80 DNA copies', *Anal. Chem.*, **77**, 4662 (2005).

(8.80) J. Q. Liu, A. Chou, W. Rahmat *et al.*, 'Achieving direct electrical connection to glucose oxidase using aligned single walled carbon nanotube arrays', *Electroanalysis*, **17**, 38 (2005).

(8.81) K. Besteman, J. O. Lee, F. G. M. Wiertz, 'Enzyme-coated carbon nanotubes as single-molecule biosensors', *Nano Lett.*, **3**, 727 (2003).

(8.82) N. W. S. Kam and H. J. Dai, 'Carbon nanotubes as intracellular protein transporters: generality and biological functionality', *J. Amer. Chem. Soc.*, **127**, 6021 (2005).

(8.83) N. W. S. Kam, M. O'Connell, J. A. Wisdom *et al.*, 'Carbon nanotubes as multifunctional biological transporters and near-infrared agents for selective cancer cell destruction', *Proc. Nat. Acad. Sci. USA*, **102**, 11 600 (2005).

(8.84) X. Yu, B. Munge, V. Patel *et al.*, 'Carbon nanotube amplification strategies for highly sensitive immunodetection of cancer biomarkers', *J. Amer. Chem. Soc.*, **128**, 11 199 (2006).

(8.85) N. Shao, S. Lu, E. Wickstrom *et al.*, 'Integrated molecular targeting of IGF1R and HER2 surface receptors and destruction of breast cancer cells using single wall carbon nanotubes', *Nanotechnology*, **18**, 315101 (2007).

(8.86) M. R. McDevitt, D. Chattopadhyay, B. J. Kappel *et al.*, 'Tumor targeting with antibody-functionalized, radiolabeled carbon nanotubes', *J. Nucl. Med.*, **48**, 1180 (2007).

(8.87) M. Zheng, A. Jagota, E. D. Semke *et al.*, 'DNA-assisted dispersion and separation of carbon nanotubes', *Nat. Mater.*, **2**, 338 (2003).

(8.88) M. Zheng, A. Jagota, M. S. Strano *et al.*, 'Structure-based carbon nanotube sorting by sequence-dependent DNA assembly', *Science*, **302**, 1545 (2003).

(8.89) M. S. Strano, M. Zheng, A. Jagota *et al.*, 'Understanding the nature of the DNA-assisted separation of single-walled carbon nanotubes using fluorescence and Raman spectroscopy', *Nano Lett.*, **4**, 543 (2004).

(8.90) N. Nakashima, S. Okuzono, H. Marakami *et al.*, 'DNA dissolves single-walled carbon nanotubes in water', *Chem. Lett.*, **32**, 456 (2003).

(8.91) S. Malik, S. Vogel, H. Rosner *et al.*, 'Physical chemical characterization of DNA-SWNT suspensions and associated composites', *Composites Sci. Technol.*, **67**, 916 (2007).

(8.92) H. Cathcart, S. Quinn, V. Nicolosi *et al.* 'Spontaneous debundling of single-walled carbon nanotubes in DNA-based dispersions', *J. Phys. Chem. C*, **111**, 66 (2007).

(8.93) D. A. Heller, E. S. Jeng, T. K. Yeung *et al.*, 'Optical detection of DNA conformational polymorphism on single-walled carbon nanotubes', *Science*, **311**, 508 (2006).

(8.94) E. S. Jeng, A. E. Moll, A. C. Roy *et al.*, 'Detection of DNA hybridization using the near-infrared band-gap fluorescence of single-walled carbon nanotubes', *Nano Lett.*, **6**, 371 (2006).

(8.95) C. Dwyer, M. Guthold, M. Falvo *et al.*, 'DNA-functionalized single-walled carbon nanotubes', *Nanotechnology*, **13**, 601 (2002).

(8.96) W. R. Yang, M. J. Moghaddam, S. Taylor *et al.*, 'Single-walled carbon nanotubes with DNA recognition', *Chem. Phys. Lett.*, **443**, 169 (2007).

(8.97) C. V. Nguyen, L. Delzeit, A. M. Cassell *et al.*, 'Preparation of nucleic acid functionalized carbon nanotube arrays', *Nano Lett.*, **2**, 1079 (2002).

(8.98) H. Cai, X. Cao, Y. Jiang *et al.*, 'Carbon nanotube-enhanced electro-chemical DNA biosensor for DNA hybridization detection', *Anal. Bioanal. Chem.*, **375**, 287 (2003).

(8.99) P. G. He, S. N. Li and L. M. Dai, 'DNA-modified carbon nanotubes for self-assembling and biosensing applications', *Synth. Met.*, **154**, 17 (2005).

(8.100) P. A. He, Y. Xu and Y. Z. Fang, 'Applications of carbon nanotubes in electrochemical DNA biosensors', *Microchimica Acta*, **152**, 175 (2006).

(8.101) K. Keren, R. S. Berman, E. Buchstab *et al.*, 'DNA-templated carbon nanotube field-effect transistor', *Science*, **302**, 1380 (2003).

(8.102) H. J. Xin and A. T. Woolley, 'DNA-templated nanotube localization', *J. Amer. Chem. Soc.*, **125**, 8710 (2003).

(8.103) C. Dwyer, V. Johri, M. Cheung *et al.*, 'Design tools for a DNA-guided self-assembling carbon nanotube technology', *Nanotechnology*, **15**, 1240 (2004).

(8.104) S. E. Stanca, R. Eritja and D. Fitzmaurice, 'DNA-templated assembly of nanoscale architectures for next-generation electronic devices', *Faraday Disc.*, **131**, 155 (2006).

(8.105) T. H. LaBean and H. Y. Li, 'Constructing novel materials with DNA', *Nano Today*, **2**, 26 (2007).

(8.106) Z. Liu, M. Winters, M. Holodniy *et al.*, 'siRNA delivery into human T cells and primary cells with carbon-nanotube transporters', *Angew. Chem. Int. Ed.*, **46**, 2023 (2007).

(8.107) B. Onoa, M. Zheng, M. S. Dresselhaus *et al.*, 'Carbon nanotubes and nucleic acids: tools and targets', *Physica Status Solidi A*, **203**, 1124 (2006).

(8.108) S. Daniel, T. P. Rao, K. S. Rao *et al.*, 'A review of DNA functionalized/grafted carbon nanotubes and their characterization', *Sensors Actuators B*, **122**, 672 (2007).

(8.109) S. K. Smart, A. I. Cassady, G. Q. Lu *et al.*, 'The biocompatibility of carbon nanotubes', *Carbon*, **44**, 1034 (2006).

(8.110) V. Stone, J. Shaw, D. M. Brown *et al.*, 'The role of oxidative stress in the prolonged inhibitory effect of ultrafine carbon black on epithelial cell function', *Toxicology In Vitro*, **12**, 649 (1998).

(8.111) K. J. Nikula, M. B. Snipes, E. B. Barr *et al.*, 'Comparative pulmonary toxicology and carcinogenicities of chronically inhaled diesel exhaust and carbon black in F344 rats', *Fund. Appl. Toxicol.*, **25**, 80 (1995).

(8.112) A. Huczko, H. Lange, E. Calko *et al.*, 'Physiological testing of carbon nanotubes: are they asbestos like?', *Fullerene Sci. Technol.*, **9**, 251 (2001).

(8.113) A. Huczko, H. Lange, M. Bystrzejewski *et al.*, 'Pulmonary toxicity of 1-D nanocarbon materials', *Fullerenes Nanotubes Carbon Nanostructures*, **13**, 141 (2005).

(8.114) H. Grubek-Jaworska, P. Nejman, K. Czuminska *et al.*, 'Preliminary results on the pathogenic effects of intratracheal exposure to one-dimensional nanocarbons', *Carbon*, **44**, 1057 (2006).

(8.115) A. A. Shvedova, E. R. Kisin, R. Mercer *et al.*, 'Unusual inflammatory and fibrogenic pulmonary responses to single-walled carbon nanotubes in mice', *Amer. J. Physiol. Lung Cellular Molec. Physiol.*, **289**, 698 (2005).

(8.116) C. C. Chou, H. Y. Hsiao, Q. S. Hong *et al.*, 'Single-walled carbon nanotubes can induce pulmonary injury in mouse model', *Nano Lett.*, **8**, 437 (2008).

(8.117) D. B. Warheit, B. R. Laurence, K. L. Reed *et al.*, 'Comparative pulmonary toxicity assessment of single-wall carbon nanotubes in rats', *Toxicol. Sci.*, **77**, 117 (2004).

(8.118) D. B. Warheit, 'What is currently known about the health risks related to carbon nanotube exposures?', *Carbon*, **44**, 1064 (2006).

(8.119) A. A. Shvedova, V. Castranova, E. R. Kisin *et al.*, 'Exposure to carbon nanotube material: assessment of nanotube cytotoxicity using human keratinocyte cells', *J. Toxicol. Environ. Health A*, **66**, 1909 (2003).

(8.120) M. Davoren, E. Herzog, A. Casey *et al.*, '*In vitro* toxicity evaluation of single walled carbon nanotubes on human A549 lung cells', *Toxicology In Vitro*, **21**, 438 (2007).

(8.121) C. M. Sayes, F. Liang, J. L. Hudson *et al.*, 'Functionalization density dependence of single-walled carbon nanotubes cytotoxicity *in vitro*', *Toxicol. Lett.*, **161**, 135 (2006).

(8.122) 'A little risky business', *Economist*, **385**(8556), 109 (2007).

(8.123) A. D. Maynard, P. A. Baron, M. Foley *et al.*, 'Exposure to carbon nanotube material: aerosol release during the handling of unrefined single-walled carbon nanotube material', *J. Toxicol. Environ. Health A*, **67**, 87 (2004).

(8.124) L. Lacerda, A. Bianco, M. Prato *et al.*, 'Carbon nanotubes as nanomedicines: from toxicology to pharmacology', *Adv. Drug Delivery Rev.*, **58**, 1460 (2006).

(8.125) A. Helland, P. Wick, A. Koehler *et al.*, 'Reviewing the environmental and human health knowledge base of carbon nanotubes', *Environ. Health Perspect.*, **115**, 1125 (2007).

(8.126) J. Boczkowski and S. Lanone, 'Potential uses of carbon nanotubes in the medical field: how worried should patients be?', *Nanomedicine*, **2**, 407 (2007).

(8.127) B. Nowack and T. D. Bucheli, 'Occurrence, behavior and effects of nanoparticles in the environment', *Environ. Poll.*, **150**, 5 (2007).

9 Carbon nanotube composites

Many of the outstanding properties of carbon nanotubes that were outlined in Chapters 6 and 7 and elsewhere in this book can be best exploited by incorporating the nanotubes into some form of matrix. The exceptional mechanical properties in particular have prompted huge interest in the production of nanotube-containing composite materials for structural applications. In many cases these composites have employed polymer matrices, but there is also interest in other matrix materials such as ceramics and metals. Preparing such composites is not without its difficulties, however, owing to the tendency of nanotubes to stick together, the challenge of forming bonds between tubes and matrix, and problems associated with the physical properties of some nanotube–matrix mixtures. The aim of this chapter is to give an overview of the very large amount of work that has been carried out on carbon nanotube composites, and to assess how successful this work has been in utilizing the full potential of nanotubes. The incorporation of nanotubes into polymer matrices is considered first.

9.1 Preparation of carbon nanotube/polymer composites

9.1.1 Solution mixing

Perhaps the simplest method for preparing nanotube/polymer composites involves mixing nanotube dispersions with solutions of the polymer and then evaporating the solvents in a controlled way. This method has been used with a range of polymers, including polyvinyl alcohol, polystyrene, polycarbonate and poly(methyl methacrylate). In order to facilitate solubilization and mixing, the nanotubes are often functionalized prior to adding to the polymer solution. In an early example of this approach, Windle and colleagues used acid treatments to disperse catalytically produced MWNTs in water (see also p. 87), and then made nanotube/PVA composites by simply mixing one of these dispersions with an aqueous solution of the polymer and casting the mixtures as films (9.1, 9.2). Subsequent studies of the effect of covalent functionalization on composite properties have been made (9.3–9.5). Problems can sometimes arise with the compatibility of functional groups with the polymer matrix. To avoid this, the nanotubes can be functionalized with polymers that are structurally similar or identical to the matrix polymers (see also p. 210). This approach has been used by Ya-Ping Sun from Clemson University and co-workers to produce SWNT/PVA composites (9.6) and by Jonathan

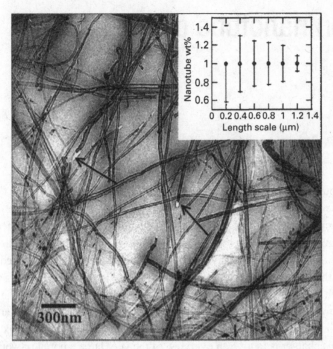

Fig. 9.1   TEM image of MWNT/polystyrene thin film prepared by solution mixing with ultrasonic agitation, from the work of Andrews *et al.* Inset shows the distribution of nanotube lengths (9.10).

Coleman of Trinity College Dublin *et al.* (9.7) to bond SWNTs with halogenated polymers. n alternative to covalent functionalization is to add a surfactant to the nanotube suspension (9.8, 9.9). The advantage here is that the structure of the tubes is not disrupted.

Functionalization of the nanotubes, or the addition of a surfactant, is not always necessary in order to prepare a polymer composite using the solution approach. Rodney Andrews and colleagues from Kentucky used a high-energy ultrasonic probe to disperse MWNTs in toluene and then mixed the dispersed suspension with a dilute solution of polystyrene in toluene, again with ultrasonic agitation (9.10, 9.11). The low viscosity of the polymer solution allowed the nanotubes to move freely through the matrix. The mixture was cast on glass and the solvent removed to yield MWNT-doped films. Specimens of the composite were prepared for TEM by placing drops of the mixed solution onto Cu TEM grids. On evaporation of the toluene, a thin film of the composite was left suspended across the grid bars. A micrograph of one of these films is shown in Fig. 9.1, and this illustrates the excellent dispersion which was achieved. Note that *in situ* polymerization has also been used to prepare nanotube/polystyrene composites, as discussed below.

A potential problem with solution mixing is that the nanotubes can agglomerate during solvent evaporation, leading to inhomogeneous distribution in the matrix. This can be alleviated by spin-casting, which reduces evaporation time (9.12). An alternative method based on coagulation has been developed by Karen Winey of the University of Pennsylvania and colleagues (9.13). This involved pouring the nanotube/PMMA suspension into an excess of non-solvent (water). The precipitating polymer chains then entrap the tubes, preventing agglomeration.

The solution mixing method can be used to make nanotube ribbons and fibres. As noted in Chapter 4 (p. 91), Ray Baughman and colleagues have infiltrated MWNT yarns with PVA by soaking in a solution of the polymer, thus enhancing their mechanical properties (9.14). In 2000, Philippe Poulin of the University of Bordeaux and co-workers described a method to continuously produce SWNT/PVA ribbons (9.15). Their method involved firstly dispersing the tubes in sodium dodecyl sulphate. The SWNT dispersion was then injected through a syringe needle into a stirred PVA solution, where it formed ribbons with diameters similar to that of the needle. Examination of the ribbons by SEM showed that the SWNT bundles were preferentially oriented along the main axis, as a consequence of flow-induced alignment. In a further development of this process, Baughman's group added a second stage, which involved unwinding the fibres then washing and drying to produce composite fibres of potentially unlimited length (9.16). The fibres, which were about 50 μm in diameter and contained approximately 60% SWNTs by weight, displayed excellent mechanical properties, as discussed in Section 9.2.1 below.

Another technique that has been used to produce nanotube/polymer fibres is electrospinning. This involves loading a polymer solution (or melt) into a syringe, and driving the liquid to the needle tip to form a droplet at the tip. The application of a large electric field between the needle and a collecting electrode leads to a jet, which is elongated and the process of solvent evaporation leads to the formation of fibres (9.17). Polymer fibres containing multiwalled (9.18) and single-walled tubes (9.19) have been produced in this way.

9.1.2   Melt processing

The solution mixing approach is limited to polymers that freely dissolve in common solvents. An alternative is to use thermoplastic polymers (i.e. polymers that soften and melt when heated), and then apply melt processing techniques. The downside is that achieving homogeneous dispersions of nanotubes in melts is generally more difficult than with solutions, and high concentrations of tubes are hard to achieve, due to the high viscosities of the mixtures. However, the dispersion of nanotubes can be improved using shear mixing. Also, melt processing lends itself to techniques such as extrusion and injection moulding.

The Kentucky group used shear mixing to disperse catalytically-produced nanotubes in a range of polymers including high-impact polystyrene, acrylonitrile-butadiene-styrene, and polypropylene (9.20). The dispersion of nanotubes was determined as a function of mixing energy and temperature and the composites were formed into fibres and thin films. The Pennsylvania group with colleagues from NIST used melt mixing methods to disperse SWNTs in polystyrene and polyethylene (9.21) and poly(methyl methacrylate) (PMMA) (9.22). In the case of the PE composites, melt-spinning was used to produce fibres with a high degree of nanotube alignment. Other groups have also used melt processing to produce nanotube/polyethylene and polypropylene composites (e.g. 9.23, 9.24). Petra Pötschke of the Polymer Research Institute in Dresden and colleagues used a screw extruder to prepare MWNT/PE composites with weight fractions ranging from 0.1 to 10 wt% (9.23). The electrical conductivity of the composites were found to be more than 16 orders of magnitude higher than that of PE. Good mechanical properties were also observed, although these began to fall off above a certain loading of tubes.

Pötschke and colleagues have studied the use of melt processing techniques to disperse nanotubes in polycarbonate (9.25–9.28). The approach they have used involves firstly preparing a nanotube/polycarbonate 'masterbatch' with a high concentration of tubes (typically 15 wt%). This is then carefully diluted with different amounts of PC in a conical twin screw extruder to obtain different tube concentrations. Good dispersions of catalytically produced MWNTs and SWNTs in PC resin were achieved in this way. Extruded strands of the composite could also be produced using a fibre-spinning apparatus. The nanotube/polycarbonate composites showed good mechanical properties (9.26).

9.1.3   *In situ* polymerization

An alternative method for preparing nanotube/polymer composites is to use the monomer rather than the polymer as a starting material, and then carry out *in situ* polymerization. Wolfgang Maser and co-workers were among the first to use this method, to prepare a MWNT/polyaniline composite (9.29). They showed that composites with MWNT loadings of up to 50 wt% could be produced in this way. Transport measurements on the composite revealed major changes in the electronic behaviour, confirming strong interaction between nanotubes and polymer, as discussed later. Since this early work, many other nanotube/polymer composites have been prepared using *in situ* polymerization, including MWNT/polystyrene (9.30), MWNT/polyurethane (9.31), MWNT/polypyrrole (9.32) and MWNT/nylon (9.33–9.35).

A large amount of work has been done on the preparation of nanotube/epoxy composites (e.g. 9.5, 9.36–9.41). This involves typically first dispersing the nanotubes in the resin followed by curing the resin with the hardener. In most cases the epoxy begins life in liquid form, facilitating mixing with nanotubes. One of the earliest studies of a nanotube-containing epoxy, by Ajayan and colleagues, was not actually carried out with the aim of producing a composite material. The objective was to obtain cross-sectional images of nanotubes by embedding purified tubes into an epoxy resin and then cutting it into thin slices with a diamond knife. However, no cross-sections were observed following this treatment; instead, the nanotubes were found to have become aligned in the direction of the knife movement, as shown in Fig. 9.2. According to Ajayan *et al.*, the alignment is primarily a consequence of extensional or shear flow of the matrix produced by the cutting, although in some cases tubes appear to come into direct contact with the knife and are pulled out of the matrix and oriented unidirectionally on the surface. In more recent work, Australian researchers (9.39) showed that entangled MWNTs in an epoxy matrix could be separated and aligned by continuous shear between steel discs.

Fig. 9.2   Alignment of nanotubes in a polymer matrix following cutting with microtome (9.36).

Some unconventional approaches to producing nanotube/epoxy composites have been reported. Ben Wang's group at the Florida Advanced Center for Composite Technologies has produced SWNT/epoxy composites by infiltrating 'buckypaper' with acetone-diluted epoxy (9.40), and reported good mechanical properties. Charles Lieber's team have produced thin films of nanotube/epoxy composite using a 'bubble-blowing' method (9.41). In this work, MWNTs were firstly modified with n-octadecylamine and then suspended in tetrahydrofuran. An epoxy was then added, mechanically mixed with the nanotube suspension, and then allowed to cure. When the mixture reached a certain viscosity range, it was blown into a bubble, as illustrated in Fig. 9.3. The films produced in this way could be transferred to substrates or open frame structures. This would seem to be a highly promising way of producing large-scale thin films of nanotube-containing polymers.

9.1.4   Effect of nanotubes on polymer structure

Before proceeding to discuss the properties of carbon nanotube/polymer composites in detail, it is worth briefly considering the effect that the inclusion of nanotubes can have on polymer microstructure. Although this topic has not yet been widely studied, a number of groups have demonstrated that the crystallization and morphology of the polymers can be strongly affected by small additions of nanotubes. Thus Jonathan Coleman and colleagues found that arc-produced MWNTs increased the crystallinity of PVA (9.42, 9.43). The same effect was observed by researchers from Georgia Tech (9.44) who recorded HRTEM images showing well resolved PVA (200) lattice fringes oriented parallel to the nanotube axes. Nanotube-induced crystallization has also been observed in a number of other polymers (e.g. 9.45–9.48), and may be an important factor in determining the properties of composites.

It has been found that that the presence of nanotubes can promote stereochemical selectivity in chiral polymers (9.49, 9.50). Wenhui Song and colleagues prepared

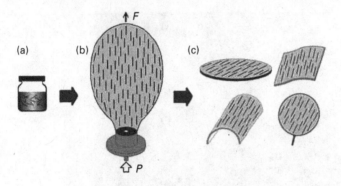

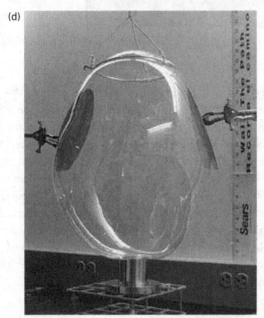

Fig. 9.3   (a)–(c) Illustrations of the blown film extrusion process developed by Lieber and colleagues for the production of nanotube/polymer thin films, (d) a photograph of an expanded bubble (9.41).

MWNT/polyaniline composite nanofibres using *in situ* polymerization (9.50) and found that the polymer contained a high proportion of one of the enantiomers. The reasons for this are not fully understood.

9.2   Properties of carbon nanotube/polymer composites

9.2.1   Mechanical properties

Much of the work on the preparation of nanotube/polymer composites has been driven by a desire to exploit the tubes' stiffness and strength. Even where the interest has been focused on other properties, the ability of nanotubes to improve the mechanical characteristics of a polymer has often been a valuable added benefit.

Shaffer and Windle were among the first to carry out a systematic study of the mechanical properties of nanotube/polymer composites. Their technique for preparing MWNT/PVA composites was described above (9.2). The tensile elastic moduli of the composite films were assessed in a dynamic mechanical thermal analyser as a function of nanotube loading and temperature. The stiffness of the composites at room temperature was relatively low. From the theory developed for short-fibre composites, a nanotube elastic modulus of 150 MPa was obtained from the room temperature experimental data. This low value may have more to do with poor stress transfer than with the weakness of the nanotubes themselves. Above the glass transition temperature of the polymer (~85°C), the nanotubes had a more significant effect on the properties of the composite.

Jonathan Coleman and colleagues achieved rather better results in 2002 with composites made from PVA and arc-produced MWNTs (9.42). The composites were prepared by solution-mixing and films were formed on glass substrates by drop casting. The presence of 1 wt% of nanotubes increased the Young's modulus and hardness of the PVA by factors of 1.8 and 1.6 respectively. This improvement over the early results of Shaffer and Windle reflects both the superior quality of the tubes and stronger interfacial bonding. Evidence for the latter was provided by TEM studies. In more recent work (9.43), the same group reported a 4.5-fold increase in the Young's modulus of a PVA with the addition of carbon nanotubes. As noted above, it was found that the enhancement of the Young's modulus was partly due to a nanotube-promoted increase in polymer crystallinity.

The mechanical properties of nanotube/polystyrene composites were studied by the Kentucky group (9.10). Two samples of catalytically-produced MWNTs were used, one with an average length of ~15 μm, the other with an average length of ~50 μm. With the addition of 1% nanotubes by weight they achieved a 36% and 42% increase in the elastic stiffness of the polymer for short and long tubes respectively (the neat polymer modulus was ~1.2 GPa). In both cases a 25% increase in the tensile strength was observed. Theoretical calculations, carried out assuming a nanotube modulus of 450 GPa, produced increases of a 48% and 62% in the stiffness of the polymer for short and long tubes. The close agreement, within ~10%, between the experimental and theoretically predicted composite moduli indicated that the external tensile loads were successfully transmitted to the nanotubes across the tube–polymer interface. At higher nanotube concentrations, the changes in the mechanical properties of the composites were more pronounced (9.20). For MWNT/polystyrene composites containing from 2.5 to 25 vol.% nanotubes, Young's modulus increased progressively from 1.9 to 4.5 GPa, with the major increases occurring when the MWNT content was at or above about 10 vol.%. However, the dependence of tensile strength on nanotube concentration was more complex. At the lower concentrations (≤10 vol.%), the tensile strength decreased from the neat polymer value of ~40 MPa, only exceeding it when the MWNT content was above 15 vol.%.

In a study of nanotube-reinforced polyethylene by a group from the University of Sydney (9.51), the Young's modulus and tensile strength were found to increase by 89% and 56%, respectively, when the nanotube loading reached 10 wt%. The Pennsylvania group (9.21) found that the tensile modulus of PE fibre was improved from 0.65 to 1.25 GPa with the addition of 5 wt% SWNT. The same group also reported excellent results with their SWNT/nylon composites (9.35): the incorporation of 2 wt% SWNTs produced a 214% improvement in elastic modulus and a 162% increase in yield strength over pure nylon. These properties were undoubtedly partly due to the fact that functionalized nanotubes were used, producing strong nanotube–polymer bonding. Perhaps the most impressive mechanical properties were observed with the SWNT/PVA fibres prepared by Baughman and colleagues (9.16). These had Young's moduli of up to 80 GPa, tensile strengths of 1.8 GPa and a very high toughness, suggesting applications such as bullet-proof vests. In 2007, Poulin's group showed that SWNT/PVA fibres prepared by the coagulation spinning process had excellent shape-memory characteristics (9.52).

The tensile fracture mechanisms of nanotube/polymer composites have been quite widely studied. In their work on MWNT/PS composites, the Kentucky workers performed deformation studies inside a transmission electron microscope (9.10). Their elegant method of preparing TEM specimens was described above. Focusing the electron beam onto the thin film resulted in local thermal stresses which initiated cracks in the composite. The propagation speed of the crack could be controlled by varying the beam flux onto the sample. These *in situ* TEM observations showed that the cracks tended to nucleate at low nanotube density areas then propagate along weak nanotube/polymer interfaces or relatively low nanotube density regions. The nanotubes became aligned perpendicular to the crack direction and bridged the crack faces in the wake, as shown in Fig. 9.4. When the crack opening displacement exceeded ~800 nm, the nanotubes began

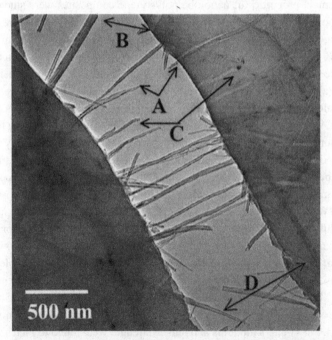

Fig. 9.4   A TEM image of a crack in a MWNT/polystyrene thin film induced by thermal stresses (9.10).

to break and/or pull out of the matrix. The fact that some of the pulled-out tubes do not appear to be coated with polymer suggests that there is room for improvement in nanotube–polymer bonding.

It is interesting to compare the work of the Kentucky group with a study by Paul Watts and Wen-Kuang Hsu of the University of Sussex (9.53). These workers also used TEM to study the fracture behaviour of a nanotube/polymer composite. Their composite consisted of arc-grown MWNTs embedded in a diblock copolymer known as MPC-DEA. To prepare specimens of the composite for TEM, the tubes were mixed with the polymer in an acidified aqueous solution and drops of this were applied to TEM grids, which were then dried in vacuum. As in the work of the Kentucky group, cracks were produced in the material by irradiation with an electron beam, but a low accelerating voltage (75 kV) was used to avoid massive disruption of the thin composite film. Individual tubes and bundles of tubes were observed to become stretched as the cracks grew wider, and pull-out was often seen. In contrast to the work of the Kentucky group, however, the tubes did not break. This was probably because high-quality arc-grown nanotubes were used, rather than catalytically produced tubes.

The fracture behaviour of SWNT/polymer composites was studied by Ajayan et al. (9.54). Two types of composite were prepared, firstly an SWNT/epoxy composite and secondly a pressed pellet containing SWNTs and carbonaceous soot material formed during nanotube synthesis. The SWNTs were used as-prepared, i.e. generally grouped into bundles rather than separated into individual tubes. The composites were loaded axially in tension until failure occurred and the fracture surfaces were then examined in detail using SEM. In some cases the tube bundles were found to have been pulled out of the matrix during the deformation and fracture of the composites, while in other cases the nanotubes were not entirely pulled out but were stretched between two fracture surfaces. The authors believed that during pull-out the nanotubes were sliding axially within the ropes and that the failure observed at large crack distances was not failure of individual tubes, but the bundles pulling apart. Studies by micro-Raman spectroscopy supported this hypothesis. It was found that little shift in the second order A_{1g} band occurred as a result of the application of axial tension, showing that the individual nanotubes were not being significantly stretched. To take full advantage of the high Young's modulus of SWNTs in polymer composites, it is clear that load must be transferred effectively from the matrix to the nanotubes. Ajayan et al. suggested that load transfer could be improved by breaking bundles down into individual tube fragments before dispersing in the matrix. Alternatively the bundles themselves could be reinforced by cross linking the tubes within bundles (e.g. by chemical treatments or irradiation).

Several reviews of the mechanical properties of carbon nanotube/polymer composites are available (9.55–9.57).

9.2.2 Electrical properties

As well as mechanical properties, the electrical properties of carbon nanotube/polymer composites have been very widely studied (e.g. 9.13, 9.32, 9.37, 9.58–9.65). These studies fall into a number of distinct categories. In some cases the nanotubes have been used to increase the conductivity of relatively low cost non-conducting polymers, as an alternative to currently used fillers such as carbon black. Other studies have involved the incorporation of nanotubes into conducting polymers such as polyaniline. In many cases, improved mechanical properties have been a valuable by-product of the inclusion of nanotubes. We consider first the addition of nanotubes to low cost, bulk polymers.

Improving the electrical conductivity of bulk polymers is important in a number of applications. For example, in some aircraft components, enhanced conductivity is required to provide electrostatic discharge and electromagnetic-radio frequency inter-ference protection. Static electrical dissipation is also needed in other applications, including computer housings and exterior automotive parts. With these kinds of applica-tions in mind, Shaffer, Windle and their colleagues investigated the electrical properties of nanotube/epoxy composites (9.37). Matrix resistivities of around 100 Ωm with filler volume fractions as low as 0.1 vol.% were achieved. These figures represented an advance on the best conductivity values previously obtained with carbon black in the same epoxy matrix. In subsequent work, the same group reported an ultra-low electrical percolation threshold in composites consisting of aligned catalytically-produced MWNTs in epoxy (9.58). The percolation threshold of a polymer containing a conduct-ing filler is the filler content required for the material to become electrically conductive. It is characterized by a sharp jump in the conductivity, usually of many orders of magnitude. The Cambridge group reported percolation thresholds as low as 0.0025 vol.% for their composites. Low conductivity thresholds have also been reported in SWNT-containing composites (e.g. 9.61, 9.62). Arjun Yodh and colleagues showed that the aspect ratios of the tubes in these composites affected the electrical properties (9.62). This group prepared SWNT/epoxy composites using tubes produced by the HiPco and laser vaporization methods. These had lengths of 167 and 516 nm and respectively. The longer tubes were found to have a significantly smaller percolation threshold.

These and other published studies suggest that carbon nanotubes have great promise in reducing the electrostatic charging of non-conducting bulk polymers. Nanotubes also have other advantages over conventional fillers such as carbon black and carbon fibres in that they are more amenable to processing and can be more easily dispersed throughout the matrix. In addition to the studies available in the open literature, there is undoubtedly a great deal of commercial work being carried out on the use of nanotubes to reduce the electrostatic charging of plastics. In fact, nanotube-containing plastics are already being used in commercial products. These include fuel lines in automobiles, where the nanotubes help to dissipate any dangerous charge which may build up. Thermoplastic polymers containing nanotubes are also used in some exterior automobile parts, so that they can be earthed during electrostatic painting.

We turn now to nanotube composites in which the matrix is a conducting polymer. Most studies in this area have involved polyaniline (PANI), a conducting polymer with many attractive characteristics such as processability and environmental stability. There has been interest in adding nanotubes to this polymer, both to provide enhanced con-ductivity and improved mechanical properties. As mentioned above, Maser and collea-gues prepared MWNT/polyaniline composites by *in situ* polymerization (9.29). The nanotubes were found to have a strong influence on the transport properties of the polymer. Thus, the room temperature resistivity of the composite was found to be an order of magnitude lower than that of pure polyaniline, while the low-temperature resistivity was much smaller than that of either PANI or of MWNTs. The temperature dependence of resistivity was also weaker than that of PANI alone. These observations were explained by assuming that *in situ* polymerization favours charge transfer between PANI and MWNTs resulting in an overall material that is more conducting than the starting components. Raman studies indicated that these charge transfer processes involved a site-selective interaction between the quinoid ring of the polymer and the MWNTs. Other groups have also demonstrated excellent conductivity in MWNT/polyaniline (9.63, 9.64) and it has been shown that SWNT/polyaniline composites can be used as printable conductors for organic electronic devices (9.59). Composites containing nanotubes dispersed in polypyrrole, another conducting polymer, have also been synthesized (9.32, 9.60, 9.65).

For general reviews on the preparation and properties of nanotube/polymer composites, references (9.56), (9.66) and (9.67) are recommended. A special issue of the journal *Composites Science and Technology* on carbon nanotube/polymer composites appeared in 2007 (9.68).

9.3   Carbon nanotube/ceramic composites

Ceramics have high stiffnesses and thermal stabilities but relatively low breaking strengths. Incorporating carbon nanotubes into a ceramic matrix might be expected to produce a composite with both toughness and high-temperature stability. However, achieving a homogeneous dispersion of tubes in an oxide, with strong bonding between tubes and matrix, presents rather more of a challenge than incorporating tubes into a polymer. Various approaches have been adopted for incorporating nanotubes into oxides, and these will now be summarized, beginning with attempts to produce nanotube/alumina composites.

Alain Peigney of the University of Toulouse and colleagues were among the pioneers in this field (e.g. 9.69–9.73). They have developed an ingenious technique that involves impregnating the ceramic with catalytic metals such as Fe, and then using these to grow nanotubes, producing the precursor for a nanotube/oxide composite. The resulting powders can then be hot-pressed to form the final composite material. The mechanical properties of these composites, however, have generally been rather disappointing. One problem is that the nanotubes can be damaged during hot-pressing. Thus the fracture strengths of the composites are frequently only marginally higher than those of the pure ceramics.

Rather better mechanical properties appear to have been achieved with nanotube/ceramic composites made using the technique of spark-plasma sintering (SPS). In this process, sample material is held in a graphite die between stainless steel electrodes, to which pressure is applied. Pulses of current (up to 20 000 A) are then passed through the die and through the sample. This heats the sample and results in sparks being produced between adjacent particles. This treatment promotes the formation of necks between the particles, and rapid sintering. An important advantage of SPS is that it allows ceramic powders to be annealed at lower temperatures and for much shorter times than in other sintering processes, leading to the fabrication of fully dense ceramics or composites with nanocrystalline microstructures under mild conditions. Unlike some of the other processes used to prepare nanotube/oxide composites, SPS does not damage the nanotubes. The use of spark-plasma sintering to make nanotube/ceramic composites has been investigated in detail by Amiya Mukherjee of the University of California, Davis, and co-workers. In 2003 this group used SPS to prepare Al_2O_3 composites containing 10 vol.% SWNTs (9.74). Prior to sintering, the powders had been mixed by ball-milling, which produced a reasonably homogeneous dispersion, again without damaging the nanotubes. The fracture toughnesses of the final composites were remarkable – about twice that of the fully densified but unreinforced Al_2O_3 matrix. In contrast to previous work on nanotube/ceramic composites, the fracture toughness was found to increase with nanotube density. This is thought to be due to the formation of entangled networks of single-walled carbon nanotubes, which may inhibit crack propagation. A subsequent study of the nanotube/Al_2O_3 composites showed that they had high electrical conductivities (9.75). Spark-plasma sintering has also been used by Korean researchers to produce nanotube-reinforced Al_2O_3 (9.76). In this case the tubes were mixed with alumina using the sol-gel process. Again good toughness was observed.

Various techniques have been used to prepare carbon nanotube/silica composites. Mauricio Terrones and colleagues described a method that involved preparing a composite gel of MWNTs with tetraethoxysilane (TEOS) and then sintering this at 1150 °C in Ar (9.77). A drawback of this technique was that the sintering process led to a partial crystallization of the SiO_2 resulting in a very inhomogeneous matrix. An alternative method, which avoids these problems has been developed (9.78). This involved using a Nd:YAG laser to rapidly heat a TEOS/nanotube mixture, resulting in partial melting of the matrix. This produced an amorphous SiO_2 matrix, with no crystallization. Spark-plasma sintering has also been used to produce nanotube/SiO_2 composites, converting insulating SiO_2 into a metallically conductive composite (9.79).

There have been a few studies on preparing nanotube composites using oxides other than alumina and silica. Several groups have produced nanotube/TiO_2 composites using sol-gel methods (9.80–9.83). Possible applications of such composites included optical wave guides, conductive films and sensors. Sol-gel methods were also used by Sakamoto and Dunn to prepare SWNT/V_2O_5 composites, with the aim of using them as electrodes in secondary lithium batteries (9.84). Nickel oxide/MWNT composites have been produced, for potential use as supercapacitors (9.85). The possibility of using hydroxyapatite/nanotube composites for biomedical applications such as bone grafts has been explored by Ian Kinloch and co-workers (9.86). Nanotubes have also been incorporated into non-oxygen containing ceramics, notably silicon carbide. In an early study, Chinese researchers fabricated nanotube/SiC composites by mixing SiC nanoparticles with 10 wt% MWNTs and hot-pressing at 2000 °C (9.87). They found that the nanotubes played a strengthening and toughening role in the composite: both bend strength and fracture toughness increased by about 10%, compared to monolithic SiC. More recent work by Tadeusz Zerda of Texas Christian University and colleagues showed that MWNT/SiC with good mechanical properties could be made by a high-pressure reactive sintering technique (9.88). Single-walled nanotubes have also been incorporated into SiC (9.89).

9.4   Carbon nanotube/carbon composites

There has been some interest in incorporating nanotubes into carbon matrices. Rodney Andrews and co-workers have produced carbon fibres containing single-walled nanotubes (9.90). These were prepared by dispersing SWNTs in petroleum pitch and then heating at high temperature. The tensile strength, modulus and electrical conductivity of the composite fibres were found to be greatly enhanced compared with the pure carbon fibres. A group from Cambridge have grown MWNTs on a carbon fibre cloth using plasma enhanced CVD, resulting in a significant increase of the bulk electrical conductivity (9.91).

Morinobu Endo and colleagues have shown that adding catalytically-produced MWNTs to the synthetic graphite used in the anodes of Li-ion batteries can greatly improve the performance of the batteries (9.92, 9.93). Nanotubes were found to have significant advantages over the conventional carbon black filler. The tubes helped to increase the conductivity of the anodes, while also enhancing their physical flexibility. Improved penetration of the electrolyte was also observed. Nanotube-containing Li-ion batteries have now been commercialized.

9.5   Carbon nanotube/metal composites

Composite materials containing conventional carbon fibres in a metal matrix such as aluminium or magnesium are used in a number of specialist applications. Such composites combine low density with high strength and modulus, making them particularly attractive to the aerospace industry. There is growing interest in the addition of carbon nanotubes to metal matrices. Pioneering work was carried out by Toru Kuzumaki and colleagues of the University of Tokyo, who described the preparation of a nanotube/aluminium composite in 1998 (9.94). Their method involved mixing a nanotube sample with a fine Al powder, mounting the mixture in a 6 mm silver sheath and then drawing and heating the wire at 700 °C in a vacuum furnace. The result was a composite wire in which the nanotubes were partially aligned along the axial direction. The tensile strengths of the as-prepared composite wires were comparable to that of pure Al, but the composite wires retained this strength after prolonged annealing at 600 °C, while the strength of pure Al decreased by about 50% after this treatment.

Quite a large amount of work has been carried out on nanotube/aluminium composites since these early studies, and in most cases significant improvements in the mechanical properties over the pure metal have been found (e.g. 9.95–9.97). A limited amount of work has been done on incorporating nanotubes into other metals such as titanium (9.98) and magnesium (9.99).

9.6   Discussion

As noted at the beginning of this chapter, incorporating carbon nanotubes into a matrix material presents a number of special challenges (9.100). The primary difficulties involve dispersing the nanotubes homogeneously throughout the matrix, and achieving good bonding between tubes and matrix. While these have not been completely overcome, notable progress has been made by a number of groups, particularly with polymer matrices. Functionalization of the nanotubes (e.g. 9.5, 9.35), or the use of surfactants (e.g. 9.8), has been quite widely used both to improve dispersion and to enhance interfacial bonding. Ultrasonic mixing has also proved to be a successful method of dispersing tubes (9.10, 9.11). As a result of these advances, nanotube/polymer composites are beginning to realize their potential. Thus, workers from the University of Pennsylvania have shown that adding 2 wt% SWNTs to nylon produces a 214% improvement in elastic modulus and a 162% increase in yield strength over the pure polymer (9.35). Some notable work has also been carried out on the production of nanotube-containing polymer fibres. Following pioneering work by Poulin and co-workers (9.15), the Baughman group have produced SWNT/PVA fibres with exceptional mechanical properties (9.16). Incorporating carbon nanotubes in polymers can also confer useful electrical properties. There has been much interest in using nanotubes to improve the conductivity of low cost bulk polymers, for possible applications in the aircraft or automobile industry. Promising results have also been achieved with nanotube/conducting polymer composites for more specialized applications.

Compared with the huge amount of research on nanotube/polymer composites, there has been less work on incorporating nanotubes into other matrices, such as ceramics. The challenges here are even greater than with polymers, but progress has been made using the technique of spark-plasma sintering (e.g. 9.74).

As far as commercial applications of carbon nanotube composites are concerned, these have been relatively limited to date. As already mentioned, nanotubes have been used to improve the anti-static properties of fuel-handling components and body panels of automobiles. Hyperion Catalysis International of Cambridge, Massachusetts were pioneers in this area. The French company Nanoledge have collaborated in the production of nanotube-containing sporting equipment such as tennis rackets, while the Swiss bicycle manufacturer BMC, in partnership with US company Easton, have incorporated nanotubes in the frames of high-performance racing bikes (see Fig. 9.5). The cost of these products places them well outside the price-range of the ordinary consumer, however. More widespread applications will have to await improvements in the low-cost production of nanotubes.

Fig. 9.5 BMC Pro Machine SLC01 racing bicycle with frame containing carbon nanotubes.

One area where nanotubes do seem to have had a significant commercial impact is in Li-ion batteries. As mentioned in Section 9.4, Endo and colleagues have demonstrated considerable improvements to the performance of such batteries by the addition of catalytically-produced MWNTs (9.92, 9.93).

References

(9.1) M. S. P. Shaffer, X. Fan and A. H. Windle, 'Dispersion and packing of carbon nanotubes', *Carbon*, **36**, 1603 (1998).

(9.2) M. S. P. Shaffer and A. H. Windle, 'Fabrication and characterization of carbon nanotube/poly(vinyl alcohol) composites', *Adv. Mater.*, **11**, 937 (1999).

(9.3) G. M. Odegard, S. J. V. Frankland and T. S. Gates, 'Effect of nanotube functionalization on the elastic properties of polyethylene nanotube composites', *AIAA J.*, **43**, 1828 (2005).

(9.4) C. Velasco-Santos, A. L. Martinez-Hernandez, F. T. Fisher *et al.*, 'Improvement of thermal and mechanical properties of carbon nanotube composites through chemical functionalization', *Chem. Mater.*, **15**, 4470 (2003).

(9.5) M. Grujicic, Y. P. Sun and K. L. Koudela, 'The effect of covalent functionalization of carbon nanotube reinforcements on the atomic-level mechanical properties of poly-vinyl-ester-epoxy', *Appl. Surface Sci.*, **253**, 3009 (2007).

(9.6) Y. Lin, M. J. Meziani and Y. P. Sun, 'Functionalized carbon nanotubes for polymeric nanocomposites', *J. Mater. Chem.*, **17**, 1143 (2007).

(9.7) R. Blake, Y. K. Gun'ko, J. Coleman *et al.*, 'A generic organometallic approach toward ultra-strong carbon nanotube polymer composites', *J. Amer. Chem. Soc.*, **126**, 10 226 (2004).

(9.8) X. Y. Gong, J. Liu, S. Baskaran *et al.*, 'Surfactant-assisted processing of carbon nanotube/polymer composites', *Chem. Mater.*, **12**, 1049 (2000).

(9.9) S. Cui, R. Canet, A. Derre *et al.*, 'Characterization of multiwall carbon nanotubes and influence of surfactant in the nanocomposite processing', *Carbon*, **41**, 797 (2003).

(9.10) D. Qian, E. C. Dickey, R. Andrews et al., 'Load transfer and deformation mechanisms in carbon nanotube-polystyrene composites', *Appl. Phys. Lett.*, **76**, 2868 (2000).

(9.11) B. Safadi, R. Andrews and E. A. Grulke, 'Multiwalled carbon nanotube polymer composites: synthesis and characterization of thin films', *J. Appl. Polym. Sci.*, **84**, 2660 (2002).

(9.12) M. L. De la Chapelle, C. Stephan, T. P. Nguyen et al., 'Raman characterization of single-walled carbon nanotubes and PMMA-nanotubes composites', *Synth. Met.*, **103**, 2510 (1999).

(9.13) F. Du, J. E. Fischer and K. I. Winey, 'Coagulation method for preparing single-walled carbon nanotube/poly(methyl methacrylate) composites and their modulus, electrical conductivity, and thermal stability', *J. Polym. Sci. B*, **41**, 3333 (2003).

(9.14) M. Zhang, K. R. Atkinson and R. H. Baughman, 'Multifunctional carbon nanotube yarns by downsizing an ancient technology', *Science*, **306**, 1358 (2004).

(9.15) B. Vigolo, A. Penicaud, C. Coulon et al., 'Macroscopic fibers and ribbons of oriented carbon nanotubes', *Science*, **290**, 1331 (2000).

(9.16) A. B. Dalton, S. Collins, E. Munoz et al., 'Super-tough carbon-nanotube fibres', *Nature*, **423**, 703 (2003).

(9.17) S. Ramakrishna, K. Fujihara, W. E. Teo et al., *An Introduction to Electrospinning and Nanofibers*, World Scientific Publishing, Singapore, 2005.

(9.18) H. Q. Hou, J. J. Ge, J. Zeng et al., 'Electrospun polyacrylonitrile nanofibers containing a high concentration of well-aligned multiwall carbon nanotubes', *Chem. Mater.*, **17**, 967 (2005).

(9.19) R. Sen, B. Zhao, D. Perea et al., 'Preparation of single-walled carbon nanotube reinforced polystyrene and polyurethane nanofibers and membranes by electrospinning', *Nano Lett.*, **4**, 459 (2004).

(9.20) R. Andrews, D. Jacques, D. Qian et al., 'Multiwall carbon nanotubes: synthesis and application', *Acc. Chem. Res.*, **35**, 1008 (2002).

(9.21) R. Haggenmueller, W. Zhou, J. E. Fischer et al., 'Production and characterization of polymer nanocomposites with highly aligned single-walled carbon nanotubes', *J. Nanosci. Nanotechnol.*, **3**, 105 (2003).

(9.22) T. Kashiwagi, J. Fagan, J. F. Douglas et al., 'Relationship between dispersion metric and properties of PMMA/SWNT nanocomposites', *Polymer*, **48**, 4855 (2007).

(9.23) T. McNally, P. Pötschke, P. Halley et al., 'Polyethylene multiwalled carbon nanotube composites', *Polymer*, **46**, 8222 (2005).

(9.24) Q. H. Zhang, S. Rastogi, D. J. Chen et al., 'Low percolation threshold in single-walled carbon nanotube/high density polyethylene composites prepared by melt processing technique', *Carbon*, **44**, 778 (2006).

(9.25) P. Pötschke, A. R. Bhattacharyya and A. Janke, 'Melt mixing of polycarbonate with multi-walled carbon nanotubes: microscopic studies on the state of dispersion', *Eur. Polymer J.*, **40**, 137 (2004).

(9.26) B. Lin, U. Sundararaj and P. Pötschke, 'Melt mixing of polycarbonate with multi-walled carbon nanotubes in miniature mixers', *Macromolecular Mater. Eng.*, **291**, 227 (2006).

(9.27) U. A. Handge and P. Pötschke, 'Deformation and orientation during shear and elongation of a polycarbonate/carbon nanotubes composite in the melt', *Rheologica Acta*, **46**, 889 (2007).

(9.28) B. K. Satapathy, R. Weidisch, P. Pötschke et al., 'Tough-to-brittle transition in multiwalled carbon nanotube (MWNT)/polycarbonate nanocomposites', *Composites Sci. Technol.*, **67**, 867 (2007).

(9.29) M. Cochet, W. K. Maser, A. M. Benito et al., 'Synthesis of a new polyaniline/nanotube composite: "*in-situ*" polymerisation and charge transfer through site-selective interaction', *Chem. Commun.*, 1450 (2001).

(9.30) S. T. Kim, H. J. Choi and S. M. Hong, 'Bulk polymerized polystyrene in the presence of multiwalled carbon nanotubes', *Colloid Polymer Sci.*, **285**, 593 (2007).

(9.31) H. J. Yoo, Y. C. Jung, N. G. Sahoo *et al.*, 'Polyurethane–carbon nanotube nanocomposites prepared by *in-situ* polymerization with electroactive shape memory', *J. Macromolecular Sci. B*, **45**, 441 (2006).

(9.32) T. M. Wu and S. H. Lin, 'Synthesis, characterization, and electrical properties of polypyrrole/multiwalled carbon nanotube composites', *J. Polymer Sci. A*, **44**, 6449 (2006).

(9.33) M. Kang, S. J. Myung and H. J. Jin, 'Nylon 610 and carbon nanotube composite by *in situ* interfacial polymerization', *Polymer*, **47**, 3961 (2006).

(9.34) R. Haggenmueller, F. Du, J. E. Fischer *et al.*, 'Interfacial *in situ* polymerization of single walled carbon nanotube/nylon 6,6 nanocomposites', *Polymer*, **47**, 2381 (2006).

(9.35) M. Moniruzzaman, J. Chattopadhyay, W. E. Billups *et al.*, 'Tuning the mechanical properties of SWNT/Nylon 6,10 composites with flexible spacers at the interface', *Nano Lett.*, **7**, 1178 (2007).

(9.36) P. M. Ajayan, O. Stephan, C. Colliex *et al.*, 'Aligned carbon nanotubes arrays formed by cutting a polymer resin-nanotubes composite', *Science*, **265**, 1212 (1994).

(9.37) J. Sandler, M. S. P. Shaffer, T. Prasse *et al.*, 'Development of a dispersion process for carbon nanotubes in an epoxy matrix and the resulting electrical properties', *Polymer*, **40**, 5967 (1999).

(9.38) X. D. Li, H. S. Gao, W. A. Scrivens *et al.*, 'Nanomechanical characterization of single-walled carbon nanotube reinforced epoxy composites', *Nanotechnology*, **15**, 1416 (2004).

(9.39) K. Q. Xiao and L. C. Zhang, 'Effective separation and alignment of long entangled carbon nanotubes in epoxy', *J. Mater. Sci.*, **40**, 6513 (2005).

(9.40) Z. Wang, Z. Y. Liang, B. Wang *et al.*, 'Processing and property investigation of single-walled carbon nanotube (SWNT) buckypaper/epoxy resin matrix nanocomposites', *Composites A*, **35**, 1225 (2004).

(9.41) G. Yu, A. Cao and C. M. Lieber, 'Large-area blown bubble films of aligned nanowires and carbon nanotubes', *Nat. Nanotech.*, **2**, 372 (2007).

(9.42) M. Cadek, J. N. Coleman, V. Barron *et al.*, 'Morphological and mechanical properties of carbon-nanotube-reinforced semicrystalline and amorphous polymer composites', *Appl. Phys. Lett.*, **81**, 5123 (2002).

(9.43) K. P. Ryan, M. Cadek, V. Nicolosi *et al.*, 'Multiwalled carbon nanotube nucleated crystallization and reinforcement in poly (vinyl alcohol) composites', *Synth. Met.*, **156**, 332 (2006).

(9.44) M. L. Minus, H. G. Chae and S. Kumar, 'Single wall carbon nanotube templated oriented crystallization of poly(vinyl alcohol)', *Polymer*, **47**, 3705 (2006).

(9.45) R. Czerw, Z. X. Guo, P. M. Ajayan *et al.*, 'Organization of polymers onto carbon nanotubes: a route to nanoscale assembly', *Nano Lett.*, **1**, 423 (2001).

(9.46) B. P. Grady, F. Pompeo, R. L. Shambaugh *et al.*, 'Nucleation of polypropylene crystallization by single-walled carbon nanotubes', *J. Phys. Chem. B*, **106**, 5852 (2002).

(9.47) L. Valentini, J. Biagiotti, J. M. Kenny *et al.*, 'Morphological characterization of single-walled carbon nanotubes-PP composites', *Composites Sci. Technol.*, **63**, 1149 (2003).

(9.48) M. C. Garcia-Gutierrez, A. Nogales, D. R. Rueda *et al.*, 'Templating of crystallization and shear-induced self-assembly of single-wall carbon nanotubes in a polymer-nanocomposite', *Polymer*, **47**, 341 (2006).

(9.49) M. in het Panhuis, R. Sainz, P. C. Innis *et al.*, 'Optically active polymer carbon nanotube composite', *J. Phys. Chem. B*, **109**, 22 725 (2005).

(9.50) X. Zhang, W. Song, P. J. F. Harris *et al.*, 'Chiral polymer–carbon-nanotube composite nanofibers', *Adv. Mater.*, **19**, 1079 (2007).

(9.51) K. Q. Xiao, L. C. Zhang and I. Zarudi, 'Mechanical and rheological properties of carbon nanotube-reinforced polyethylene composites', *Composites Sci. Technol.*, **67**, 177 (2007).

(9.52) P. Miaudet, A. Derre, M. Maugey *et al.*, 'Shape and temperature memory of nanocomposites with broadened glass transition', *Science*, **318**, 1294 (2007).

(9.53) P. C. P. Watts and W. K. Hsu, 'Behaviours of embedded carbon nanotubes during film cracking', *Nanotechnology*, **14**, L7 (2003).

(9.54) P. M. Ajayan, L. S. Schadler, C. Giannaris *et al.*, 'Single-walled carbon nanotube-polymer composites: strength and weakness', *Adv. Mater.*, **12**, 750 (2000).

(9.55) D. Hui, M. Chipara, M. Sankar *et al.*, 'Mechanical properties of carbon nanotubes composites', *J. Comput. Theor. Nanosci.*, **1**, 204 (2004).

(9.56) M. Shaffer and I. A. Kinloch, 'Prospects for nanotubes and nanofiber composites', *Composites Sci. Technol.*, **64**, 2281 (2004).

(9.57) J. N. Coleman, U. Khan, W. J. Blau *et al.*, 'Small but strong: a review of the mechanical properties of carbon nanotube-polymer composites', *Carbon*, **44**, 1624 (2006).

(9.58) J. K. W. Sandler, J. E. Kirk, I. A. Kinloch *et al.*, 'Ultra-low electrical percolation threshold in carbon-nanotube–epoxy composites', *Polymer*, **44**, 5893 (2003).

(9.59) G. B. Blanchet, C. R. Fincher and F. Gao, 'Polyaniline nanotube composites: a high-resolution printable conductor', *Appl. Phys. Lett.*, **82**, 1290 (2003).

(9.60) Y. Z. Long, Z. J. Chen, X. T. Zhang *et al.*, 'Electrical properties of multi-walled carbon nanotube/polypyrrole nanocables: percolation-dominated conductivity', *J. Phys. D*, **37**, 1965 (2004).

(9.61) J. C. Grunlan, A. R. Mehrabi, M. V. Bannon *et al.*, 'Water-based single-walled-nanotube-filled polymer composite with an exceptionally low percolation threshold', *Adv. Mater.*, **16**, 150 (2004).

(9.62) M. B. Bryning, M. F. Islam, J. M. Kikkawa *et al.*, 'Very low conductivity threshold in bulk isotropic single-walled carbon nanotube–epoxy composites', *Adv. Mater.*, **17**, 1186 (2005).

(9.63) R. Sainz, A. M. Benito, M. T. Martinez *et al.*, 'A soluble and highly functional polyaniline–carbon nanotube composite', *Nanotechnology*, **16**, S150 (2005).

(9.64) M. Ginic-Markovic, J. G. Matisons, R. Cervini *et al.*, 'Synthesis of new polyaniline/nanotube composites using ultrasonically initiated emulsion polymerization', *Chem. Mater.*, **18**, 6258 (2006).

(9.65) J. Wang, Y. L. Xu, X. Chen *et al.*, 'Capacitance properties of single wall carbon nanotube/polypyrrole composite films', *Composites Sci. Technol.*, **67**, 2981 (2007).

(9.66) M. Moniruzzaman and K. I. Winey, 'Polymer nanocomposites containing carbon nanotubes', *Macromolecules*, **39**, 5194 (2006).

(9.67) N. Grossiord, J. Loos, O. Regev *et al.*, 'Toolbox for dispersing carbon nanotubes into polymers to get conductive nanocomposites', *Chem. Mater.*, **18**, 1089 (2006).

(9.68) K. Schulte and A. H. Windle (eds.), *Composites Sci. Technol.*, special issue on carbon nanotube–polymer composites, **67**(5) (April 2007).

(9.69) A. Peigney, C. Laurent, F. Dobigeon *et al.*, 'Carbon nanotubes grown *in situ* by a novel catalytic method', *J. Mater. Res.*, **12**, 613 (1997).

(9.70) E. Flahaut, A. Peigney, C. Laurent *et al.*, 'Carbon nanotube–metal-oxide nanocomposites: microstructure, electrical conductivity and mechanical properties', *Acta Mater.*, **48**, 3803 (2000).

(9.71) A. Peigney, E. Flahaut, C. Laurent, *et al.*, 'Aligned carbon nanotubes in ceramic-matrix nanocomposites prepared by high-temperature extrusion' *Chem. Phys. Lett.*, **352**, 20 (2002).

(9.72) A. Cordier, E. Flahaut, C. Viazzi *et al.*, '*In situ* CCVD synthesis of carbon nanotubes within a commercial ceramic foam', *J. Mater. Chem.*, **15**, 4041 (2005).

(9.73) A. Peigney, S. Rul, F. Lefevre-Schlick *et al.*, 'Densification during hot-pressing of carbon nanotube–metal–magnesium aluminate spinel nanocomposites', *J. Eur. Ceramic Soc.*, **27**, 2183 (2007).

(9.74) G. D. Zhan, J. D. Kuntz, J. L. Wan *et al.*, 'Single-wall carbon nanotubes as attractive toughening agents in alumina-based nanocomposites', *Nat. Mater.*, **2**, 38 (2003).

(9.75) G. D. Zhan, J. D. Kuntz, J. E. Garay *et al.*, 'Electrical properties of nanoceramics reinforced with ropes of single-walled carbon nanotubes', *Appl. Phys. Lett.* **83**, 1228 (2003).

(9.76) C. B. Mo, S. I. Cha, K. T. Kim *et al.*, 'Fabrication of carbon nanotube reinforced alumina matrix nanocomposite by sol-gel process', *Mater. Sci. Eng. A*, **395**, 124 (2005).

(9.77) T. Seeger, T. Köhler, T. Frauenheim *et al.*, 'Nanotube composites: novel SiO_2 coated carbon nanotubes', *Chem. Commun.*, 34 (2002).

(9.78) T. Seeger, G. De La Fuente, W. K. Maser *et al.*, 'Evolution of multiwalled carbon-nanotube/SiO_2 composites via laser treatment', *Nanotechnology*, **14**, 184 (2003).

(9.79) S. Q. Guo, R. Sivakumar, H. Kitazawa *et al.*, 'Electrical properties of silica-based nano-composites with multiwall carbon nanotubes', *J. Amer. Ceram. Soc.*, **90**, 1667 (2007).

(9.80) P. Vincent, A. Brioude, C. Journet *et al.*, 'Inclusion of carbon nanotubes in a TiO_2 sol-gel matrix', *J. Non-Cryst. Sol.*, **311**, 130 (2002).

(9.81) A. Jitianu, T. Cacciaguerra, R. Benoit *et al.*, 'Synthesis and characterization of carbon nanotubes–TiO_2 nanocomposites', *Carbon*, **42**, 1147 (2004).

(9.82) X. B. Yan, B. K. Tay and Y. Yang, 'Dispersing and functionalizing multiwalled carbon nanotubes in TiO_2 sol', *J. Phys. Chem. B*, **110**, 25 844 (2006).

(9.83) M. Sanchez, R. Guirado and M. E. Rincon, 'Multiwalled carbon nanotubes embedded in sol-gel derived TiO_2 matrices and their use as room temperature gas sensors', *J. Mater. Sci. Materials Electron.*, **18**, 1131 (2007).

(9.84) J. S. Sakamoto and B. Dunn, 'Vanadium oxide–carbon nanotube composite electrodes for use in secondary lithium batteries', *J. Electrochem. Soc.*, **149**, A26 (2002).

(9.85) Y. Z. Zheng, M. L. Zhang and P. Gao, 'Preparation and electrochemical properties of multiwalled carbon nanotubes-nickel oxide porous composite for supercapacitors', *Mater. Res. Bull.*, **42**, 1740 (2007).

(9.86) A. A. White, S. M. Best and I. A. Kinloch, 'Hydroxyapatite–carbon nanotube composites for biomedical applications: a review', *Int. J. Appl. Ceramic Technol.*, **4**, 1 (2007).

(9.87) R. Z. Ma, J. Wu, B. Q. Wei *et al.*, 'Processing and properties of carbon nanotubes–nano-SiC ceramic', *J. Mater. Sci.*, **33**, 5243 (1998).

(9.88) Y. Wang, G. A. Voronin, T. W. Zerda *et al.*, 'SiC–CNT nanocomposites: high pressure reaction synthesis and characterization', *J. Phys. Cond. Matter*, **18**, 275 (2006).

(9.89) Y. Wang, Z. Iqbal and S. Mitra, 'Rapid, low temperature microwave synthesis of novel carbon nano tube–silicon carbide composite', *Carbon*, **44**, 2804 (2006).

(9.90) R. Andrews, D. Jacques, A. M. Rao *et al.*, 'Nanotube composite carbon fibers', *Appl. Phys. Lett.*, **75**, 1329 (1999).

(9.91) B. O. Boskovic, V. B. Golovko, M. Cantoro *et al.*, 'Low temperature synthesis of carbon nanofibres on carbon fibre matrices', *Carbon*, **43**, 2643 (2005).

(9.92) M. Endo, Y. A. Kim, T. Hayashi *et al.*, 'Vapor-grown carbon fibers (VGCFs) – basic properties and their battery applications', *Carbon*, **39**, 1287 (2001).

(9.93) M. Endo, T. Hayashi and Y. A. Kim, 'Large-scale production of carbon nanotubes and their applications', *Pure Appl. Chem.*, **78**, 1703 (2006).

(9.94) T. Kuzumaki, K. Miyazawa, H. Ichinose *et al.*, 'Processing of carbon nanotube reinforced aluminum composite', *J. Mater. Res.*, **13**, 2445 (1998).

(9.95) S. M. Zhou, X. B. Zhang, Z. P. Ding *et al.*, 'Fabrication and tribological properties of carbon nanotubes reinforced Al composites prepared by pressureless infiltration technique', *Composites A*, **38**, 301 (2007).

(9.96) C. F. Deng, D. Z. Wang, X. X. Zhang *et al.*, 'Processing and properties of carbon nanotubes reinforced aluminum composites', *Mater. Sci. Eng. A*, **444**, 138 (2007).

(9.97) S. R. Bakshi, V. Singh, K. Balani *et al.*, 'Carbon nanotube reinforced aluminium coating via cold spraying', *Surface Coatings Technol.*, **202**, 5162 (2008).

(9.98) T. Kuzumaki, O. Ujiie, H. Ichinose *et al.*, 'Mechanical characteristics and preparation of carbon nanotube fiber-reinforced Ti composite', *Adv. Eng. Mater.*, **2**, 416 (2000).

(9.99) C. S. Goh, J. Wei, L. C. Lee *et al.*, 'Effect of fabrication techniques on the properties of carbon nanotubes reinforced magnesium', *Solid State Phenomena*, **111**, 179 (2006).

(9.100) P. M. Ajayan and J. M Tour, 'Nanotube composites', *Nature*, **447**, 1066 (2007).

10 Filled and heterogeneous nanotubes

Attempts to introduce foreign materials into the empty central cavities of carbon nanotubes began soon after the appearance of Iijima's paper in 1991. This work was driven partly by curiosity – the empty cavities represented ideal 'nano-test-tubes' for the study of matter in confined spaces – and partly by the idea of using nanotubes as templates for nanowires. There was also interest in the possibility of filling opened tubes with catalytic metals, to produce new size-selective catalysts. Multiwalled nanotubes were first successfully filled in 1993, using a technique that involved carrying out arc-evaporation in the usual way, but with an anode containing some of the material to be encapsulated. This method generally seems to favour the formation of filled nanoparticles rather than nanotubes, and is only applicable to materials that can survive the extreme conditions of the electric arc. A more generally applicable method, in which the tubes were opened and filled by chemical means, was introduced a short time later. This has now been applied to a wide range of materials, including biological molecules. Single-walled tubes were first opened and filled in 1997, and some fascinating work has been carried out on the effect of confinement in SWNTs on the structure of crystalline materials. There is also great interest in filling SWNTs with fullerenes. In the first part of this chapter the methods used to open and fill carbon nanotubes, and the new science and possible applications that are emerging from this work, will be summarized.

The second part of this chapter gives a brief overview of heterogeneous nanotubes, defined as nanotubes whose carbon atoms are partially substituted with hetero-atoms, typically nitrogen and/or boron. Since this book is solely concerned with carbon-containing structures, pure boron nitride nanotubes are not covered.

10.1 Filling by arc-evaporation

The first attempts to put foreign material inside nanotubes used the arc-evaporation technique to vaporize a mixture of graphite and lanthanum oxide. Thus, instead of using pure graphite electrodes, the anode was drilled out and a mixture of La_2O_3 and graphite powder inserted. Arc discharge was carried out in the usual way, and the carbon deposited on the cathode collected. Experiments of this kind were carried out independently in 1993 by Rodney Ruoff's group in the USA (10.1) and Yahachi Saito's group in Japan (10.2) and produced identical results: instead of filled nanotubes, the cathodic soot contained significant numbers of filled nanoparticles.

Although arc-evaporation with modified electrodes usually seems to favour the formation of filled nanoparticles (10.3, 10.4) rather than nanotubes, a French group has studied the preparation of filled nanotubes in this way (10.5–10.7). They have successfully used arc-evaporation to produce nanotubes filled with a range of transition elements, rare earth elements and non-metals such as S, Se and Ge. A few other groups have also experimented with filling using arc-evaporation (e.g. 10.8). However, this method of filling carbon nanotubes has not been widely taken up since chemical methods, discussed in the next section, are much more widely applicable and offer a greater degree of control. It should be noted, however, that experiments on filling nanotubes using arc-evaporation led to the discovery of single-walled nanotubes (see p. 27).

10.2   Opening and filling of multiwalled nanotubes using chemical methods

10.2.1   Early work

Pulickel Ajayan and Sumio Iijima were the first to use chemical methods and capillarity to open and fill carbon nanotubes (10.9). Earlier theoretical work had suggested that opened nanotubes should act as 'nanopipettes', sucking liquid inside by capillary action (10.10). Ajayan and Iijima tested this idea by treating a sample of tubes with molten lead, hoping that some of the lead might be drawn inside open tubes. Their method involved depositing particles of lead onto the tubes in a vacuum using electron-beam evaporation and then heating in air at 400 °C, a temperature sufficient to melt the lead. When they examined the resulting samples using TEM they found that a small proportion of the nanotubes had clearly been filled; examples are shown in Fig. 10.1. The fillings extended for distances up to a few hundred nm, but were frequently blocked by the presence of internal caps. The proportion of nanotubes filled with lead was estimated to be 1%. The proportion of open tubes in a fresh sample is only around one in a million, so it is clear that the lead was not entering the tubes through existing holes. It seems that the tubes were opened by lead-catalysed oxidation.

Ajayan and Iijima's work demonstrated that oxidation, apparently catalysed by a metal, could be used to selectively remove the tips of nanotubes, while leaving the tube walls unaffected. Work carried out a short time later by Iijima's group in Japan and by Edman Tsang, Malcolm Green and the present author in Oxford showed that nanotubes could be opened with a reasonable degree of selectivity simply by heating in a mildly oxidizing environment, with no catalyst being necessary (10.11, 10.12). The Oxford group used carbon dioxide as the oxidizing agent, making use of the 'reverse Boudouard reaction':

$$C_{(s)} + CO_{2(g)} \rightarrow 2CO_{(g)}$$

Samples of nanotubes were heated at a range of temperatures up to 950 °C in CO_2 for periods up to 24 hours. When the oxidized samples were examined using TEM, it was found that in a small, but significant, number of cases the tube caps had been selectively attacked, frequently being thinned at the extremity, and sometimes completely opened. The Japanese group heated the tubes in oxygen at temperatures up to 850 °C and observed

Fig. 10.1   Nanotubes filled with lead or lead oxide, prepared using capillarity by Ajayan and Iijima (10.9).

similar behaviour. At temperatures below about 700 °C, little oxidation was observed, but at higher temperatures oxidation occurred rapidly: at 850 °C the entire sample was consumed after 15 minutes. Samples containing opened tubes could be prepared by heating in oxygen at 700 °C for short periods (typically 10 minutes). Using oxygen rather than carbon dioxide as the oxidizing agent resulted in a higher proportion of open nanotubes, but the treatment was considerably more destructive, with many tubes being massively corroded.

Having succeeded in preparing opened nanotubes, both groups then endeavoured to fill the tubes with inorganic materials, but this proved far more difficult than expected. The Oxford group attempted to introduce solutions of metal salts into the tubes using 'incipient wetness' techniques of the kind used in preparing supported catalysts, with the aim of drying off the solvent and reducing the salt to leave metal crystallites inside the tube. This met with little success, as did experiments by the Japanese group aimed at filling opened tubes with molten lead. It was not immediately clear why the two-step approach to filling nanotubes failed where the one-step method used by Ajayan and Iijima to fill tubes with molten lead succeeded so well. One possibility is that tubes become blocked by amorphous carbon as soon as they are opened, making subsequent filling difficult.

10.2.2   Opening by treatment with acid

Following the difficulties with the carbon dioxide oxidation method, Tsang, Green and their colleagues in Oxford looked for alternative ways to open nanotubes. Previous work on the treatment of other fullerene-related carbons with nitric acid had suggested that this might be an effective way of selectively attacking pentagonal rings, so a similar treatment

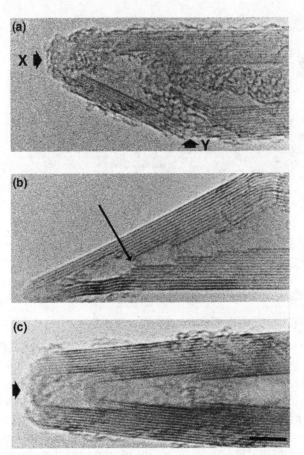

Fig. 10.2   Typical nanotube caps following treatment with boiling nitric acid. (a) Micrograph showing selective attack at points X and Y, where non-six-membered rings are present. (b) and (c) Micrographs showing the destruction of multiple internal caps (10.13). Scale bar 5 nm.

was applied to nanotubes. This proved to be highly successful (10.13). Not only did the nitric acid treatment result in a high yield of opened tubes, but the selectivity with which the tube tips were attacked was extraordinary. This is illustrated in Fig. 10.2, which shows micrographs of tubes from a sample treated with boiling nitric acid for 4.5 hours. In Fig. 10.2(a) the tube cap has been attacked at two points, labelled X and Y, both of which are at positions where pentagonal rings would have been present. In Fig. 10.2(b) the tube cap has been opened, and the internal caps have been selectively removed, so that a passage exists to the central cavity, although the remainder of the tube remains intact. This indicates that the acid has reacted only with the pentagonal rings; even where the edges of graphite sheets have been exposed, these have not been attacked, and very little thinning or stripping of the outer layers is observed. A similar effect is shown in Fig. 10.2(c). Tsang and colleagues also showed that nanotubes opened using the nitric acid method, unlike those opened by the less selective gas-phase oxidation methods, did not appear to be blocked with amorphous material. Tubes opened in this way were therefore much more amenable to filling.

The question arises of why the nitric acid method is so exquisitely selective. Once the reactive edge of the basal planes have been exposed, why are they not rapidly consumed, as is the case with gaseous oxidants? The answer is probably that a reaction occurs between the exposed edges and the nitric acid, resulting in the formation of surface carboxyl and other groups that act as a barrier to further reaction. These hydrophilic groups would be expected to facilitate the filling of tubes with aqueous solutions and with biological material as discussed below.

10.2.3 Filling opened tubes

In their original *Nature* paper (10.13), the Oxford team showed that the tubes could be opened and filled in a one-stage process. Thus, when nickel nitrate was added to the nitric acid used in the oxidation treatment, tubes containing crystalline nickel oxide resulted. Further treatment with H_2 at 400 °C reduced the oxide to nickel metal, showing that chemical reactions could be effected inside the opened tubes. A micrograph showing a tube containing metallic nickel is shown in Fig. 10.3(a). In a similar way, tubes were filled with samarium oxide (Sm_2O_3) by treating closed tubes with $Sm(NO_3)_3/HNO_3$, as shown in Fig. 10.3(b) (10.14). Alternatively a two-stage process can be used in which opened tubes are subsequently filled simply by treating them with a solution of the substance, or with a molten material. The use of molten materials has the advantage that it enables complete filling of the tubes. Deposition from solution can never result in

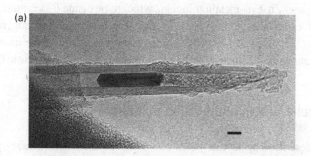

(a)

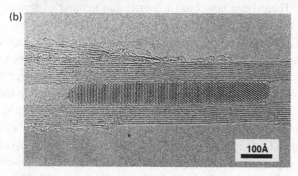

(b)

100Å

Fig. 10.3 Nanotubes opened and filled using acid treatments. (a) Tube filled with metallic nickel (10.13). Scale bar 5 nm. (b) Samarium oxide crystallite inside nanotube (10.14).

complete filling, since the precipitated material will always occupy far less space than the solution. Ajayan and colleagues prepared a sample of tubes by treatment with oxygen or nitric acid and then mixed the sample with V_2O_5 powder (m.pt. 690 °C) and heated to a temperature of 750 °C (10.15). The Oxford group showed that treatment of opened nanotubes with molten MoO_3, followed by heating in H_2, resulted in tubes completely filled with single-crystal MoO_2 (10.16).

In the course of these studies, it became clear that some materials entered the opened tubes much more readily than others. Whether or not a liquid will enter the central core of a nanotube depends to a large extent on the interfacial energy between two. If the liquid–solid contact angle is less than 90°, liquid will enter the tube spontaneously, while if the angle is greater than 90° it will not. Dujardin, Ebbesen and colleagues studied the wetting of nanotubes with a range of materials in an attempt to determine the 'critical' surface tension, below which wetting would occur (10.17). They concluded that this cutoff surface tension lies somewhere between 100 and 200 mN m^{-1}. Thus water, with a surface tension of ~72 mN m^{-1} would be expected to enter nanotubes spontaneously, as would most organic solvents, which have lower surface energies than water. This is consistent with the observation that aqueous solutions enter opened tubes.

As well as inorganic materials, biological molecules have also been introduced into multiwalled nanotubes. Green and Tsang, together with bioinorganic chemists from Oxford and from Birkbeck College, London, showed that small enzyme molecules could be inserted into opened MWNTs (10.18, 10.19). The technique was simply to suspend samples of opened nanotubes in aqueous solutions of the proteins for 24 hours and then evaporate off the volatile water under reduced pressure. In this way they successfully introduced the enzymes Zn_2Cd_5-metallothionein, cytochrome c_3 and β-lactamase I into the nanotubes' central cavities in high yield. The enzyme molecules also adhered to the outsides of the tubes. This work probably represents the first study of the interaction of biomolecules with nanotubes, a subject that has since grown rapidly, as we saw in Chapter 8.

10.3   Filling catalytically-grown multiwalled nanotubes

All of the work described up to this point has involved multiwalled tubes produced by arc-evaporation. There have now been a substantial number of studies on filling catalytically-grown multiwalled nanotubes. Partially filled nanotubes are quite often seen in samples of CVD-grown tubes, but complete filling requires an excess of catalyst to be present. It has been found that this can be achieved in a one-stage process by pyrolysing precursors that contain the catalytic metals and a source of carbon. C. N. R. Rao and colleagues from Bangalore showed in 1998 that heating ferrocene, or ferrocene-acetylene mixtures, to approximately 1100 °C produced aligned MWNTs, most of which were partially or completely filled with Fe (10.20). At about the same time, Nicole Grobert and colleagues prepared MWNTs filled with Ni by pyrolysing thin films of C_{60} and Ni deposited on a silica plate (10.21). The approach has since become quite widely used to produce tubes filled with Fe, Ni and Co (e.g. 10.22–10.25) as well as with alloys of the metals (10.26). Micrographs showing some typical filled tubes

Fig. 10.4   Images of MWNTs filled by an *in situ* CVD process, from the work of Watts *et al.* The filling material is α-Fe (10.24, 10.121).

produced in this way by Paul Watts of the University of Sussex and colleagues (10.24) are shown in Fig. 10.4. Tubes filled with these ferromagnetic metals might be useful in magnetic storage applications.

A fascinating study of filled MWNTs produced by pyrolysis was described in 2006 by an international collaboration which included Terrones, Ajayan and Banhart (10.27). This demonstrated the effect of irradiating filled tubes with an electron beam. Previous work had shown that simultaneous annealing and irradiation of carbon onions could produce immense pressures inside the onion, which resulted in the formation of diamond (10.28). It had also been shown that irradiating pure MWNTs with electrons could result in the shrinkage of the tubes by a loss of atoms and diffusion of interstitials through the inner cavity (10.29). In the 2006 work, MWNTs were filled with Fe, Fe_3C and Co using metallocene-based pyrolytic methods. The filled tubes were then irradiated in the TEM, while being held at a temperature of 600 °C with a heating holder. As with the pure tubes, this resulted in shrinkage of the tubes, exerting intense pressures on the contents. The contraction caused by the tube restructuring can be so strong that the crystals inside are squeezed and extruded, as shown in Fig. 10.5. Calculations showed that pressures could reach more than 40 GPa in the radial direction. Therefore, this method offers a way of studying the effect of high pressures on nanoscale materials.

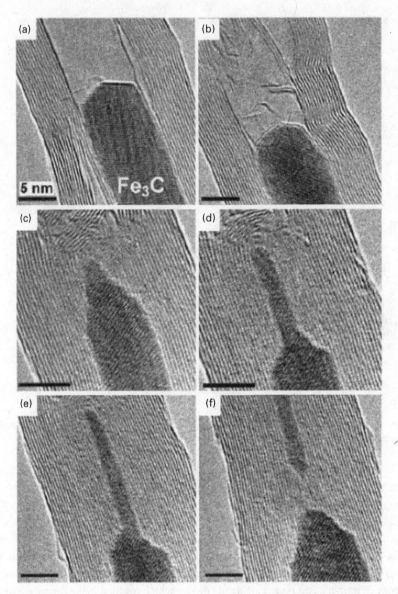

Fig. 10.5   Deformation and extrusion of Fe$_3$C crystal inside a MWNT (10.27). The nanotube was exposed
to intense electron irradiation at a specimen temperature of 600 °C.

10.4 Water in multiwalled nanotubes

Some remarkable studies of water inside multiwalled nanotubes have been described by Yury Gogotsi of Drexel University and his co-workers (10.30–10.32). The first of these studies involved preparing tubes using a hydrothermal catalytic process, which resulted in liquid water and gases (e.g. CO and CH_4) being trapped inside the tubes' central cavities (10.30). Images of the partially filled tubes showed a good wettability of carbon with water, indicating that the formation of functional groups during synthesis has rendered the carbon surface hydrophilic. In a later study (10.32), the interaction of the water with the nanotube walls was examined in more detail. Liquid was observed to penetrate between the layers of the tube, and some dissolution of hydrated carbon layers was seen. When the tubes were locally heated with the electron beam, a more drastic dissolution of the tube walls was induced. As well as producing water-filled tubes using hydrothermal synthesis, Gogotsi and colleagues have introduced water into tubes previously produced by the catalytic and arc methods (10.32). This was accomplished by treating them at high pressures in an autoclave.

The flow of water through membranes consisting of aligned carbon nanotubes was investigated by a group led by Olgica Bakajin of Lawrence Livermore National Laboratory (10.33). The flows observed were several orders of magnitude higher than those of commercial polycarbonate membranes. This was despite the fact that the tube diameters were smaller than the pore sizes of the commercial membranes. These results, which are consistent with theoretical predictions (10.34), suggest that nanotube membranes may be useful in applications such as removing salt from water.

10.5 Filling single- and double-walled nanotubes

With their extremely small diameters, single-walled nanotubes are even more difficult to fill than their multiwalled counterparts. However, in an exceptional piece of work, Jeremy Sloan and colleagues from Oxford demonstrated in 1998 that SWNTs can be opened and filled using techniques similar to those employed for MWNTs (10.35). A short time later, Brian Smith, Marc Monthioux and David Luzzi made the amazing discovery that fullerene-filled SWNTs formed spontaneously during the purification and annealing treatments applied to raw SWNT material produced by the laser vaporization technique (10.36, 10.37). These studies stimulated great interest, and much beautiful work has now been done on filled SWNTs. Filling with inorganic materials will be discussed first.

10.5.1 Filling with inorganic materials

In their initial study (10.35), the Oxford group introduced Ru crystallites and other materials into the tubes using a solution method. Subsequent work showed that almost complete filling of single-walled tubes could be achieved using molten materials (10.38). A typical filling procedure would entail heating as-made nanotubes with the molten salt or oxide to a temperature 100 °C above its melting point. Using this technique, Sloan

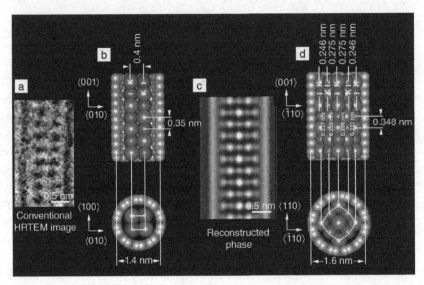

Fig. 10.6 (a) Conventional HRTEM image of a 2 × 2 KI crystal formed within a 1.4 nm diameter single-walled nanotube, (b) structure model derived from (a), (c) super-resolved HRTEM image of a 3 × 3 KI crystal inside a 1.6 nm diameter SWNT, (d) corresponding structure (10.39, 10.120).

and colleagues filled SWNTs with a range of metals, metal salts and oxides (10.38–10.43). In some cases chemical reactions have been carried out on material inside the tubes. For example, metallic Ag was produced by photolytic reduction of AgBr (10.38). Other groups have also experimented with putting inorganic materials into SWNTs. Monthioux and colleagues showed that SWNTs could be filled with CrO_3 by soaking as-prepared SWNT materials in a mixture of the oxide and HCl (10.44), while Luzzi's group produced nanowires of magnetic metals (Fe, Co, Ho, Gd) by filling SWNTs with precursor metal chlorides and subsequent reduction (10.45).

The structure of crystals inside SWNTs has been studied using electron diffraction and HRTEM, with the Oxford group again leading the way. Some of their work on determining tube structure using HRTEM was discussed in Section 5.5.3. In 2000 they described HRTEM imaging of KI crystals inside SWNTs (10.39). One of their images of a 2 × 2 KI crystal, recorded at optimum Scherzer defocus is shown in Fig. 10.6(a). In this image, only the strongly scattering I atoms contribute significantly to the contrast, with the much lighter K atoms making a negligible contribution. Higher quality images, with some of the aberrations removed, can be obtained by digitally combining a tilt or focal series of images. An example of such a restored image can be seen in Fig. 10.6(c), which shows a 3 × 3 KI crystal inside a 1.6 nm diameter SWNT. Here the contribution of the K atoms as well as the I atoms is visible. It was found that both 1D crystals displayed considerable lattice distortions compared with their bulk structures. In the 2 × 2 case, a lattice expansion of ~17% occurred across the nanotube, whereas in the 3 × 3 case, a differential expansion was observed, with the I columns being more compressed than the K columns.

Sometimes twisted 1D crystals are observed inside single-walled tubes (10.42). Figure 10.7(a) shows a CoI_2 crystal inside a SWNT, with Fig. 10.7(b) illustrating its

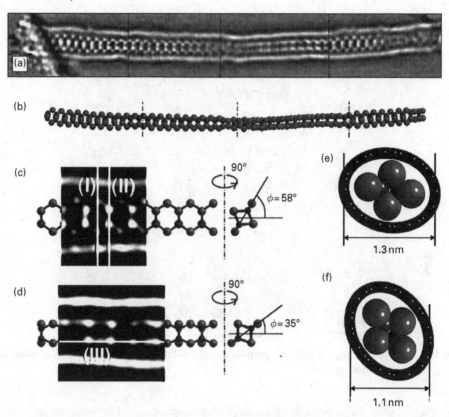

Fig. 10.7   (a) A super-resolved HRTEM image of a one-dimensional (1D) CoI₂ crystal inside a distorted single-walled nanotube, (b) derived structure model, (c), (d) details from the left and right middle sections of (a), showing the microstructure of the 1D crystal in two of the different projections, (e), (f) end-on views of Co₂I₄ units in two different orientations, showing distortion of nanotube (10.42).

twisted structure. The contrast of the encapsulated crystal was interpreted in terms of Co_2I_4 repeating units. A remarkable feature of this system was that the asymmetric Co_2I_4 subunits caused a distortion of the tube's cross-section, as shown in Figs. 10.7(e) and (f).

As well as inorganic crystals, Sloan and colleagues have inserted molecular species into SWNTs. For example, ortho-carborane was introduced into SWNTs by sublimation, and individual o-carborane molecules were imaged, (10.46). Studies such as this, and the work on fullerene-filled tubes discussed below, show that placing molecules inside single-walled nanotubes represents an excellent way of stabilizing them for HRTEM imaging.

10.5.2   Filling with fullerenes: 'nano-peapods'

Turning now to single-walled tubes filled with fullerenes, or 'peapods', some extraordinary work has been done on these materials since the initial studies of Smith, Monthioux and Luzzi (10.36, 10.37). As noted above, the fullerene-filled tubes were first produced accidentally during the purification and annealing of as-produced SWNT

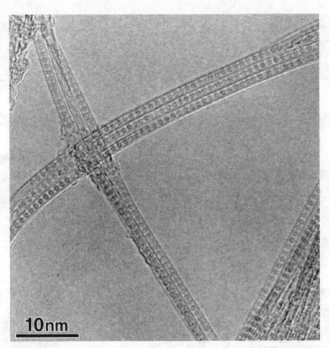

10nm

Fig. 10.8   Typical image of fullerene 'peapods': single-walled nanotubes filled with C_{60} molecules. Courtesy Kazu Suenaga.

soot. Subsequent work revealed that peapods also formed spontaneously during the synthesis of SWNTs using arc discharge, although with low yield (10.47, 10.48). More controlled methods of synthesis have now been developed, in which the tubes are firstly opened using acids and then vacuum-annealed in the presence of fullerenes (e.g. 10.49–10.51). A typical image of peapods is shown in Fig. 10.8. Smith, *et al.* measured the average C_{60}–C_{60} separation and found a value of ~0.97 nm (10.50). This is slightly smaller than the intermolecular separation in FCC crystalline C_{60}, but this may be due to a slight tilt of the tube. Selected-area diffraction measurements produced separations very close to that in crystalline C_{60}.

It is thought that the fullerenes enter the SWNTs through opened ends, although the possibility that they might enter through side-wall defects has also been considered (10.52). Raman measurements by Kataura and colleagues (10.51) showed that significant numbers of C_{70}, as well as C_{60} molecules had been encapsulated. It has also been demonstrated by Smith and colleagues that electron irradiation of peapods within the microscope can induce diffusion and coalescence of the encapsulated fullerenes (10.37, 10.49). Moreover, high-temperature heat treatments of peapods can result in the coalescence of many adjacent fullerenes, to produce extremely narrow inner tubes (10.37, 10.53, 10.54). This is illustrated in Fig. 10.9, taken from work by Kazu Suenaga and colleagues in collaboration with workers from the University of Vienna (10.54).

A fascinating study of fullerenes inside both double- and single-walled nanotubes was described by Andrei Khlobystov of the Materials Department at Oxford and

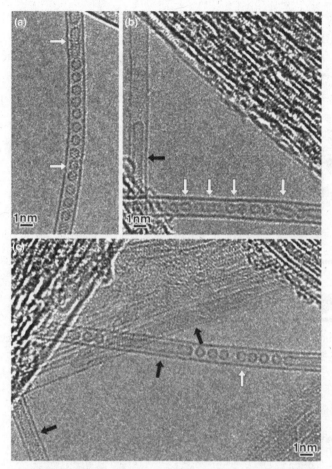

Fig. 10.9 Images showing the coalescence of fullerenes inside SWNTs at a temperature of 1250 °C, after (a) 5 min, (b) 15 min and (c) 25 min (10.54).

colleagues in 2004 (10.55). One of their images of a DWNT with a single chain of fullerenes inside is shown in Fig. 10.10(a). Larger-diameter DWNTs were also filled, and in this case some novel packings were seen, including zigzag arrangements (Fig. 10.10b) and double-helical phases.

The electronic properties of peapods were probed by Ali Yazdani of the University of Illinois, with workers from Pennsylvania in 2002 (10.56). Using a low-temperature scanning tunnelling microscope they found that the C_{60} molecules induced periodic modifications in the nanotube's local electronic structure. This was interpreted in terms of a mixing of the nanotubes' electronic states and the C_{60} orbitals (10.57). It is conceivable that encapsulated molecules could be used to deliberately 'tune' the electronic properties of carbon nanotubes although, needless to say, the degree of control required to achieve this is beyond current capabilities.

The encapsulation of metallofullerenes in single-walled nanotubes was reported by Iijima, Suenaga and colleagues from Japan and France in 2000 (10.58). In this study,

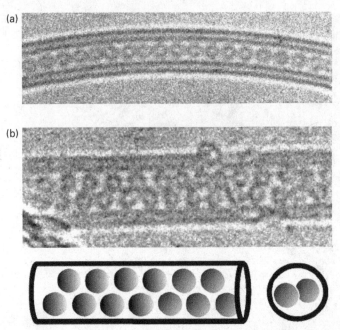

Fig. 10.10 C_{60} molecules inside double-walled nanotubes (a) single chain of molecules, (b) 'zigzag' phase of molecules with schematic illustration of arrangement (10.55).

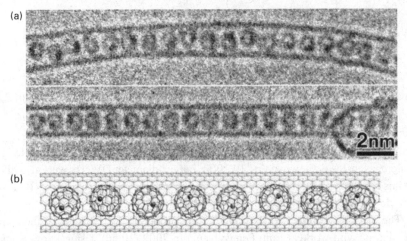

Fig. 10.11   Images of encapsulated metallofullerenes (a) and computer-generated (b) images of $Gd@C_{82}$ peapods (10.58).

gadolinium metallofullerenes were inserted into SWNTs by heating a mixture of tubes and $Gd@C_{82}$ molecules in a sealed ampoule at 500 °C. The metallofullerenes were found to pack tightly into the tubes, as shown in Fig. 10.11(a). Remarkably, the individual encapsulated Gd atoms were visible as dark spots on many of the C_{82} molecules. Figure 10.11(b) is a schematic representation of the metallofullerene-containing SWNTs.

This was the first time that atoms inside metallofullerenes had been seen directly, showing that encapsulating molecular species inside nanotubes is an excellent way of imaging them. It was also shown that the individual encapsulated Gd atoms could be detected by electron energy-loss spectroscopy (10.59). In subsequent work, $Gd_2@C_{92}$ molecules were encapsulated in SWNTs using similar methods, and rapid movement of the Gd atoms inside the fullerene cages was observed (10.60). The most spectacular images of nanotube-encapsulated fullerenes were published in 2007 by Iijima, Suenaga and co-workers (10.61, 10.62). In this work, aberration-corrected TEM was used to directly image the atomic structures of both the fullerene molecules and the encapsulating nanotubes. One of the superb images from these studies is shown in Fig. 10.12.

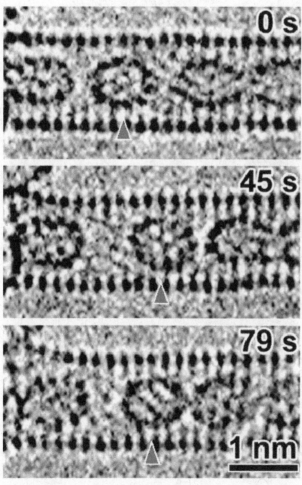

Fig. 10.12   Aberration-corrected TEM images of D_{5d}-C_{80} fullerene molecules inside an (18, 1) SWNT. Triangles indicate orientational changes of encapsulated molecule (10.62).

As well as obtaining ultra-high-resolution TEM images of encapsulated fullerenes, the Iijima–Suenaga group have also shown that it is possible to image small molecules attached to fullerenes inside nanotubes (10.63). The molecule they studied was retinal, which is found in photoreceptor cells. This was attached to C_{60} molecules that were then inserted into SWNTs. Some of the images they obtained are shown in Fig. 10.13. The

(a)

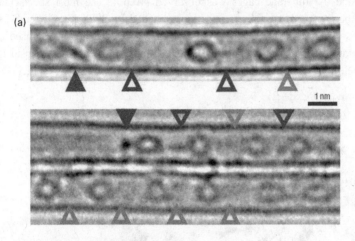

1 nm

(b)

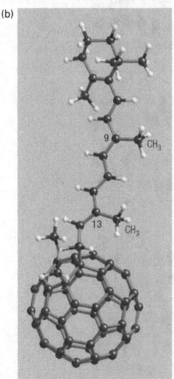

Fig. 10.13   (a) HRTEM images of retinal-C_{60} molecules inside a single-walled nanotube, (b) structure of the *trans* version of retinal-C_{60} (10.63).

retinyl groups were found to appear in 'strong' or 'weak' contrast next to the C_{60} 'markers', apparently due to conformational changes.

10.6   Gases in nanotubes

10.6.1   Hydrogen

In early 1997, a group led by Michael Heben of the National Renewable Energy Laboratory (NREL) in Golden, Colorado, claimed in a Letter to *Nature* (10.64) that single-walled carbon nanotubes could store up to 8% by weight of H_2 at room temperature and moderate pressure. These remarkable results attracted great publicity and initiated the most controversial episode in nanotube science (10.65, 10.66).

The experimental method used by Heben's team involved exposing the tube-containing material to H_2 at 300 torr, cooling to 90 K and then using temperature programmed desorption (TPD) spectroscopy to observe the desorption behaviour. The as-prepared SWNT soot displayed desorption behaviour similar to that of an activated carbon sample. However, soot that had been heated in vacuum at 970 K showed an extra desorption peak that was apparently consistent with adsorption of H_2 within the cavities of the tubes. The authors suggested that the tubes had been opened by the heat treatment in vacuum, and thus made accessible to the H_2.

The high uptakes observed suggested that single-walled tubes might make useful candidates as H_2-storage materials; hence the huge interest in the results. One of the major obstacles preventing the use of H_2 fuel cells to power automobiles is the lack of a suitable method for storing H_2. The US Department of Energy has set 6.5 wt% as the target capacity for a practical H_2-storage material for use in vehicles, a target comfortably exceeded by the NREL group. Following the publication of the Heben *Nature* paper, other groups reported high H_2 uptakes on SWNT-based materials (10.67–10.69). For example, a group led by Mildred Dresselhaus reported a storage capacity of 4.2 wt% for large-diameter SWNTs at room temperature under a modestly high pressure (10.67). Jianyi Lin and colleagues from the National University of Singapore (10.69) reported even more spectacular uptakes for MWNTs doped with alkali metals (up to 14 wt% at room temperature). Unfortunately, however, many more groups were unable to reproduce these results (10.70–10.75). Thus, Michael Hirscher, from the MPI in Stuttgart, and his co-workers, found uptakes of less than 1 wt% at room temperature and ambient pressure (10.74). These researchers suggested that the uptake observed by Heben may be due to a titanium contaminant in the nanotube samples. Workers from the University of Utrecht measured the adsorption of a number of different carbon adsorbents for H_2 at 77 K and 1 bar, and found that the storage capacities depended simply on surface area, with carbon nanotubes showing no special properties (10.75). It is now generally believed that any uptake of H_2 by carbon materials is primarily due to physisorption and therefore only occurs to any appreciable extent at low temperatures. As far as the work of Lin and co-workers is concerned, the observed mass increase has been attributed to hydroxide formation (10.72).

Even more controversial than the work of Heben and colleagues were the claims of Nelly Rodriguez and co-workers at Northeastern University, who reported that 'graphite nanofibres' (see Fig. 3.2) could store H_2 at levels exceeding 50 wt% at room temperature (10.76). The results were widely seen as incredible and have not been reproduced by other groups.

It is interesting to consider whether the story of hydrogen storage in carbon nanotubes is an example of 'pathological science'. This term was coined by Irving Langmuir (10.77) to describe the process whereby 'people are tricked into false results ... by subjective effects, wishful thinking or threshold interactions'. Cold fusion, a phenomenon that also involves hydrogen, was seen by many to be an example of pathological science. In the case of hydrogen storage in nanotubes, however, it would probably be unfair to attach such a label. One characteristic of pathological science is that 'The effect is of a magnitude that remains close to the limit of detectability, or many measurements are necessary because of the very low statistical significance of the results.' In the case of the experiments of Heben, Rodriguez, Lin and others it was the very size of the effects that seemed incredible. Nevertheless, one aspect of pathological science does seem to apply: 'The ratio of supporters to critics rises and then falls gradually to oblivion'.

10.6.2   Other gases

Compared with the massive interest in storing hydrogen in nanotubes, there has been relatively little work on introducing other gases into the tubes. A few studies have been reported however. In 1997, shortly after the Heben work on hydrogen, Australian researchers described the trapping of argon inside graphitic tubes, in a paper entitled 'The world's smallest gas cylinders' (10.78). These tubes were prepared by catalytic reduction of CO_2 and were much larger than typical carbon nanotubes, having diameters in the range 20–150 nm. The argon was introduced by hot isostatically pressing (HIP) the carbon material for 48 hours at 650 °C under a pressure of 170 megapascals, and could be detected inside the tubes using energy dispersive X-ray spectroscopy. The argon appeared to enter the tubes through defects in the relatively imperfect structures, and may have become sealed inside as a result of amorphization during the HIPing process. The pressure inside the tubes was estimated at 60 megapascals, and appeared to change little over several months at room temperature, so it seems possible to store gases inside these tubes for long periods.

In 2004, Nicole Grobert, Mauricio Terrones and colleagues described the encapsulation of gaseous nitrogen inside bamboo-like MWNTs (10.79). The filled tubes were prepared by heating aerosols of ferrocene/benzylamine solutions at 850 °C. The trapping of SF_6 and CO_2 in opened SWNT bundles was reported by US workers in 2005 (10.80). The gases were cryogenically adsorbed into opened SWNTs and then locked inside by functionalizing the sample with a low-temperature ozone treatment. This had the effect of closing the tube entrances with an impermeable barrier. The samples were stable under vacuum for periods of at least 24 hours and the trapped gases could be released by vacuum heating to 430 °C.

As well as studies of gases encapsulated inside nanotubes, the flow of gases through opened nanotubes has been investigated. It was mentioned above that rapid flows of water through aligned nanotube membranes was observed by workers from Lawrence Livermore Laboratory (10.33). In the same study the flow of gases through the membranes was also found to be exceptionally high. This is in agreement with a number of theoretical studies (e.g. 10.81, 10.82).

10.7 Heterogeneous nanotubes

Boron nitride exists in a graphite-like layered form, as shown in Fig. 10.14, and in the 1980s various groups showed that graphite hybrids containing C, B and N could be prepared (e.g. 10.84). Following the discovery of carbon nanotubes, theoreticians predicted that BN and BCN nanotubes should be stable (10.85, 10.86). This was confirmed a short time later when both types of nanotube were successfully produced using variations of the Krätschmer–Huffman technique (10.87, 10.88). At about the same time a 'catalytic' synthesis of BN nanotubes was also demonstrated (10.89), and subsequently various other methods of producing BN nanostructures have been developed. As already mentioned, we are only concerned with carbon-containing structures here, so pure boron nitride nanotubes are not covered. In the following sections, the synthesis of BCN, CN and CB nanotubes is summarized, and an outline of their properties given.

10.7.1 Boron–carbon–nitrogen tubes

Nanotubes containing boron, carbon and nitrogen have been prepared using all the major techniques that have been used for pure carbon tubes, i.e. arc-evaporation, laser ablation and catalytic CVD. The first arc-evaporation synthesis was reported by researchers from the Université de Paris-Sud and the Université de Montpellier in France in 1994 (10.87). Their method involved placing a mixture of boron and graphite powder into a

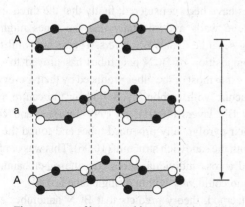

Fig. 10.14 The structure of hexagonal boron nitride (10.83).

hollowed-out graphite anode and carrying out the arc-evaporation in an atmosphere of nitrogen. As is the case for pure carbon nanotubes, the BCN tubes were found inside the deposit that formed on the cathode. They were accompanied by BN and BCN sheets, disordered carbon and pure carbon nanotubes and nanoparticles; elemental analysis of these structures was carried out using electron energy loss spectroscopy (EELS). Two types of B- and N-containing nanotubes were present: relatively large-diameter fibres (100–500 nm) with irregular thickening and narrower tubes up to 100 μm in length. In both cases, the caps of the tubes were found to be poorly formed in comparison with the caps of pure carbon nanotubes, presumably because five-membered rings are less easy to form in BCN networks than in pure carbon networks. The difficulty in forming caps also probably explains why in some cases the BCN tubes grew to much greater lengths than usually observed for carbon nanotubes. A short time after the French work appeared, Marvin Cohen's group from the University of California at Berkeley also described the synthesis of BCN nanotubes, this time using a graphite anode that had been drilled out to contain a BN rod, and arc-evaporating in helium (10.90). In 1997, a French team led by Christian Colliex (10.91) synthesized BCN nanotubes and nanoparticles by arc-evaporating a hafnium diboride rod with graphite in a nitrogen atmosphere.

Laser ablation was first used to produce BCN nanotubes by Iijima and colleagues in 1997 (10.92). A composite target containing BN, C, Ni and Co was laser-ablated at 1000 °C under N_2, resulting in the formation of multiwalled nanotubes containing B, C and N, as well as pure carbon tubes. A catalytic pyrolysis method was described by Mauricio Terrones and co-workers in 1996 (10.93). This involved pyrolysis of $CH_3CN.BCl_3$ at 900–1000 °C over Co powder, and resulted in the formation of graphitic BCN nanofibres and nanotubes possessing a range of morphologies. More recently, Renzhi Ma and Yoshio Bando have produced BCN nanotubes by pyrolysing dimethylamine borane $((CH_3)_2NH$ $BH_3)$ at 1050 °C in the presence of Fe or Ni nanoparticles under N_2 (10.94). Chemical vapour deposition techniques have also been used to produce boron–carbon–nitrogen nanotubes. Xuedong Bai and colleagues employed this method to prepare highly oriented BCN tubes on Ni substrates from a gas mixture of N_2, H_2, CH_4 and B_2H_6 (10.95).

Most boron–carbon–nitrogen nanotubes have rather imperfect bamboo-type structures. There is some uncertainty about the distribution of boron and nitrogen within the carbon lattice. Two possibilities have been considered: firstly that the three species are randomly distributed in the tube walls and secondly that the tubes might have a 'sandwich' structure containing separate C and BN layers. The most useful technique for determining the atomic composition of BCN nanotubes has proved to be EELS. Colliex *et al.* used EELS to show that most of the tubes produced by their arc-evaporation technique had a sandwich structure with carbon layers both in the centre and at the periphery, separated by a few BN layers (10.91). In 1999 Terrones and colleagues described an EELS study of their pyrolytically-produced tubes and found the stoichiometry to be BC_2N, consistent with the sandwich structure (10.96). This was confirmed by concentration profiles recorded across individual tubes. On the other hand, Ma and Bando found that the B:C:N ratio could vary within a single tube (10.94).

As far as properties are concerned, theory predicts that BCN nanotubes should be semiconducting. Marvin Cohen and co-workers calculated the electronic properties of crystalline BC_2N in 1988 (10.97), and predicted a 2.0 eV band gap. Experimental measurements on bundles of BCN tubes by Bando and colleagues led to a value of ~1.0 eV (10.98). Xuedong Bai and co-workers have reported promising field emission (10.95) and photoluminescence (10.99) behaviour with BCN tubes.

The above discussion refers exclusively to multiwalled BCN nanotubes. The synthesis of BCN SWNTs was reported in 2006 by an international team including Xuedong Bai, Enge Wang and Dmitri Golberg (10.100). A CVD method was used, with an Fe–Mo/MgO catalyst and CH_4, B_2H_6, and ethylenediamine vapour as the reactant gases. The tubes appeared to be of high quality and may be expected to have unique properties.

Several reviews of boron–carbon–nitrogen nanotubes have been given (10.101–10.103).

10.7.2 Carbon–nitrogen tubes

It has been speculated that the carbon–nitrogen compounds CN and C_3N_4 might represent a new class of superhard materials (10.104, 10.105), and this has prompted interest in preparing nitrogen-doped carbon nanotubes. If we consider multiwalled tubes first, most attempts to prepare these have involved pyrolytic or catalytic methods. In 1997, in probably the first such study, Japanese researchers grew N-doped MWNTs on a quartz substrate, by decomposition of Ni phthalocyanine (10.106). Terrones and co-workers produced similar arrays of CN multiwalled tubes by pyrolysing mixtures of ferrocene and melamine (10.107). The tubes produced in this way had relatively low nitrogen contents (~2%). Marianne Glerup and co-workers have incorporated higher concentrations of N (up to 20%) into multiwalled nanotubes by aerosol assisted CVD methods (10.108). However, even a concentration of 20% falls far short of a stoichiometry of CN or C_3N_4, and it appears that these levels of nitrogen doping have not yet been achieved. Structurally, multiwalled CN nanotubes tend to have bamboo or stacked-cone morphologies. This has been explained in terms of the way the nitrogen interacts with the catalyst particles (10.109).

As far as nitrogen doped single-walled tubes are concerned, these have been produced by arc-evaporation. A group from Brazil (10.110) prepared CN SWNTs in 2002 by evaporating a metal-containing graphite rod in a N_2–He atmosphere. Glerup and colleagues used a slightly different approach, introducing a nitrogen-rich precursor into the anode rods together with graphite and the catalysts (10.111). The N-doped SWNTs exhibited morphologies similar to their undoped counterparts.

10.7.3 Carbon–boron tubes

Boron interacts with carbon materials in interesting ways. For example, it is well established that boron can act as a graphitization 'catalyst' (10.112), while boron doping can improve the oxidation resistance of carbons (10.113). Doping carbon nanotubes with boron has also produced some novel results. In an early study, a group from Stuttgart prepared boron-doped carbon nanotubes using arc-discharge with an anode made of BC_4N (10.114). The effect of B doping was to increase the tube lengths to ~100 μm and to improve graphitization. This is believed to be because the B acts as a surfactant, preventing tube closure. In 1999 a multinational group led by Xavier Blase produced evidence that that B doping not only increased the length of MWNTs but also led to 'near zigzag' chiralities (10.115). Again the tubes were prepared using arc-discharge with anodes filled with B or BN powders. In an attempt to understand these results, they compared the energetics of B-doped zigzag and armchair nanotubes. They found that B atoms were more stabilized at zigzag edges than at armchair ones, suggesting that the atoms would preferentially remain on zigzag edges, acting as surfactants during the growth. If correct, this represents one of the very few examples in the literature of a method that can preferentially produce tubes with a given structure.

As discussed in Chapter 2 (p. 25), Robert Chang and colleagues have shown that boron can promote the growth of MWNTs by high-temperature heat treatment (10.116). They found that MWNTs could be formed by the annealing of fullerene soot and other carbons to 2200–2400 °C in a graphite resistance furnace, but that the yield was greatly enhanced by the addition of boron. It is not clear whether the tubes produced in this way were pure carbon or B-doped.

There are a few reports of the synthesis of B-doped multiwalled tubes by CVD (e.g. 10.117, 10.118). It appears that boron–carbon single-walled tubes have not yet been prepared.

10.8   Discussion

Research into the filling of carbon nanotubes has not yet led to any major commercial applications. On the other hand, inserting material into nanotubes has enabled some intriguing experiments to be conducted into the behaviour of confined matter on the nanoscale. The work of Terrones, Ajayan, Banhart and colleagues on filled MWNTs produced by pyrolysis provides one example (10.27). By irradiating the filled tubes with an intense electron beam, these workers caused the tubes to contract, enabling them to study the effect of high pressures on the contained material. Similarly, the experiments of Gogotsi and co-workers on the behaviour of water inside MWNTs (10.32) may provide new insights into the behaviour of fluids at the nanoscale. Single-walled nanotubes, with their very small and well-defined diameters, have even more potential for studying confined materials. Crystals inside SWNTs are forced to adopt a genuinely 1D morphology, often adopting configurations not seen in the bulk, as demonstrated by the Oxford group (10.39). Filled SWNTs have proved to be superb 'nano-test-tubes' for the imaging of encapsulated molecules using HRTEM. Particularly notable here are the studies of retinal attached to C_{60} by Suenaga, Iijima and their colleagues (10.63). Normally, such molecules would be highly unstable under an electron beam, making imaging extremely difficult, but placing them inside SWNTs enables incredible images such as those shown in Fig. 10.13 to be recorded. For a more detailed discussion of the filling of carbon nanotubes than has been possible in this chapter, a number of reviews have been given (10.40, 10.119–10.122), with those by Sloan *et al.* (10.120) and by Monthioux and colleagues (10.121) being particularly recommended. Mattia and Gogotsi have recently given a useful overview of the static and dynamic behaviour of liquids inside carbon nanotubes (10.123)

Heterogeneous nanotubes have been briefly discussed. In this area, most interest has focused on boron nitride tubes, rather than tubes containing carbon, and a detailed discussion of these structures would lie outside the scope of this book. For up-to-date reviews of this subject, see references (10.124) and (10.125).

References

(10.1) R. S. Ruoff, D. C. Lorents, B. Chan *et al.*, 'Single crystal metals encapsulated in carbon nanoparticles', *Science*, **259**, 346 (1993).

(10.2) M. Tomita, Y. Saito and T. Hayashi, 'LaC$_2$ encapsulated in graphite nanoparticle', *Jpn. J. Appl. Phys.*, **32**, L280 (1993).

(10.3) Y. Saito, 'Nanoparticles and filled nanocapsules', *Carbon*, **33**, 979 (1995).

(10.4) S. Seraphin, D. Zhou and J. Jiao, 'Filling the carbon nanocages', *J. Appl. Phys.*, **80**, 2097 (1996).

(10.5) C. Guerret-Piécourt, Y. Le Bouar, A. Loiseau *et al.*, 'Relation between metal electronic structure and morphology of metal compounds inside carbon nanotubes', *Nature*, **372**, 761 (1994).

(10.6) A. Loiseau and H. Pascard, 'Synthesis of long carbon nanotubes filled with Se, S, Sb and Ge by the arc method', *Chem. Phys. Lett.*, **256**, 246 (1996).

(10.7) A. Loiseau and F. Willaime, 'Filled and mixed nanotubes: from TEM studies to the growth mechanism within a phase-diagram approach', *Appl. Surface Sci.*, **164**, 227 (2000).

(10.8) Z. Y. Wang, Z. B. Zhao and J. S. Qiu, '*In situ* synthesis of super-long Cu nanowires inside carbon nanotubes with coal as carbon source', *Carbon*, **44**, 1845 (2006).

(10.9) P. M. Ajayan and S. Iijima, 'Capillarity-induced filling of carbon nanotubes', *Nature*, **361**, 333, (1993).

(10.10) M. R. Pederson and J. Q. Broughton, 'Nanocapillarity in fullerene tubules', *Phys. Rev. Lett.*, **69**, 2689 (1992).

(10.11) S. C. Tsang, P. J. F. Harris and M. L. H. Green, 'Thinning and opening of carbon nanotubes by oxidation using carbon dioxide', *Nature*, **362**, 520 (1993).

(10.12) P. M. Ajayan, T. W. Ebbesen, T. Ichihashi *et al.*, 'Opening carbon nanotubes with oxygen and implications for filling', *Nature*, **362**, 522 (1993).

(10.13) S. C. Tsang, Y. K. Chen, P. J. F. Harris *et al.*, 'A simple chemical method of opening and filling carbon nanotubes', *Nature*, **372**, 159 (1994).

(10.14) J. Sloan, J. Cook, M. L. H. Green *et al.*, 'Crystallisation inside fullerene related structures', *J. Mater. Chem.*, **7**, 1089 (1997).

(10.15) P. M. Ajayan, O. Stephan, P. Redlich *et al.*, 'Carbon nanotubes as removable templates for metal oxide nanocomposites and nanostructures', *Nature*, **375**, 564 (1995).

(10.16) Y. K. Chen, M. L. H. Green and S. C. Tsang, 'Synthesis of carbon nanotubes filled with long continuous crystals of molybdenum oxides', *Chem. Commun.*, 2489 (1996).

(10.17) E. Dujardin, T. W. Ebbesen, H. Hiura *et al.*, 'Capillarity and wetting of carbon nanotubes', *Science*, **265**, 1850 (1994).

(10.18) S. C. Tsang, J. J. Davis, M. L. H. Green *et al.*, 'Immobilisation of small proteins in carbon nanotubes: high resolution transmission electron microscopy study and catalytic activity', *Chem. Commun.*, 1803 (1995).

(10.19) J. J. Davis, M. L. H. Green, H. A. O. Hill *et al.*, 'The immobilisation of proteins in carbon nanotubes', *Inorganica Chimica Acta*, **272**, 261 (1998).

(10.20) C. N. R. Rao, R. Sen, B. C. Satishkumar *et al.*, 'Large aligned-nanotube bundles from ferrocene pyrolysis', *Chem. Commun.*, 1525 (1998).

(10.21) N. Grobert, M. Terrones, O. J. Osborne *et al.*, 'Thermolysis of C_{60} thin films yields Ni-filled tapered nanotubes', *Appl. Phys. A*, **67**, 595 (1998).

(10.22) N. Grobert, W. K. Hsu, Y. Q. Zhu *et al.*, 'Enhanced magnetic coercivities in Fe nanowires', *Appl. Phys. Lett.*, **75**, 3363 (1999).

(10.23) N. Grobert, M. Mayne, M. Terrones *et al.*, 'Alloy nanowires: Invar inside carbon nanotubes', *Chem. Commun.*, 471 (2001).

(10.24) P. C. P. Watts, W. K. Hsu, V. Kotzeva *et al.*, 'Fe-filled carbon nanotube–polystyrene: RCL composites', *Chem. Phys. Lett.*, **366**, 42 (2002).

(10.25) A. Leonhardt, A. Ritschel, R. Kozhuharova *et al.*, 'Synthesis and properties of filled carbon nanotubes', *Diamond Related Mater.*, **12**, 790 (2003).

(10.26) A. L. Elías, J. A. Rodriguez-Manzo, M. R. McCartney *et al.*, 'Production and characterization of single-crystal FeCo nanowires inside carbon nanotubes', *Nano Lett.*, **5**, 467 (2005).

(10.27) L. Sun, F. Banhart, A. V. Krasheninnikov *et al.*, 'Carbon nanotubes as high-pressure cylinders and nanoextruders', *Science*, **312**, 1199 (2006).

(10.28) F. Banhart and P. M. Ajayan, 'Carbon onions as nanoscopic pressure cells for diamond formation', *Nature*, **382**, 433 (1996).

(10.29) F. Banhart, J. X. Li and A. V. Krasheninnikov, 'Carbon nanotubes under electron irradiation: stability of the tubes and their action as pipes or atom transport', *Phys. Rev. B*, **71**, 241408 (2005).

(10.30) Y. Gogotsi, J. A. Libera and M. Yoshimura, 'Hydrothermal synthesis of multiwall carbon nanotubes', *J. Mater. Res.*, **15**, 2591 (2000).

(10.31) Y. Gogotsi, N. Naguib and J. A. Libera , '*In situ* chemical experiments in carbon nanotubes', *Chem. Phys. Lett.*, **365**, 354 (2002).

(10.32) N. Naguib, H. Ye, Y. Gogotsi *et al.*, 'Observation of water confined in nanometer channels of closed carbon nanotubes', *Nano Lett.*, **4**, 2237 (2004).

(10.33) J. K. Holt, H. G. Park, Y. M. Wang *et al.*, 'Fast mass transport through sub-2-nanometer carbon nanotubes', *Science*, **312**, 1034 (2006).

(10.34) M. Whitby and N. Quirke, 'Fluid flow in carbon nanotubes and nanopipes', *Nat. Nanotech.*, **2**, 87 (2007).

(10.35) J. Sloan, J. Hammer, M. Zwiefka-Sibley *et al.*, 'The opening and filling of single walled carbon nanotubes (SWTs)', *Chem. Commun.*, 347 (1998).

(10.36) B. W. Smith, M. Monthioux and D. E. Luzzi, 'Encapsulated C_{60} in carbon nanotubes', *Nature*, **396**, 323 (1998).

(10.37) B. W. Smith, M. Monthioux and D. E. Luzzi, 'Carbon nanotube encapsulated fullerenes: a unique class of hybrid materials', *Chem. Phys. Lett.*, **315**, 31 (1999).

(10.38) J. Sloan, D. M. Wright, S. R. Bailey *et al.*, 'Capillarity and silver nanowire formation observed in single walled carbon nanotubes', *Chem. Commun.*, 699 (1999).

(10.39) J. Sloan, M. C. Novotny, S. R. Bailey *et al.*, 'Two layer 4:4 co-ordinated KI crystals grown within single walled carbon nanotubes', *Chem. Phys. Lett.*, **329**, 61 (2000).

(10.40) J. Sloan, A. I. Kirkland, J. L. Hutchison *et al.*, 'Integral atomic layer architectures of 1D crystals inserted into single walled carbon nanotubes', *Chem. Commun.*, 1319 (2002).

(10.41) S. Friedrichs, J. Sloan, M. L. H. Green *et al.*, 'Simultaneous determination of inclusion crystallography and nanotube conformation for a Sb_2O_3/SWNT single-walled nanotube composite', *Phys. Rev. B*, **64**, 0454061 (2001).

(10.42) E. Philp, J. Sloan, A. I. Kirkland *et al.*, 'An encapsulated helical one-dimensional cobalt iodide nanostructure', *Nat. Mater.*, **2**, 788 (2003).

(10.43) E. Flahaut, J. Sloan, S. Friedrichs *et al.*, 'Crystallization of 2H and 4H PbI_2 in carbon nanotubes of varying diameters and morphologies', *Chem. Mater.*, **18**, 2059 (2006).

(10.44) J. Mittal, M. Monthioux, H. Allouche *et al.*, 'Room temperature filling of single-wall carbon nanotubes with chromium oxide in open air', *Chem. Phys. Lett.*, **339**, 311 (2001).

(10.45) B. C. Satishkumar, A. Taubert and D. E. Luzzi, 'Filling single-wall carbon nanotubes with d- and f-metal chloride and metal nanowires', *J. Nanosci. Nanotech.*, **3**, 159 (2003).

(10.46) D. A. Morgan, J. Sloan, and M. L. H. Green, 'Direct imaging of o-carborane molecules within single walled carbon nanotubes', *Chem. Commun.*, 2442 (2002).

(10.47) Y. Zhang, S. Iijima, Z. Shi *et al.*, 'Defects in arc-discharge-produced single-walled carbon nanotubes', *Phil. Mag. Lett.*, **79**, 473 (1999).

(10.48) J. Sloan, R. E. Dunin-Borkowski, J. L. Hutchison *et al.*, 'The size distribution, imaging and obstructing properties of C_{60} and higher fullerenes formed within arc-grown single walled carbon nanotubes', *Chem. Phys. Lett.*, **316**, 191 (2000).

(10.49) D. E. Luzzi and B. W. Smith, 'Carbon cage structures in single wall carbon nanotubes: a new class of materials', *Carbon*, **38**, 1751 (2000).

(10.50) B. W. Smith, R. M. Russo, S. B. Chikkannanavar *et al.*, 'High-yield synthesis and one-dimensional structure of C_{60} encapsulated in single-wall carbon nanotubes', *J. Appl. Phys.*, **91**, 9333 (2002).

(10.51) H. Kataura, Y. Maniwa, T. Kodama *et al.*, 'High-yield fullerene encapsulation in single-wall carbon nanotubes', *Synth. Met.*, **121**, 1195 (2001).

(10.52) M. Monthioux, 'Filling single-wall carbon nanotubes', *Carbon*, **40**, 1809 (2002).

(10.53) S. Bandow, T. Hiraoka, T. Yumura *et al.*, 'Raman scattering study on fullerene derived intermediates formed within single-wall carbon nanotube: from peapod to double-wall carbon nanotube', *Chem. Phys. Lett.*, **384**, 320 (2004).

(10.54) R. Pfeiffer, M. Holzweber, H. Peterlik *et al.*, 'Dynamics of carbon nanotube growth from fullerenes', *Nano Lett.*, **7**, 2428 (2007).

(10.55) A. N. Khlobystov, D. A. Britz, A. Ardavan *et al.*, 'Observation of ordered phases of fullerenes in carbon nanotubes', *Phys. Rev. Lett.*, **92**, 245507 (2004).

(10.56) D. J. Hornbaker, S. J. Kahng, S. Misra *et al.*, 'Mapping the one-dimensional electronic states of nanotube peapod structures', *Science*, **295**, 828 (2002).

(10.57) A. Yazdani and E. J. Mele, 'Probing the electronic structure of nanotube peapods with the scanning tunneling microscope', *Appl. Phys. A*, **76**, 469 (2003).

(10.58) K. Hirahara, K. Suenaga, S. Bandow *et al.*, 'One-dimensional metallofullerene crystal generated inside single-walled carbon nanotubes', *Phys. Rev. Lett.*, **85**, 5384 (2000).

(10.59) K. Suenaga, M. Tencé, C. Mory *et al.*, 'Element selective single atom imaging', *Science*, **290**, 2280 (2000).

(10.60) K. Suenaga, R. Taniguchi, T. Shimada *et al.*, 'Evidence for the intramolecular motion of Gd atoms in a $Gd_2@C_{92}$ nanopeapod', *Nano Lett.*, **3**, 1395 (2003).

(10.61) Z. Liu, K. Suenaga and S. Iijima, 'Imaging the structure of an individual C_{60} fullerene molecule and its deformation process using HRTEM with atomic sensitivity', *J. Amer. Chem. Soc.*, **129**, 6666 (2007).

(10.62) Y. Sato, K. Suenaga, S. Okubo *et al.*, 'Structures of D_{5d}-C_{80} and I_h-$Er_3N@C_{80}$ fullerenes and their rotation inside carbon nanotubes demonstrated by aberration-corrected electron microscopy', *Nano Lett.*, **7**, 3704 (2007).

(10.63) Z. Liu, K. Yanagi, K. Suenaga *et al.*, 'Imaging the dynamic behaviour of individual retinal chromophores confined inside carbon nanotubes', *Nat. Nanotech.*, **2**, 422 (2007).

(10.64) A. C. Dillon, K. M. Jones, T. A. Bekkedahl *et al.*, 'Storage of hydrogen in single-walled carbon nanotubes', *Nature*, **386**, 377 (1997).

(10.65) C. Zandonella, 'Is it all just a pipe dream?', *Nature*, **410**, 734 (2001).

(10.66) R. Dagani, 'Tempest in a tiny tube', *Chem. Eng. News*, **80**(2), 25 (2002).

(10.67) C. Liu, Y. Y. Fan, M. Liu *et al.*, 'Hydrogen storage in single-walled carbon nanotubes at room temperature', *Science*, **286**, 1127 (1999).

(10.68) Y. Ye, C. C. Ahn, C. Witham *et al.*, 'Hydrogen adsorption and cohesive energy of single-walled carbon nanotubes', *Appl. Phys. Lett.*, **74**, 2307 (1999).

(10.69) P. Chen, X. Wu, J. Lin *et al.*, 'High H_2 uptake by alkali-doped carbon nanotubes under ambient pressure and moderate temperatures', *Science*, **285**, 91 (1999).

(10.70) L. Schlapbach and A. Züttel, 'Hydrogen-storage materials for mobile applications', *Nature*, **414**, 353 (2001).

(10.71) F. L. Darkrim, P. Malbrunot and G. P. Tartaglia, 'Review of hydrogen storage by adsorption in carbon nanotubes', *Int. J. Hydrogen Energy*, **27**, 193 (2002).

(10.72) M. Hirscher, M. Becher, M. Haluska *et al.*, 'Hydrogen storage in carbon nanostructures', *J. Alloys Compounds*, **330**, 654 (2002).

(10.73) M. Becher, M. Haluska, M. Hirscher *et al.*, 'Hydrogen storage in carbon nanotubes', *Comptes Rendus Physique*, **4**, 1055 (2003).

(10.74) M. Hirscher, M. Becher, M. Haluska *et al.*, 'Hydrogen storage in sonicated carbon materials', *Appl. Phys. A*, **72**, 129 (2001).

(10.75) M. G. Nijkamp, J. E. M. J. Raaymakers, A. J. Van Dillen *et al.*, 'Hydrogen storage using physisorption – materials demands. *Appl. Phys. A*, **72**, 619 (2001).

(10.76) A. Chambers, C. Park, R. T. K. Baker *et al.*, 'Hydrogen storage in graphite nanofibers', *J. Phys. Chem. B*, **102**, 4253 (1998).

(10.77) I. Langmuir, 'Pathological science', *Phys. Today*, **42**(10), 36 (1989).

(10.78) G. E. Gadd, M. Blackford, D. Moricca *et al.*, 'The world's smallest gas cylinders', *Science*, **277**, 933 (1997).

(10.79) M. Reyes-Reyes, N. Grobert, R. Kamalakaran *et al.*, 'Efficient encapsulation of gaseous nitrogen inside carbon nanotubes with bamboo-like structure using aerosol thermolysis', *Chem. Phys. Lett.*, **396**, 167 (2004).

(10.80) C. Matranga and B. Bockrath, 'Controlled confinement and release of gases in single-walled carbon nanotube bundles', *J. Phys. Chem. B*, **109**, 9209 (2005).

(10.81) V. P. Sokhan, D. Nicholson and N. Quirke, 'Transport properties of nitrogen in single walled carbon nanotubes', *J. Chem. Phys.*, **120**, 3855 (2004).

(10.82) D. M. Ackerman, A. I. Skoulidas, D. S. Sholl *et al.*, 'Diffusivities of Ar and Ne in carbon nanotubes', *Molec. Simul.*, **29**, 677 (2003).

(10.83) L. Boulanger, B. Andriot, M. Cauchetier *et al.*, 'Concentric shelled and plate-like graphitic boron nitride nanoparticles produced by CO_2 laser pyrolysis', *Chem. Phys. Lett.*, **234**, 227 (1995).

(10.84) R. B. Kaner, J. Kouvetakis, C. E. Warble *et al.*, 'Boron–carbon–nitrogen materials of graphite-like structure', *Mat. Res. Bull.*, **22**, 399 (1987).

(10.85) A. Rubio, J. L. Corkill and M. L Cohen, 'Theory of graphitic boron nitride nanotubes', *Phys. Rev. B*, **49**, 5081 (1994).

(10.86) Y. Miyamoto, A. Rubio, M. L Cohen *et al.*, 'Chiral tubules of hexagonal BC_2N', *Phys. Rev. B*, **50**, 4976 (1994).

(10.87) O. Stéphan, P. M. Ajayan, C. Colliex *et al.*, 'Doping graphitic and carbon nanotube structures with boron and nitrogen', *Science*, **266**, 1683 (1994).

(10.88) N. G. Chopra, R. J. Luyken, K. Cherrey *et al.*, 'Boron nitride nanotubes', *Science*, **269**, 966 (1995).

(10.89) P. Gleize, M. C. Schouler, P. Gadelle *et al.*, 'Growth of tubular boron nitride filaments', *J. Mater. Sci.*, **29**, 1575 (1994).

(10.90) Z. Weng-Sieh, K. Cherrey, N. G. Chopra *et al.*, 'Synthesis of $B_xC_yN_z$ nanotubules', *Phys. Rev. B*, **51**, 11 229 (1995).

(10.91) K. Suenaga, C. Colliex, N. Demoncy *et al.*, 'Synthesis of nanoparticles and nanotubes with well separated layers of boron nitride and carbon', *Science*, **278**, 653 (1997).

(10.92) Y. Zhang, H. Gu, K. Suenaga *et al.*, 'Heterogeneous growth of B–C–N nanotubes by laser ablation', *Chem. Phys. Lett.*, **279**, 264 (1997).

(10.93) M. Terrones, A. M. Benito, C. Manteca-Diego, *et al.*, 'Pyrolytically grown $B_xC_yN_z$ nanomaterials: nanofibres and nanotubes', *Chem. Phys. Lett.*, **257**, 576 (1996).

(10.94) R. Z. Ma and Y. Bando, 'Pyrolytic-grown B–C–N and BN nanotubes', *Sci. Technol. Adv. Mater.*, **4**, 403 (2004).

(10.95) X. D. Bai, J. D. Guo, J. Yu *et al.*, 'Synthesis and field-emission behavior of highly oriented boron carbonitride nanofibers', *Appl. Phys. Lett.*, **76**, 2624 (2000).

(10.96) P. Kohler-Redlich, M. Terrones, C. Manteca-Diego *et al.*, 'Stable BC_2N nanostructures: low-temperature production of segregated C/BN layered materials', *Chem. Phys. Lett.*, **310**, 459 (1999).

(10.97) A. Y. Liu, R. M. Wetzcovitch and M. L. Cohen, 'Atomic arrangement and electronic structure of BC_2N', *Phys. Rev. B*, **39**, 1760 (1988).

(10.98) D. Golberg, P. Dorozhkin, Y. Bando *et al.*, 'Semiconducting B–C–N nanotubes with few layers', *Chem. Phys. Lett.*, **359**, 220 (2002).

(10.99) X. D. Bai, E. G. Wang, J. Yu *et al.*, 'Blue–violet photoluminescence from large-scale highly aligned boron carbonitride nanofibers', *Appl. Phys. Lett.*, **77**, 67 (2000).

(10.100) W. L. Wang, X. D. Bai, K. H. Liu *et al.*, 'Direct synthesis of B–C–N single-walled nanotubes by bias-assisted hot filament chemical vapor deposition', *J. Amer. Chem. Soc.*, **128**, 6530 (2006).

(10.101) M. Terrones, N. Grobert and H. Terrones, 'Synthetic routes to nanoscale $B_xC_yN_z$ architectures', *Carbon*, **40**, 1665 (2002).

(10.102) R. Z. Ma, D. Goldberg, Y. Bando *et al.*, 'Syntheses and properties of $B_xC_yN_z$ and BN nanostructures', *Phil. Trans. Roy. Soc. A*, **362**, 2161 (2004).

(10.103) C. Y. Zhi, X. D. Bai and E. G. Wang, 'Boron carbonitride nanotubes', *J. Nanosci. Nanotech.*, **4**, 35 (2004).

(10.104) M. L. Cohen, 'Calculation of bulk moduli of diamond and zincblende solids', *Phys. Rev. B*, **32**, 7988 (1985).

(10.105) D. M. Teter and R. J. Hemley, 'Low-compressibility carbon nitrides', *Science*, **271**, 53 (1996).

(10.106) M. Yudasaka, R. Kikuchi, Y. Ohki *et al.*, 'Nitrogen-containing carbon nanotube growth from Ni phthalocyanine by chemical vapor deposition', *Carbon*, **35**, 195 (1997).

(10.107) M. Terrones, H. Terrones, N. Grobert *et al.*, 'Efficient route to large arrays of CN_x nanofibers by pyrolysis of ferrocene/melamine mixtures', *Appl. Phys. Lett.*, **75**, 3932 (1999).

(10.108) M. Glerup, M. Castignolles, M. Holzinger *et al.*, 'Synthesis of highly nitrogen-doped multi-walled carbon nanotubes', *Chem. Commun.*, 2542 (2003).

(10.109) C. P. Ewels and M. Glerup, 'Nitrogen doping in carbon nanotubes', *J. Nanosci. Nanotech.*, **5**, 1345 (2005).

(10.110) R. Droppa, P. Hammer, A. C. M. Carvalho *et al.*, 'Incorporation of nitrogen in carbon nanotubes', *J. Non-Cryst. Solids*, **299**, 874 (2002).

(10.111) M. Glerup, J. Steinmetz, D. Samaille *et al.*, 'Synthesis of N-doped SWNT using the arc-discharge procedure', *Chem. Phys. Lett.*, **387**, 193 (2004).

(10.112) A. Oya and H. Marsh, 'Phenomena of catalytic graphitization', *J. Mater. Sci.*, **17**, 309 (1982).

(10.113) X. X. Wu and L. R. Radovic, 'Inhibition of catalytic oxidation of carbon/carbon composites by boron-doping', *Carbon*, **43**, 1768 (2005).

(10.114) P. Redlich, J. Loeffler, P. M. Ajayan *et al.*, 'B–C–N nanotubes and boron doping of carbon nanotubes', *Chem. Phys. Lett.*, **260**, 465 (1996).

(10.115) X. Blase, J. C. Charlier, A. De Vita *et al.*, 'Boron-mediated growth of long helicity-selected carbon nanotubes', *Phys. Rev. Lett.*, **83**, 5078 (1999).

(10.116) A. A. Setlur, S. P. Doherty, J. Y. Dai *et al.*, 'A promising pathway to make multiwalled carbon nanotubes', *Appl. Phys. Lett.*, **76**, 3008 (2000).

(10.117) Z. Wang, C. H. Yu, D. C. Ba *et al.*, 'Influence of the gas composition on the synthesis of boron-doped carbon nanotubes by ECR-CVD', *Vacuum*, **81**, 579 (2007).

(10.118) K. C. Mondal, A. M. Strydom, R. M. Erasmus *et al.*, 'Physical properties of CVD boron-doped multiwalled carbon nanotubes', *Mater. Chemi. Phys.*, **111**, 386 (2008).

(10.119) F. Banhart, N. Grobert, M. Terrones *et al.*, 'Metal atoms in carbon nanotubes and related nanoparticles', *Int. J. Mod. Phys. B*, **15**, 4037 (2001).

(10.120) J. Sloan, D. E. Luzzi, A. I. Kirkland *et al.*, 'Imaging and characterization of molecules and one-dimensional crystals formed within carbon nanotubes', *MRS Bull.*, **29**, 265 (2004).

(10.121) M. Monthioux, E. Flahaut and J. P. Cleuziou, 'Hybrid carbon nanotubes: strategy, progress, and perspectives', *J. Mater. Res.*, **21**, 2774 (2006).

(10.122) Z. Y. Wang, Z. B. Zhao and J. S. Qiu, 'Development of filling carbon nanotubes', *Progr. Chem.*, **18**, 563 (2006).

(10.123) D. Mattia and Y. Gogotsi, 'Static and dynamic behavior of liquids inside carbon nanotubes', *Microfluidics Nanofluidics*, **5**, 289 (2008).

(10.124) D. Golberg, Y. Bando, C. C. Tang *et al.*, 'Boron nitride nanotubes', *Adv. Mater.*, **19**, 2413 (2007).

(10.125) M. Terrones, J. M. Romo-Herrera, E. Cruz-Silva *et al.*, 'Pure and doped boron nitride nanotubes', *Materials Today*, **10**(5), 30 (2007).

11   Probes and sensors

The development of new methods for imaging, measurement and sensing is an important theme in modern research, and the unique properties of carbon nanotubes give them great potential in these areas. For example, nanotubes' outstanding mechanical properties and unique geometry suggest that they should be ideal tips for atomic force microscopy (AFM). Currently, AFM tips typically consist of microfabricated pyramids of silicon or silicon nitride mounted on cantilevers. These probes can be relatively 'blunt' on the scale of the features that are being imaged, and are thus often unable to probe narrow crevices on the specimen surface. Carbon nanotubes, with their elongated shape and tiny diameter not only offer the possibility of much higher resolution imaging, but are also capable of probing the narrowest of fissures. The potential advantages of carbon nanotube tips are illustrated in Fig. 11.1, from the work of Arvind Raman of Purdue University and colleagues (11.1), which shows a nanotube attached to a conventional microfabricated probe. Preparing nanotube AFM tips is not straightforward, however. Two methods can be used: attaching previously produced tubes to the probes, or growing the tubes *in situ*. The first part of this chapter summarizes the methods available for preparing nanotube AFM tips. The performance of nanotube AFM tips is then discussed.

Gas sensing is another area where the properties of carbon nanotubes can be exploited. The discovery by Alex Zettl's group that the electronic properties of carbon nanotubes are highly sensitive to the presence of oxygen (11.2) was mentioned in Chapter 6 (p. 167). This extreme oxygen sensitivity appeared to be bad news for the application of nanotubes in electronic devices but, on the positive side, it suggested that they might be very useful in gas sensors. Zettl's paper, and another published a short time earlier by Hongjie Dai and colleagues (11.3) attracted much attention, and interest in the gas sensing properties of nanotubes continues to grow. A brief review of this field is given, followed by a discussion of the use of nanotubes in biosensors, another rapidly growing area. Finally, some of the ways in which nanotubes can be used as physical sensors are considered.

11.1   Nanotube tips for atomic force microscopy

11.1.1   Preparing nanotube tips: mechanical assembly

The earliest attempts to use carbon nanotubes as AFM tips involved attaching ready-made tubes to the tips of commercial Si or Si_3N_4 pyramids. In a 1996 paper, Smalley and

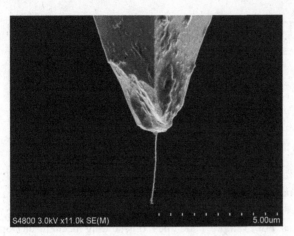

Fig. 11.1   Carbon nanotube attached to a conventional AFM tip (11.1).

colleagues described a method for attaching multiwalled nanotubes to the tips of commercial silicon pyramids (11.4). This was achieved as follows. Firstly the bottom part of a pyramid was coated with an acrylic adhesive, by lowering it onto an adhesive-coated carbon tape. This tip was then brought into contact with a bundle of 5–10 MWNTs, while under direct view of an optical microscope, and then a single tube was drawn out from the bundle to act as the imaging probe. The nanotubes tips were then used to obtain AFM images of a patterned film.

Following this pioneering work, Lieber and colleagues also prepared carbon nanotube AFM probes using mechanical assembly (11.5–11.8). They attached open-ended MWNTs to silicon pyramids using an acrylic glue under an optical microscope. The tips were used for imaging biological systems, as discussed below. In addition to multi-walled tubes, single-walled nanotube AFM tips were also prepared by Lieber's group. The SWNTs were grown on oxidized silicon substrates by CCVD using an iron catalyst. The nanotube-covered wafers were then imaged in tapping mode with silicon tips. Isolated, vertically aligned SWNTs were located and were then 'picked up' on silicon pyramids. For probes to be used for imaging under fluids, the pyramids were coated with a thin layer of a UV-cure adhesive prior to picking up the nanotubes. A method was also described for shortening the tube tips in a controlled way by applying voltage pulses. It was found that SWNT probes need to be very short (c. 10 nm) to be used for imaging. As well as examining the nanotube tips using SEM or TEM, Lieber and colleagues have also used gold nanoparticles to characterize the tips. These nanoparticles represent appropriate imaging standards for AFM since they are essentially incompressible and can be prepared with very uniform diameters. The effective tip radius is calculated from the particle image using the two-sphere model. Manually assembled MWNT tips were found to have diameters as small as 12 nm, which is typical for arc-produced MWNT tubes. For the mechanically assembled SWNT tips the measured radii were larger than would be expected if the tips were individual tubes, suggesting the tubes were bundled together.

Neil Wilson and Julie Macpherson of the University of Warwick, and their colleagues have also prepared SWNT tips using the 'pick-up' technique, and used them for imaging in non-contact (tapping) mode (11.9–11.11). Both individual SWNTs and bundles of SWNTs were attached to metal-coated (gold and platinum) silicon tips in this way. The SWNT bundles were found to adhere much better to the tips than single SWNTs, meaning that they could be used in solution and for extensive periods of time (several months). Also, the bundles could be used without shortening: tips of greater than 1 μm in length could be used for imaging. This group is particularly interested in producing electrically connecting nanotube-based probes for electrochemical, conducting AFM and electrostatic force microscopy applications. In this connection they have described a method for sputter coating SWNT-AFM probes with AuPd or Au, to produce metal probes as small as 30 nm in diameter (11.10, 11.11).

Mechanical assembly of nanotube AFM probes can also be performed in a rather more controlled way inside a scanning electron microscope. Japanese workers described a method that involved firstly aligning nanotubes on the edge of a razor using an AC electrophoresis technique and then transferring one of the tubes to a conventional Si tip inside an SEM equipped with two independent translation stages (11.12). The nanotube tips were then used to image DNA strands in tapping mode, and showed superior resolution to conventional tips. A group from Korea used piezoelectric nanomanipulators inside an SEM to attach nanotubes to AFM tips (11.13). The tubes were 'welded' to the tips using the electron beam (a similar method was used by Ruoff and colleagues in their work on mechanical properties – see p. 186). More recently, researchers from the University of Nottingham have used mechanically assembled nanotube tips to image biomolecules at very high resolution (11.14, 11.15).

11.1.2 Preparing nanotube tips: chemical vapour deposition

An alternative method for attaching carbon nanotubes to AFM pyramids is to grow the tubes directly on a cantilever using catalytic chemical vapour deposition (CVD). This can be achieved in several different ways. One approach, pioneered by Lieber's group in 1999 (11.16), involves producing a porous surface, depositing catalytic metal particles in the pores and then exposing the particles to a carbon-containing gas at high temperature (e.g. ethylene at 800 °C). In this way, the growth of aligned nanotubes out of the pores can be achieved. Lieber and colleagues created a porous Si surface by firstly producing a flattened area of 1–5 μm on Si AFM tips (this was done by 'hard scanning' the tips on a diamond surface) and then anodizing the Si in hydrofluoric acid to create 100 nm diameter pores in the surface. Observation by TEM revealed thin individual multiwalled nanotubes protruding from the ends of the silicon tips, with typical diameters ranging from 6 to 10 nm. Subsequently, the same group has demonstrated the growth of thin SWNT bundles of 1–3 nm in diameter from pores made at the silicon tip ends (11.17). The pore-growth method clearly has potential for producing nanotube AFM probes. However, there are drawbacks: the preparation of a porous layer can be time consuming, and the pores may not be in the correct orientation to promote growth of nanotubes in the optimal direction.

Instead of growing the nanotubes in porous Si, Lieber and colleagues have also shown that they can be grown on the surface of Si pyramids (11.18). This approach involves simply depositing catalytic metal particles on the surface of the pyramids and then exposing them to ethylene under conditions that promote nanotube growth. As the tubes grow, it is found that surface energy tends to guide the tubes towards the apex of the tip.

The direct growth of nanotubes has become quite a popular method for the preparation of AFM tips (e.g. 11.19–11.21). In principle, this approach lends itself to the large-scale production of nanotube AFM tips. The first demonstration of this was given by the Stanford group in 2002 (11.21). They started with a commercially available wafer that contained 375 prefabricated Si cantilevers with pyramidal AFM tips. Polymethylmethacrylate was then spin-coated and baked onto the wafer in such a way that the tips of the pyramids were left exposed. Catalyst particles were deposited onto the wafer from a solution, and the PMMA coating was removed. This left the catalyst particles deposited only on the pyramids, and these were then used to grow nanotubes. Finally the tubes were shortened by applying voltage pulses. It is not clear whether these tips were successfully used for imaging.

11.1.3   Imaging using nanotube AFM tips

Smalley and colleagues used their mechanically assembled SWNT probes to obtain tapping-mode AFM images of a patterned film on a silicon wafer (11.4). It was found that the tips could reach to the bottom of deep trenches in the film, and thus produce much more realistic images than those obtained using pyramidal tips, as can be seen in Fig. 11.2

As well as the experiments with gold nanoparticles mentioned above, the Lieber group has also carried out studies of isolated proteins using nanotube probes. Pore-grown MWNT CCVD tips were used to image isolated immunoglobulin-G (IgG) and immunoglobulin-M (IgM) antibody proteins. Immunoglobulin-G has a molecular weight of ~180 000, with a characteristic Y-shaped structure approximately 15 nm across. In previous AFM studies using conventional tips, the Y-shape was only seen at cryogenic temperatures. Using MWNT CCVD tips, Lieber *et al.* were readily able to resolve the Y-shaped structure at room temperature (11.17). The pentameric structure of IgM was also resolved using this type of probe. In order to test the performance of the smaller diameter CCVD SWNT tips, Lieber and colleagues have used a smaller protein, GroES, as a test specimen, and obtained higher resolution images than had been achieved with conventional tips, as shown in Fig. 11.3.

The Lieber group has also pioneered the use of functionalized carbon nanotube tips to sense specific interactions with functional groups on substrates (11.5, 11.6). The tubes were initially oxidized, which removed the caps and resulted in the formation of carboxyl surface groups. The carboxyl-terminated tubes were then used to carry out chemically sensitive imaging of surfaces patterned with different molecules. Tubes terminated with amine groups were used in a similar way. In addition to chemical imaging, Lieber and co-workers used nanotube tips to investigate interactions between biological molecules, specifically the ligand-receptor interaction of biotin with streptavidin. Biotin ligand was

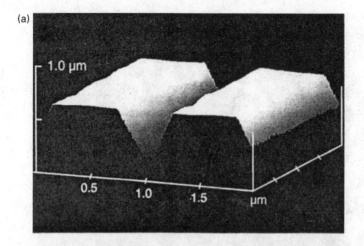

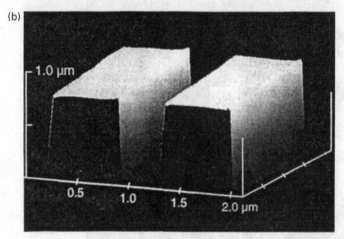

Fig. 11.2   AFM images of patterned film obtained using (a) conventional tip and (b) nanotube tip (11.4).

covalently linked to nanotube tips by the formation of amide bonds, and the modified tips were then used to probe immobilized streptavidin molecules adsorbed on mica. In this way it proved possible to measure the binding forces between the biotin–streptavidin pairs.

Most of the studies using nanotube AFM tips have involved biological samples. There have been fewer studies of the potential of nanotube tips for high-resolution imaging of non-biological materials. The early work of Smalley and colleagues on imaging a patterned silicon film has already been mentioned. The Lieber group have used mechanically assembled SWNT tips to resolve substructure within SWNTs deposited on surfaces (11.7). Also, Japanese workers have used carbon nanotubes as probes for friction force microscopy (11.22). Periodic hexagonal images corresponding to the atomic structure of the mica surface were obtained.

A useful review of the use of carbon nanotubes as AFM tips has been given by Cattien Nguyen and colleagues of NASA (11.23).

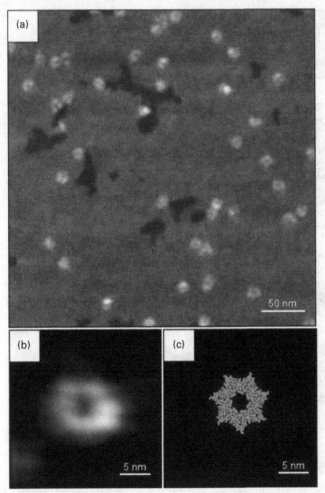

Fig. 11.3   GroES protein imaged by a CVD nanotube tip, from the work of the Lieber group (11.17).
(a) Large area scan (b) higher resolution image showing heptameric symmetry (c) crystal
structure of protein.

11.2   Gas sensors

In their *Science* paper published in 2000 (11.2), Zettl and colleagues described measurements of the DC electrical resistance and the thermoelectric power of bundles and thin
films of SWNTs. The samples were mounted in a vacuum test chamber with provisions
for heating and cooling and injecting different gases. Resistance measurements were
made at room temperature using a four-probe contact configuration. It was found that the
presence of air resulted in a 10–15% drop in resistance, which could be reversed by
evacuating the chamber once again. A similar effect was observed when pure oxygen was
used. Measurements were also made of the thermoelectric power, S, of the SWNT

samples i.e. the voltage generated when two ends of the samples were held at different temperatures. A positive S was observed in the presence of oxygen, indicating p-type behaviour, while a smaller, negative S, occurred in vacuum, consistent with n-type behaviour.

Although theoretical work by Marvin Cohen's group (11.24) provided support for the Zettl results, subsequent studies, both experimental and theoretical, have cast doubt on its validity. In 2002, Tobias Hertel and colleagues showed that the binding energies for oxygen on SWNTs could be attributed to van der Waals interactions, i.e. physisorption, rather than the chemisorption assumed by the Zettl group (11.25). A paper by Goldoni and co-workers in 2003 suggested that the observed oxygen sensitivity might have been due to contamination (11.26). This group studied purified 'buckypaper', and used photoemission spectroscopy to study the adsorption of oxygen on the carbon. They found evidence that oxygen was chemisorbed on the nanotubes, which would be consistent with the view that oxygen can strongly affect the tubes' electronic properties. However, the photoemission spectra also showed the presence of several contaminants in the nominally purified buckypaper. In particular, there was a significant amount of Na, probably a residue of the purification process, together with Ni from the catalyst particles and other minor contaminants. It appears, therefore, that there is some doubt about the results originally reported by Zettl *et al.*

The work by Dai and colleagues, published slightly before the Zettl paper (11.3) seems to be less controversial. This study showed that the conductivity of individual SWNTs was extremely sensitive to nitrogen dioxide and ammonia. An arrangement was used in which each end of a single semiconducting SWNT was connected to titanium and/or gold pads. Using the pads as electrodes, they found that the conductivity of the SWNT changed rapidly over several orders of magnitude upon exposure to nitrogen dioxide and ammonia. Thus, an increase in the conductivity by up to three orders of magnitude was observed within 10 s after exposing the tube to 200 ppm NO_2. With 1% NH_3, the conductance decreased by two orders of magnitude within two minutes. An important aspect of the work was that the sensors worked at room temperature. Conventional sensors for NO_2 and NH_3 based on semiconducting metal oxides need to operate at temperatures of up to 600 °C for high sensitivity, while those that are based on conducting polymers have limited sensitivity. The SWNT sensors therefore appear to have considerable promise. One drawback, however, is that they take several hours to release the analyte at room temperature, although this can be speeded up by heating.

Several other groups have explored the use of SWNTs to sense NH_3, often using functionalized tubes (e.g. 11.27–11.29) Robert Haddon and colleagues found that functionalized SWNTs experienced a much greater change of resistance upon exposure to NH_3 than did pristine tubes, giving them greater sensitivity as sensors (11.28). This was attributed to electron transfer between the attached molecules and the valence band of semiconducting SWNTs (11.29). Functionalized tubes have also been used in NO_2 sensors (11.30). In 2007, Alexander Star and colleagues from the University of Pittsburgh described a sensor for nitric oxide that employed SWNTs functionalized with poly(ethylene imine) (11.31). The NO was firstly oxidized to NO_2, which was then passed over a network of SWNTs in a field-effect transistor device, inducing changes in the conductance of the tubes. The authors suggested that the device could be useful in diagnosing asthma, by monitoring NO in exhaled breath.

Nanotubes decorated with metal particles have also been used in sensing. The first demonstration of this was given by Dai and co-workers, who showed in 2001 that SWNTs coated with Pd nanoparticles can selectively sense hydrogen in a flow of air (11.32). In this work, both individual SWNTs and nanotube bundles were sputter-coated with Pd nanoparticles. Decreases in conductivity up to 50% and 33% were observed for the Pd-coated individual SWNT and nanotube bundles, respectively, upon exposure to a flow of air mixed with 400 ppm of hydrogen. This was followed by a rapid recovery of the conductivity after the hydrogen flow was turned off. The sensing mechanism is believed to involve the dissolution of atomic hydrogen into Pd, leading to a decrease in the work function for Pd. This in turn, causes electron transfer from Pd to the SWNTs to reduce the hole-carriers in the p-type nanotube, and hence a decreased conductivity. Workers from NASA have used Pd-decorated SWNTs as sensors for methane (11.33), while the Pittsburgh group have fabricated arrays of SWNTs decorated with Pd, Pt, Rh and Au, and used them to detect H_2, CO, CH_4 and H_2S (11.34).

In an interesting study published in *Science* in 2005, workers from the USA and Sweden showed that electronic transport in metallic SWNTs was sensitive to collisions with inert gas atoms or small molecules, including He, Ar, Ne, Kr, Xe and N_2 (11.35). These gases are difficult to detect with current measurement technologies.

11.3   Biosensors

Carbon nanotubes have been used as biosensors in a variety of ways. There has been much interest in using 'bulk' quantities of nanotubes in macroscopic electrodes to replace, or complement, existing electrode materials such as glassy carbon or precious metals. At the other extreme, it has been demonstrated that individual single-walled tubes can be used as sensors. We begin by considering macroscale electrochemical biosensors employing nanotube electrodes.

Electrochemical biosensors usually contain three electrodes, a reference electrode, an active electrode and a sink electrode. The analyte reacts with the active electrode surface, and the ions produced create a potential that is subtracted from that of the reference electrode to give a signal. Carbon nanotubes have a number of qualities that suggest they might make attractive electrode materials, including their high surface areas and con-ductivities. The use of functionalized nanotubes in electrodes is also of interest, as previously discussed in Chapter 8. Many studies have now been carried out using nanotube-containing electrodes (e.g. 11.36–11.46), and in many cases they have dis-played characteristics that are equal or superior to that of conventional electrodes (11.47). Various methods have been used to make the electrodes, including simply mixing the nanotubes with a binder (11.36, 11.40, 11.45), or drop coating onto a glassy carbon electrode (11.38, 11.41). Aligned nanotubes have also been used in electrodes. These have sometimes been prepared using self-asembly, as described in Chapter 4 (p. 93) (11.37, 11.43), and sometimes by direct growth (e.g. 11.39). Among the biomolecules

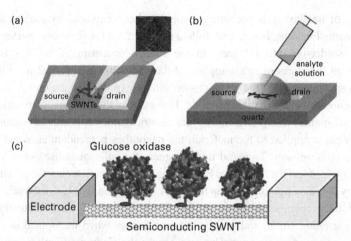

(a)

(b)

analyte
solution

source   drain
SWNTs

source   drain

quartz

(c)   Glucose oxidase

Electrode

Semiconducting SWNT

Fig. 11.4   (a), (b) Using nanotubes as electronic devices for sensing in aqueous solutions, from the work
of Dai *et al.* (11.48). (c) Two electrodes connecting a semiconducting SWNT with GOx enzymes
immobilized on its surface, from work by the Dekker group (11.49).

that have been analysed using nanotube-containing electrodes are DNA (11.39, 11.42,
11.44), enzymes (11.40, 11.46), proteins (11.43) and glucose (11.45).

Turning now to nanoscale biosensing using nanotubes. This has generally involved
using SWNTs in a field effect transistor (FET) configuration, similar to that used in many
of the chemical sensors described above. The work of Hongjie Dai's group on the
development of protein sensors was mentioned in Section 8.4.1 (11.48). This involved
coating SWNTs with a surfactant, and then attaching specific receptors to the coated
tubes. The tubes in these devices were grown *in situ* on quartz substrates. Metal
evaporation through a shadow mask then formed the source and drain electrodes, as
shown in Fig. 11.4(a). Sensing in solution was carried out by monitoring electrical
current through the device during protein additions (Fig. 11.4b). The glucose sensors
developed by Dekker's group (11.49) worked in a similar way, but this time using
individual tubes, as in Fig. 11.4(c). Again, the tubes were grown *in situ* on SiO$_2$, and
metal electrodes were deposited on top of the SWNTs using electron-beam lithography.
Glucose oxidase (GOx) was then immobilized on the tubes via a pyrenyl group linking
molecule. The conductivity of the GOx-coated tubes was found to change with pH and
with addition of glucose, demonstrating that single nanotubes can act as sensors.

For further information on the enormous amount of work done on nanotube biosensors, several excellent reviews are available (11.47, 11.50–11.52).

11.4   Physical sensors

The application of carbon nanotubes in atomic force microscopy was discussed above.
Here we are concerned with the use of nanotubes to measure physical phenomena such as
pressure and flow rate, and as 'nanobalances'.

The use of nanotubes as pressure sensors has been explored by Daniel Wagner of the Weizmann Institute, Israel, and colleagues (11.53–11.55). These workers showed that the disorder-induced D* band in the Raman spectrum of SWNTs is strongly dependent on the stress or strain applied to the nanotubes (see p. 192 for a discussion of Raman spectroscopy). This phenomenon can be exploited to determine the strain in nanotube-containing composite materials. However, if the nanotubes are oriented randomly in the matrix, interpretation may not be straightforward. This is because when a uniaxial stress is applied to the material, the nanotubes perpendicular to the load may experience compression. To avoid this, Wagner and colleagues use polarized Raman spectroscopy to select the nanotubes lying along the polarization direction. Using Raman microscopy it is also possible to map the strain distribution in these materials. There are other potential problems with this technique. The shifts in the Raman frequencies caused by strain are rather small, and will be reduced further when the optical signal passes through the surrounding medium. This might make the method difficult to apply in some situations.

The idea that it might be possible to generate electricity by passing a liquid over carbon nanotubes, and that the tubes could therefore be used as flow sensors was first put forward by Petr Král and Moshe Shapiro, also of the Weizmann Institute, in 2001 (11.56). Their calculations suggested that passing a polar liquid such as water past a conducting nanotube should induce an electron flow along the walls of the tubes in the same direction as the liquid flow. In early 2003, Ajay Sood of the Indian Institute of Science in Bangalore and colleagues claimed to have demonstrated this effect experimentally (11.57). They described a device consisting of a bundle of randomly oriented SWNTs packed between two metal electrodes and suspended in a metre-long glass tube. Water was pumped through the tube and the voltage across the nanotubes measured. It was found that even very small flow velocities produced relatively large voltages across the nanotube bundle. Thus, a flow velocity of $5\,\mathrm{m\,s^{-1}}$ induced a voltage of $0.65\,\mathrm{mV}$. Liquids with greater ionic strength than water were found to induce higher voltages, while non-polar liquids such as methanol had a much smaller effect. The one-dimensional nature of the SWNTs appeared to be crucial for the generation of an electrical signal: experiments with graphite did not produce any measurable signal. Multi walled nanotubes did generate a voltage, but this was about ten times smaller than that produced by SWNTs. It is not clear whether the results of Sood and colleagues have been repeated by other groups.

Another interesting idea that does not seem to have been widely taken up is the use of nanotubes to weigh nanoscale particles. This was demonstrated in 1999 by de Heer and colleagues (11.58). These workers applied alternating voltages to nanotubes inside an electron microscope, causing them to vibrate. By adjusting the frequency of the applied potential they were able to excite the nanotubes resonantly at their fundamental frequency and at higher harmonics. In this way they could determine not only the tubes' moduli, but could also measure the masses of carbon nanoparticles that were attached to the tubes. They suggest that this 'nanobalance' approach could be applied to other particles of similar dimensions such as viruses.

11.5   Discussion

The potential of carbon nanotubes as AFM tips was recognized quite early on by Smalley and his colleagues (11.4). Important early work in this area was also carried out by the Lieber group (11.5–11.8). Following these pioneering studies, notable progress has been made in the production of nanotube AFM tips, and many impressive images have been obtained using these probes. Nanotube tips can now be purchased from a number of AFM companies. Currently, the price of these tips is relatively high (typically around \$400 each compared with about \$50 for conventional tips). It remains to be seen whether nanotube tips will eventually become as widely used as Si and Si_3N_4 probes.

Interest in gas sensing using nanotubes has also grown steadily since the publication of the *Science* papers by the Zettl and Dai groups in 2000 (11.2, 11.3). Nanotubes have the advantage over alternative materials of extremely large surface areas (single-walled nanotubes are essentially 'all surface'), leading to high sensitivity. In addition, nanotubes can be tailored to sense certain gases by functionalization and doping with catalysts. A further advantage over conventional solid-state sensors, which typically operate at temperatures over 400 °C, is that nanotube-based sensors can operate at room temperature. It appears, however, that there are still issues with the 'recovery time' after exposure to the gas.

The use of nanotubes in biosensors has attracted just as much interest as their application in gas sensing. A huge amount of work has been done on macroscale electrochemical biosensors employing nanotube electrodes, and it seems that these offer significant advantages over conventional electrodes. Some impressive demonstrations of the use of individual tubes as biosensors have also been given.

References

(11.1) M. C. Strus, A. Raman, C. S. Han *et al.*, 'Imaging artefacts in atomic force microscopy with carbon nanotube tips', *Nanotechnology*, **16**, 2482 (2005).

(11.2) P. G. Collins, K. Bradley, M. Ishigami *et al.*, 'Extreme oxygen sensitivity of electronic properties of carbon nanotubes', *Science*, **287**, 1801 (2000).

(11.3) J. Kong, N. R. Franklin, C. W. Zhou *et al.*, 'Nanotube molecular wires as chemical sensors', *Science*, **287**, 622 (2000).

(11.4) H. J. Dai, J. H. Hafner, A. G. Rinzler *et al.*, 'Nanotubes as nanoprobes in scanning probe microscopy', *Nature*, **384**, 147 (1996).

(11.5) S. S. Wong, E. Joselevich, A. T. Woolley *et al.*, 'Covalently functionalized nanotubes as nanometre-sized probes in chemistry and biology', *Nature*, **394**, 52 (1998).

(11.6) S. S. Wong, A. T. Woolley, E. Joselevich *et al.*, 'Covalently-functionalized single-walled carbon nanotube probe tips for chemical force microscopy', *J. Amer. Chem. Soc.*, **120**, 8557 (1998).

(11.7) S. S. Wong, A. T. Woolley, T. W. Odom *et al.*, 'Single-walled carbon nanotube probes for high-resolution nanostructure imaging', *Appl. Phys. Lett.*, **73**, 3465 (1998).

(11.8) J. Hafner, C. L. Cheung, T. H. Oosterkamp *et al.*, 'High yield assembly of individual single-walled carbon nanotube tips for scanning probe microscopies', *J. Phys. Chem. B*, **105**, 743 (2001).

(11.9) N. R. Wilson, D. H. Cobden and J. V. Macpherson, 'Single-wall carbon nanotube conducting probe tips', *J. Phys. Chem. B*, **106**, 13 102 (2002).

(11.10) N. R. Wilson and J. V. Macpherson, 'Single-walled carbon nanotubes as templates for nanowire conducting probes', *Nano Lett.*, **3**, 1365 (2003).

(11.11) D. P. Burt, N. R. Wilson, J. M. R. Weaver *et al.*, 'Nanowire probes for high resolution combined scanning electrochemical microscopy – atomic force microscopy', *Nano Lett.*, **5**, 639 (2005).

(11.12) H. Nishijima, S. Kamo, S. Akita *et al.*, 'Carbon-nanotube tips for scanning probe microscopy: preparation by a controlled process and observation of deoxyribonucleic acid', *Appl. Phys. Lett.*, **74**, 4061 (1999).

(11.13) K. S. Kim, S. C. Lim, I. B. Lee *et al.*, '*In situ* manipulation and characterizations using nanomanipulators inside a field emission-scanning electron microscope', *Rev. Sci. Instrum.*, **74**, 4021 (2003).

(11.14) C. T. Gibson, S. Carnally and C. J. Roberts, 'Attachment of carbon nanotubes to atomic force microscope probes', *Ultramicroscopy*, **107**, 1118 (2007).

(11.15) S. Carnally, K. Barrow, M. R. Alexander *et al.*, 'Ultra-resolution imaging of a self-assembling biomolecular system using robust carbon nanotube AFM probes', *Langmuir*, **23**, 3906 (2007).

(11.16) J. H. Hafner, C. L. Cheung and C. M. Lieber, 'Growth of nanotubes for probe microscopy tips', *Nature*, **398**, 761 (1999).

(11.17) C. L. Cheung, J. H. Hafner and C. M. Lieber, 'Carbon nanotube atomic force microscopy tips: direct growth by chemical vapor deposition and application to high-resolution imaging', *Proc. Nat. Acad. Sci. USA*, **97**, 3809 (2000).

(11.18) J. H. Hafner, C. L. Cheung and C. M. Lieber, 'Direct growth of single-walled carbon nanotube scanning probe microscopy tips', *J. Amer. Chem. Soc.*, **121**, 9750 (1999).

(11.19) C. L. Cheung, J. H. Hafner, T. W. Odom *et al.*, 'Growth and fabrication with single-walled carbon nanotube probe microscopy tips', *Appl. Phys. Lett.*, **76**, 3136 (2000).

(11.20) H. Cui, S. V. Kalinin, X. Yang *et al.*, 'Growth of carbon nanofibers on tipless cantilevers for high resolution topography and magnetic force imaging', *Nano. Lett.*, **4**, 2157 (2004).

(11.21) E. Yenilmez, Q. Wang, R. J. Chen *et al.*, 'Wafer scale production of carbon nanotube scanning probe tips for atomic force microscopy', *Appl. Phys. Lett.*, **80**, 2225 (2002).

(11.22) M. Ishikawa, M. Yoshimura and K. Ueda, 'Carbon nanotube as a probe for friction force microscopy', *Physica B*, **323**, 184 (2002).

(11.23) C. V. Nguyen, Q. Ye and M. Meyyappan, 'Carbon nanotube tips for scanning probe microscopy: fabrication and high aspect ratio nanometrology', *Measurement Sci. Technol.*, **16**, 2138 (2005).

(11.24) S. H. Jhi, S. G. Louie and M. L. Cohen, 'Electronic properties of oxidized carbon nanotubes', *Phys. Rev. Lett*, **85**, 1710 (2000).

(11.25) H. Ulbricht, G. Moos and T. Hertel, 'Physisorption of molecular oxygen on single-wall carbon nanotube bundles and graphite', *Phys. Rev. B*, **66**, 075404 (2002).

(11.26) A. Goldoni, R. Larciprete, L. Petaccia *et al.*, 'Single-wall carbon nanotube interaction with gases: sample contaminants and environmental monitoring', *J. Amer. Chem. Soc.*, **125**, 11 329 (2003).

(11.27) X. Feng, S. Irle, H. Witek *et al.*, 'Sensitivity of ammonia interaction with single-walled carbon nanotube bundles to the presence of defect sites and functionalities', *J. Amer. Chem. Soc.*, **127**, 10 533 (2005).

(11.28) E. Bekyarova, M. Davis, T. Burch *et al.* 'Chemically functionalized single-walled carbon nanotubes as ammonia sensors', *J. Phys. Chem. B*, **108**, 19 717 (2004).

(11.29) E. Bekyarova, I. Kalinina, M. E. Itkis *et al.*, 'Mechanism of ammonia detection by chemically functionalized single-walled carbon nanotubes: *in situ* electrical and optical study of gas analyte detection', *J. Amer. Chem. Soc.*, **129**, 10 700 (2007).

(11.30) Q. F. Pengfei, O. Vermesh, M. Grecu *et al.*, 'Toward large arrays of multiplex functionalized carbon nanotube sensors for highly sensitive and selective molecular detection', *Nano. Lett.*, **3**, 347 (2003).

(11.31) O. Kuzmych, B. L. Allen and A. Star, 'Carbon nanotube sensors for exhaled breath components', *Nanotechnology*, **18**, 375502 (2007).

(11.32) J. Kong, M. G. Chapline and H. J. Dai, 'Functionalized carbon nanotubes for molecular hydrogen sensors', *Adv. Mater.*, **13**, 1384 (2001).

(11.33) Y. Lu, J. Li, J. Han *et al.* 'Room temperature methane detection using palladium loaded single-walled carbon nanotube sensors', *Chem. Phys. Lett.*, **391**, 344 (2004).

(11.34) A. Star, V. Joshi, S. Skarupo *et al.* 'Gas sensor array based on metal-decorated carbon nanotubes', *J. Phys. Chem. B*, **110**, 21 014 (2006).

(11.35) H. E. Romero, K. Bolton, A. Rosen *et al.*, 'Atom collision-induced resistivity of carbon nanotubes', *Science*, **307**, 89 (2005).

(11.36) P. J. Britto, K. S. V. Santhanam and P. M. Ajayan, 'Carbon nanotube electrode for oxidation of dopamine', *Bioelectrochem. Bioenergetics*, **41**, 121 (1996).

(11.37) Z. F. Liu, Z. Y. Shen, T. Zhu *et al.*, 'Organizing single-walled carbon nanotubes on gold using a wet chemical self-assembling technique', *Langmuir*, **16**, 3569 (2000).

(11.38) J. X. Wang, M. X. Li, Z. J. Shi *et al.*, 'Electrocatalytic oxidation of norepinephrine at a glassy carbon electrode modified with single wall carbon nanotubes', *Electroanalysis*, **14**, 225 (2002).

(11.39) C. V. Nguyen, L. Delzeit, A. M. Cassell *et al.*, 'Preparation of nucleic acid functionalized carbon nanotube arrays', *Nano Lett.*, **2**, 1079 (2002).

(11.40) F. Valentini, A. Amine, S. Orlanducci *et al.*, 'Carbon nanotube purification: preparation and characterization of carbon nanotube paste electrodes', *Anal. Chem.*, **75**, 5413 (2003).

(11.41) J. Wang, R. P. Deo and M. Musameh, 'Stable and sensitive electrochemical detection of phenolic compounds at carbon nanotube modified glassy carbon electrodes', *Electroanalysis*, **15**, 1830 (2003).

(11.42) H. Cai, X. Cao, Y. Jiang *et al.*, 'Carbon nanotube-enhanced electro-chemical DNA biosensor for DNA hybridization detection', *Anal. Bioanal. Chem.*, **375**, 287 (2003).

(11.43) J. J. Gooding, R. Wibowo, J. Q. Liu *et al.*, 'Protein electrochemistry using aligned carbon nanotube arrays', *J. Amer. Chem. Soc.*, **125**, 9006 (2003).

(11.44) P. G. He, S. N. Li and L. M. Dai, 'DNA-modified carbon nanotubes for self-assembling and biosensing applications', *Synthetic Metals*, **154**, 17 (2005).

(11.45) G. L. Luque, N. F. Ferreyra and G. A. Rivas, 'Glucose biosensor based on the use of a carbon nanotube paste electrode modified with metallic particles', *Microchimica Acta*, **152**, 277 (2006).

(11.46) A. Arvinte, F. Valentini, A. Radoi *et al.*, 'The NADH electrochemical detection performed at carbon nanofibers modified glassy carbon electrode', *Electroanalysis*, **19**, 1455 (2007).

(11.47) J. J. Gooding, 'Nanostructuring electrodes with carbon nanotubes: a review on electro-chemistry and applications for sensing', *Electrochim. Acta*, **50**, 3049 (2005).

(11.48) R. J. Chen, S. Bangsaruntip, K. A. Drouvalakis *et al.*, 'Noncovalent functionalization of carbon nanotubes for highly specific electronic biosensors', *Proc. Nat. Acad. Sci. USA*, **100**, 4984 (2003).

(11.49) K. Besteman, J. O. Lee, F. G. M. Wiertz *et al.*, 'Enzyme-coated carbon nanotubes as single-molecule biosensors', *Nano Lett.*, **3**, 727 (2003).

(11.50) K. Balasubramanian and M. Burghard, 'Biosensors based on carbon nanotubes', *Anal. Bioanal. Chem.*, **385**, 452 (2006).

(11.51) P. He and L. M. Dai, 'Carbon nanotube biosensors', in *BioMEMS and Biomedical Nanotechnology*, vol. 1, ed. M. Ferrari, Springer, Berlin, 2006, p. 171.

(11.52) P. Pandey, M. Datta and B. D. Malhotra, 'Prospects of nanomaterials in biosensors', *Anal. Lett.*, **41**, 159 (2008).

(11.53) J. R. Wood and H. D. Wagner, 'Single-wall carbon nanotubes as molecular pressure sensors', *Appl. Phys. Lett.*, **76**, 2883 (2000).

(11.54) J. R. Wood, Q. Zhao and H. D. Wagner, 'Orientation of carbon nanotubes in polymers and its detection by Raman spectroscopy', *Composites A*, **32**, 391 (2001).

(11.55) Q. Zhao, M. D. Frogley and H. D. Wagner, 'Direction-sensitive strain-mapping with carbon nanotube sensors', *Composites Sci. Technol.*, **62**, 147 (2002).

(11.56) P. Král and M. Shapiro, 'Nanotube electron drag in flowing liquids', *Phys. Rev. Lett.*, **86**, 131 (2001).

(11.57) S. Ghosh, A. K. Sood and N. Kumar, 'Carbon nanotube flow sensors', *Science*, **299**, 1042 (2003).

(11.58) P. Poncharal, Z. L. Wang, D. Ugarte *et al.*, 'Electrostatic deflections and electromechanical resonances of carbon nanotubes', *Science*, **283**, 1513 (1999).

12 Conclusions

Few materials can have been as intensively studied as carbon nanotubes. As a result of the enormous amount of work carried out on these structures since 1991, we now have a wealth of information on their electronic, mechanical, thermal, optical and other physical properties. We know how to attach things to the outside of tubes and to put things inside. We can manipulate nanotubes into defined arrangements and incorporate them into polymer, ceramic or metal matrices. At present, however, there are still relatively few nanotube-containing products on the market. The few commercial products that have emerged are in highly specialized areas such as AFM tips. Perhaps the most commercially significant application of nanotubes to date has been as components of lithium ion batteries, although these employ relatively poor quality catalytically-produced MWNTs. It would be fair to say, then, that carbon nanotubes have not yet fulfilled their potential. The main reason for this is quite simple: they are still far too expensive. The price of single-walled nanotubes is currently around $100 per gram, compared with about $30 per gram for gold. Clearly there is still much work to be done in improving the quality and yield of nanotube production.

In this final chapter, a brief and subjective summary is given of some of the highlights of carbon nanotube research. Some comments are then made on the areas where progress still needs to be made.

12.1 Highlights of carbon nanotube research

Any list of classic papers on carbon nanotubes would probably have to begin with Sumio Iijima's 1991 letter to *Nature* (12.1). It should be recognized however that important work on catalytically-produced carbon fibres was carried out in the 1970s and 1980s, most notably by Terry Baker (e.g. 12.2) and Morinobu Endo (12.3) and their co-workers. The fact that this work was not widely recognized until after the appearance of Iijima's paper does not lessen its significance. Following Iijima's 1991 paper, the next major event was the discovery of single-walled nanotubes in 1993 (12.4, 12.5). These were initially produced using arc-evaporation, a technique that is difficult to scale up. The demonstration in 1996 that single-walled tubes could be produced catalytically (12.6) was therefore of great importance. Catalytic processes are readily scaleable, and a number of methods for the bulk synthesis of SWNTs have been developed, notably those known by the acronyms HiPco (12.7) and CoMoCAT (12.8). The production of single-walled tubes in this way has greatly increased their availability (although, as noted above, the commercial price remains stubbornly high). It is important to note that the quality of single-walled nanotubes produced using catalytic processes is similar to those made using arc- or laser-vaporization.

Producing carbon nanotubes catalytically enables them to be grown in defined arrangements. A huge amount of research has been done on the growth of vertical arrays of multiwalled tubes on substrates, following the pioneering work of Zhifeng Ren and colleagues in 1998 (12.9). A primary aim of this work has been to manufacture field emission display devices, although it is not clear whether this has led to any commercial products. There has also been great interest in the directed growth of SWNTs across substrates, to produce nanoscale circuits. Hongjie Dai's group showed in 2001 that this can be achieved using electric fields (12.10). The same group also used catalytic growth to integrate nanotube transistors into a silicon circuit in 2004 (12.11). On a much larger scale, catalytic methods have been successfully used in the continuous production of aligned MWNT aerogels (12.12), which can be processed to produce fibres with exceptional strength and stiffness (12.13).

The alternative to growing nanotubes in defined arrangements is to process ready-made tubes. Some fascinating work has been done by the Ralph Krupke group on the assembly of single-walled nanotubes into defined patterns using dielectrophoresis (12.14). They claim that millions of nanotube devices can be constructed in this way. Equally remarkable is the work of Ray Baughman and colleagues on spinning MWNT yarns from arrays grown catalytically on substrates (12.15). A discussion of nanotube processing should also mention the development of 'buckypaper', i.e. thin sheets of purified single-walled tubes. This was first described by Andrew Rinzler and co-workers in 1998 (12.16). By refining his original process, Rinzler has been able to prepare ultra-thin buckypaper films which are transparent to visible light and electrically conductive. It seems certain that this extraordinary material will find important applications.

Turning now to the electronic properties of carbon nanotubes, this is an area where theory led experiment. As noted in the opening chapter, Carter White's group, from the Naval Research Laboratory (12.17), and Mildred Dresselhaus and co-workers from MIT (12.18) both submitted papers on the electronic properties of fullerene tubes prior to the appearance of Iijima's paper. A short time later, the MIT group, and Noriaki Hamada and colleagues from Tsukuba, carried out band structure calculations which demonstrated that electronic properties were a function of both tube structure and diameter (12.19, 12.20). In other words, nanotubes could be either semiconducting or metallic depending on their size and structure. Experimental confirmation of these predictions required extraordinary skill, involving the precise positioning of nanotubes on substrates, and amazingly sensitive conductivity measurements. Key achievements here were the demonstration by Ebbesen and colleagues in 1996 that different MWNTs within a single sample can display different electronic properties (12.21); the direct correlation of electronic properties with SWNT structure, by Dekker and co-workers in 1998 (12.22); and the demonstration in 1997 of quantum transport in SWNTs (12.23) and in MWNTs one year later (12.24). Other significant achievements include the construction in 2001 of a SWNT-based single-electron transistor which operated at room temperature (12.25), and of a logic gate based on a single nanotube bundle (12.26). Of course, translating these laboratory demonstrations into practical electronic devices will not only require progress to be made in preparing nanotubes with a defined structure, but also the development of techniques for reliably growing or arranging them in a defined manner. Some workers have recently argued that graphene offers a better prospect than nanotubes for the construction of nanoscale devices. Graphene is certainly a fascinating and important material, but its use in nanoelectronics is still at a very early stage, and nanotubes possess important advantages in terms of chemical and physical stability. It would be premature, therefore, to abandon research on the electronics of nanotubes in favour of graphene.

The first indication that nanotubes possessed outstanding mechanical properties came in the 1996 study by Michael Treacy and colleagues (12.27), who carried out *in situ* measurements of the intrinsic thermal vibrations of multiwalled carbon nanotubes in a TEM. The lack of precision in this approach led to a wide spread in values for the Young's modulus, with an average of 1.8 TPa. Subsequent, more accurate, measurements (e.g. 12.28–12.30) have quite consistently produced values close to 1 TPa for the modulus. This makes carbon nanotubes the stiffest materials known: significantly stiffer than the best carbon fibres and about 5 times stiffer than steel. But what makes nanotubes truly exceptional is that they combine a high modulus with other outstanding mechanical properties. Their tensile strength, as measured by Rodney Ruoff's group, for example (12.31), can be as high as 63 GPa, around 50 times higher than steel. In addition, nanotubes have the remarkable ability to recover from deformations apparently undamaged. It is important to appreciate, however, that not all nanotubes exhibit these exceptional properties. When produced catalytically, most multiwalled tubes have much lower stiffnesses and strengths, as shown by the Forró group (12.29).

The extraordinary mechanical properties of carbon nanotubes have prompted a major research effort into the production of nanotube-containing composite materials, usually with a polymer matrix. However, successfully incorporating the tubes into a matrix in a way which fully exploits their properties has not been easy, and some of the early results in this area were disappointing. More recently, significant progress has been made in dispersing nanotubes homogeneously in a variety of matrices, and achieving good bonding between the tubes and the matrix. As a result, several groups have produced nanotube/polymer composites with impressive mechanical properties. Two examples are the SWNTs/nylon composites prepared by Karen Winey and colleagues (12.32) and the SWNT/PVA fibres produced by the Baughman group (12.33). Nanotubes can also be used to improve the conductivity of polymers, and have been used commercially in this way for a number of years.

The production of better nanotube-containing composites has been enhanced by progress in nanotube functionalization, and the chemistry of nanotubes has developed into a flourishing field. As well as facilitating the preparation of composite materials, work on nanotube functionalization has helped in the development of techniques for solubilizing, purifying and processing nanotubes. Notable work in this area has been done by the groups of Robert Haddon (12.34), Maurizio Prato (12.35) and Andreas Hirsch (12.36). The functionalization of nanotubes with biomolecules is still a relatively new area, but one that is creating great excitement, partly because of the potential therapeutic applications. In this connection, a very important development has been the demonstration by Bianco, Kostarelos, Prato and colleagues (12.37) that functionalized tubes can pass through cell membranes.

If the science of attaching things to the outside of nanotubes is attracting increasing interest, research on inserting things inside has also advanced greatly since the early studies by Iijima and colleagues (12.38, 12.39) and the Oxford group (12.40, 12.41). Remarkable work on the structure of inorganic crystals inside single-walled nanotubes has been carried out by Jeremy Sloan and co-workers (12.42), and there is great interest in SWNTs filled with fullerenes, or 'nano-peapods', first prepared by Smith, Monthioux and Luzzi (12.43). Some amazing images of individual molecules inside SWNTs have also been recorded by Sloan *et al.* (12.44) and by the Suenaga group (12.45). Studies of small organic molecules in this way represent something quite new for transmission electron microscopy.

The application of nanotubes as AFM tips and sensors is an important growth area. Pioneering work by the Smalley (12.46) and Lieber (12.47) groups demonstrated that nanotubes had significant advantages over conventional silicon or silicon nitride AFM probes. The idea of using nanotubes in this way has been quite widely taken up, and nanotube tips are now commercially available. An area with greater commercial potential is the use of nanotubes as sensors. Two studies published in 2000 (12.48, 12.49) revealed the potential of nanotubes as gas sensors, and prompted intense interest in this area. The use of nanotubes in biosensors is also attracting a major research effort. Both bulk amounts of nanotubes (12.50) and individual tubes (12.51) can be used in this way, and the prospects for commercial exploitation appear to be good.

12.2   Final thoughts

At a conference in 1993 (12.52) Richard Smalley summarized his opinion about the most important issue in nanotube research by paraphrasing Bill Clinton: 'It's the mechanism, stupid'. At this early stage in the development of the subject, Smalley realized that a fully mature nanotube science could only be built on the firm foundation of an understanding of the growth mechanism. Without such an understanding there would be little hope of developing techniques for preparing nanotubes with defined structures, or for the bulk synthesis of high-quality nanotubes.

Sixteen years later, the central problem in nanotube science is still the mechanism. Although progress has been made, the growth mechanisms of both single- and multi-walled nanotubes remain controversial. In particular, the detailed mechanism of multi-walled tube formation in the arc remains obscure. This is despite the fact that some of the most spectacular nanotube properties have been demonstrated using arc-grown MWNTs. In some ways, then, nanotube science has got ahead of itself (12.53). We know how to make single-electron transistors from individual carbon nanotubes, but we do not know how to prepare nanotubes with a defined structure. Perhaps it is time for a shift in the emphasis of nanotube research away from some of these more spectacular areas towards the fundamental issue of understanding how these amazing structures actually grow.

References

(12.1) S. Iijima, 'Helical microtubules of graphitic carbon', *Nature*, **354**, 56 (1991).

(12.2) R. T. K. Baker and P. S. Harris, 'The formation of filamentous carbon', *Chem. Phys. Carbon*, **14**, 83 (1978).

(12.3) M. Endo, 'Grow carbon fibres in the vapor phase', *Chemtech*, **18**, 568 (1988) (September issue).

(12.4) S. Iijima and T. Ichihashi, 'Single-shell carbon nanotubes of 1-nm diameter', *Nature*, **363**, 603 (1993).

(12.5) D. S. Bethune, C. H. Kiang, M. S. deVries *et al.*, 'Cobalt-catalysed growth of carbon nanotubes with single-atomic-layer walls', *Nature*, **363**, 605 (1993).

(12.6) H. J. Dai, A. G. Rinzler, P. Nikolaev *et al.*, 'Single-wall nanotubes produced by metal-catalyzed disproportionation of carbon monoxide', *Chem. Phys. Lett.*, **260**, 471 (1996).

(12.7) P. Nikolaev, M. J. Bronikowski, R. K. Bradley *et al.*, 'Gas-phase catalytic growth of single-walled carbon nanotubes from carbon monoxide', *Chem. Phys. Lett.*, **313**, 91 (1999).

(12.8) D. E. Resasco, W. E. Alvarez, F. Pompeo *et al.*, 'A scalable process for production of single-walled carbon nanotubes (SWNTs) by catalytic disproportionation of CO on a solid catalyst', *J. Nanoparticle Res.*, **4**, 131 (2002).

(12.9) Z. F. Ren, Z. P. Huang, J. W. Xu *et al.*, 'Synthesis of large arrays of well-aligned carbon nanotubes on glass', *Science*, **282**, 1105 (1998).

(12.10) Y. G. Zhang, A. L. Chang, J. Cao *et al.*, 'Electric-field-directed growth of aligned single-walled carbon nanotubes', *Appl. Phys. Lett.*, **79**, 3155 (2001).

(12.11) Y. C. Tseng, P. Q. Xuan, A. Javey *et al.*, 'Monolithic integration of carbon nanotube devices with silicon MOS technology', *Nano Lett.*, **4**, 123 (2004).

(12.12) Y. L. Li, I. A. Kinloch and A. H. Windle, 'Direct spinning of carbon nanotube fibers from chemical vapor deposition synthesis', *Science*, **304**, 276 (2004).

(12.13) K. Koziol, J. Vilatela, A. Moisala *et al.*, 'High-performance carbon nanotube fiber', *Science*, **318**, 1892 (2007).

(12.14) A. Vijayaraghavan, S. Blatt, D. Weissenberger *et al.*, 'Ultra-large-scale directed assembly of single-walled carbon nanotube devices', *Nano Lett.*, **7**, 1556 (2007).

(12.15) M. Zhang, K. R. Atkinson, R. H. Baughman, 'Multifunctional carbon nanotube yarns by downsizing an ancient technology', *Science*, **306**, 1358 (2004).

(12.16) A. G. Rinzler, J. Liu, H. J. Dai *et al.*, 'Large-scale purification of single-wall carbon nanotubes: process, product, and characterization', *Appl. Phys. A*, **67**, 29 (1998).

(12.17) J. W. Mintmire, B. I. Dunlap and C. T. White, 'Are fullerene tubules metallic?', *Phys. Rev. Lett.*, **68**, 631 (1992).

(12.18) M. S. Dresselhaus, G. Dresselhaus and R. Saito, 'Carbon fibers based on C_{60} and their symmetry', *Phys. Rev. B*, **45**, 6234 (1992).

(12.19) R. Saito, M. Fujita, G. Dresselhaus *et al.*, 'Electronic structure of graphene tubules based on C_{60}', *Phys. Rev. B*, **46**, 1804 (1992).

(12.20) N. Hamada, S. Sawada and A. Oshiyama, 'New one-dimensional conductors: graphitic microtubules', *Phys. Rev. Lett.*, **68**, 1579 (1992).

(12.21) T. W. Ebbesen, H. J. Lezec, H. Hiura *et al.*, 'Electrical conductivity of individual carbon nanotubes', *Nature*, **382**, 54 (1996).

(12.22) J. W. G. Wildöer, L. C. Venema, A. G. Rinzler *et al.*, 'Electronic structure of atomically resolved carbon nanotubes', *Nature*, **391**, 59 (1998).

(12.23) S. J. Tans, M. H. Devoret, H. Dai *et al.*, 'Individual single-wall carbon nanotubes as quantum wires', *Nature*, **386**, 474 (1997).

(12.24) S. Frank, P. Poncharal, Z. L. Wang *et al.*, 'Carbon nanotube quantum resistors', *Science*, **280**, 1744 (1998).

(12.25) H. W. C. Postma, T. Teepen, Z. Yao *et al.*, 'Carbon nanotube single-electron transistors at room temperature', *Science*, **293**, 76 (2001).

(12.26) V. Derycke, R. Martel, J. Appenzeller *et al.*, 'Carbon nanotube inter- and intramolecular logic gates', *Nano Lett.*, **1**, 453 (2001).

(12.27) M. M. J. Treacy, T. W. Ebbesen and J. M. Gibson, 'Exceptionally high Young's modulus observed for individual carbon nanotubes', *Nature*, **381**, 678 (1996).

(12.28) E. W. Wong, P. E. Sheehan and C. M. Lieber, 'Nanobeam mechanics: elasticity, strength, and toughness of nanorods and nanotubes', *Science*, **277**, 1971 (1997).

(12.29) J. P. Salvetat, A. J. Kulik, J. M. Bonard *et al.*, 'Elastic modulus of ordered and disordered multiwalled carbon nanotubes', *Adv. Mater.*, **11**, 161 (1999).

(12.30) B. G. Demczyk, Y. M. Wang, J. Cumings *et al.*, 'Direct mechanical measurement of the tensile strength and elastic modulus of multiwalled carbon nanotubes', *Mater. Sci. Eng. A*, **334**, 173 (2002).

(12.31) M. F. Yu, O. Lourie, M. J. Dyer *et al.*, 'Strength and breaking mechanism of multiwalled carbon nanotubes under tensile load', *Science*, **287**, 637 (2000).

(12.32) M. Moniruzzaman, J. Chattopadhyay, W. E. Billups *et al.*, 'Tuning the mechanical properties of SWNT/Nylon 6,10 composites with flexible spacers at the interface', *Nano Lett.*, **7**, 1178 (2007).

(12.33) A. B. Dalton, S. Collins, E. Munoz *et al.*, 'Super-tough carbon-nanotube fibres', *Nature*, **423**, 703 (2003).

(12.34) J. Chen, M. A. Hamon, H. Hu *et al.*, 'Solution properties of single-walled carbon nanotubes', *Science*, **282**, 95 (1998).

(12.35) D. Tasis, N. Tagmatarchis, A. Bianco *et al.*, 'Chemistry of carbon nanotubes', *Chem. Rev.*, **106**, 1105 (2006).

(12.36) A. Hirsch and O. Vostrowsky, 'Functionalization of carbon nanotubes', *Topics Current Chem.*, **245**, 193 (2005).

(12.37) D. Pantarotto, R. Singh, D. McCarthy *et al.*, 'Functionalized carbon nanotubes for plasmid DNA gene delivery', *Angew. Chem. Int. Ed.*, **43**, 5242 (2004).

(12.38) P. M. Ajayan and S. Iijima, 'Capillarity-induced filling of carbon nanotubes', *Nature*, **361**, 333 (1993).

(12.39) P. M. Ajayan, T. W. Ebbesen, T. Ichihashi *et al.*, 'Opening carbon nanotubes with oxygen and implications for filling', *Nature*, **362**, 522 (1993).

(12.40) S. C. Tsang, P. J. F. Harris and M. L. H. Green, 'Thinning and opening of carbon nanotubes by oxidation using carbon dioxide', *Nature*, **362**, 520 (1993).

(12.41) S. C. Tsang, Y. K. Chen, P. J. F. Harris *et al.*, 'A simple chemical method of opening and thinning carbon nanotubes', *Nature*, **372**, 159 (1994).

(12.42) J. Sloan, D. E. Luzzi, A. I. Kirkland *et al.*, 'Imaging and characterization of molecules and one-dimensional crystals formed within carbon nanotubes', *MRS Bull.*, **29**, 265 (2004).

(12.43) B. W. Smith, M. Monthioux and D. E. Luzzi, 'Encapsulated C_{60} in carbon nanotubes', *Nature*, **396**, 323 (1998).

(12.44) D. A. Morgan, J. Sloan and M. L. H. Green, 'Direct imaging of o-carborane molecules within single walled carbon nanotubes', *Chem. Commun.*, 2442 (2002).

(12.45) Z. Liu, K. Yanagi, K. Suenaga *et al.*, 'Imaging the dynamic behaviour of individual retinal chromophores confined inside carbon nanotubes', *Nat. Nanotech.*, **2**, 422 (2007).

(12.46) H. J. Dai, J. H. Hafner, A. G. Rinzler *et al.*, 'Nanotubes as nanoprobes in scanning probe microscopy', *Nature*, **384**, 147 (1996).

(12.47) S. S. Wong, E. Joselevich, A. T. Woolley *et al.*, 'Covalently functionalized nanotubes as nanometre-sized probes in chemistry and biology', *Nature*, **394**, 52 (1998).

(12.48) P. G. Collins, K. Bradley, M. Ishigami *et al.*, 'Extreme oxygen sensitivity of electronic properties of carbon nanotubes', *Science*, **287**, 1801 (2000).

(12.49) J. Kong, N. R. Franklin, C. W. Zhou *et al.*, 'Nanotube molecular wires as chemical sensors', *Science*, **287**, 622 (2000).

(12.50) J. J. Gooding, 'Nanostructuring electrodes with carbon nanotubes: a review on electrochemistry and applications for sensing', *Electrochim. Acta*, **50**, 3049 (2005).

(12.51) R. J. Chen, S. Bangsaruntip, K. A. Drouvalakis *et al.*, 'Noncovalent functionalization of carbon nanotubes for highly specific electronic biosensors', *Proc. Nat. Acad. Sci. USA*, **100**, 4984 (2003).

(12.52) *Fullerenes '93*, Santa Barbara, California, June 27–July 1, 1993.

(12.53) N. Grobert, 'Nanotubes – grow or go?', *Materials Today*, **9**(10), 64 (2006).

Name Index

Subject Index